SOCIETY OF GENERAL PHYSIOLOGISTS SERIES
Volume 38

Electrogenic Transport: Fundamental Principles and Physiological Implications

Society of General Physiologists Series

Published by Raven Press

Vol. 38. Electrogenic Transport: Fundamental Principles and Physiological Implications
Mordecai P. Blaustein and Melvyn Lieberman, editors. 416 pp., 1984.

Vol. 37. Basic Biology of Muscles: A Comparative Approach
B.M. Twarog, R.J.C. Levine, and M.M. Dewey, editors. 424 pp., 1982.

Vol. 36. Ion Transport by Epithelia
Stanley G. Schultz, editor. 288 pp., 1981.

Vol. 35. Peptides: Integrators of Cell and Tissue Function
Floyd E. Bloom, editor. 257 pp., 1980.

Vol. 34. Membrane-Membrane Interactions
Norton B. Gilula, editor. 238 pp., 1980.

Vol. 33. Membrane Transduction Mechanisms
Richard A. Cone and John E. Dowling, editors. 248 pp., 1979.

Vol. 32. Cell and Tissue Interactions
J.W. Lash and M.M. Burger, editors. 324 pp., 1977.

Vol. 31. Biogenesis and Turnover of Membrane Macromolecules
J.S. Cook, editor. 276 pp., 1976.

Vol. 30. Molecules and Cell Movement
S. Inoué and R.E. Stephens, editors. 460 pp., 1975.

Vol. 29. Cellular Selection and Regulation in the Immune Response
G.M. Edelman, editor. 299 pp., 1974.

Vol. 28. Synaptic Transmission and Neuronal Interaction
M.V.L. Bennett, editor. 401 pp., 1974.

SOCIETY OF GENERAL PHYSIOLOGISTS SERIES
Volume 38

Electrogenic Transport
Fundamental Principles and Physiological Implications

Editors

Mordecai P. Blaustein, M.D.
Department of Physiology
University of Maryland School of Medicine
Baltimore, Maryland

Melvyn Lieberman, Ph.D.
Department of Physiology
Duke University Medical Center
Durham, North Carolina

Raven Press ■ New York

Raven Press, 1140 Avenue of the Americas, New York, New York 10036

© 1984 by Raven Press Books, Ltd. All rights reserved. This book is protected by copyright. No part of it may be reproduced, stored in a retrieval system, or transmitted, in any form or by any means, electronic, mechanical, photocopying, recording, or otherwise, without the prior written permission of the publisher.

Made in the United States of America

Library of Congress Cataloging In Publication Data
Main entry under title:

Electrogenic transport.

   (Society of General Physiologists series ; v. 38)
   Includes index.
   1. Biological transport, Active.   2. Ion channels.
3. Action potentials (Electrophysiology)   I. Blaustein, Mordecai P.   II. Lieberman, Melvyn.   III. Series.
[DNLM: 1. Biological transport—Congresses. 2. Cell membrane permeability—Congresses.   W1 S0872G v.38 / QH601 E375 1982]
QH509.E44   1984   591.1   83-23091
ISBN 0-89004-959-9

The material contained in this volume was submitted as previously unpublished material, except in the instances in which credit has been given to the source from which some of the illustrative material was derived.

Great care has been taken to maintain the accuracy of the information contained in the volume. However, Raven Press cannot be held responsible for errors or for any consequences arising from the use of the information contained herein.

Materials appearing in this book prepared by individuals as part of their official duties as U.S. Government employees are not covered by the above-mentioned copyright.

Preface

Since its inception in 1946, the Society of General Physiologists has sponsored symposia on timely and important topics of interest to physiologists. The goal of the 1982 Symposium was to bring together distinguished scientists identified with fundamental contributions in electrogenic transport. Exciting developments in the field of electrogenic pumps provided the rationale for discussing these newer aspects and merging them with some of the older observations. As a result, these discussions have expanded our knowledge of the theoretical basis of electrogenic pump mechanisms, and have drawn attention to the physiological consequences of electrogenic transport. The importance of electrogenic transport in the function of both excitable and nonexcitable cells has been confirmed with the recent application of ion-selective electrode and simultaneous tracer flux and voltage-clamp methods, and with the study of transport in membrane vesicles, lipid bilayer systems, and cultured cells. A number of these findings are reported here for the first time.

In 1955, Hodgkin and Keynes first suggested that the sodium (Na-K exchange) pump might be electrogenic. Experimental evidence adduced since these pioneering studies has amply supported their conjecture. Nevertheless, arguments about pump stoichiometry and the nature of the pump mechanism continue. Some of these critical unresolved questions are considered in the first section of this volume, which addresses the fundamental principles that underly electrogenic transport systems.

Another important area that has emerged in recent years concerns ion-gradient coupled transport systems, or so-called secondary active transport. In these transport systems, the uphill transport of a solute is linked to the dissipation of energy from an established ion gradient—for example, the sodium electrochemical gradient. The basic principles of secondary active transport, and the resultant changes in membrane potential (which may appropriately be referred to as secondary electrogenic transport), are considered in the second section of this volume.

Secondary active transport depends on both the membrane potential and the chemical gradients for the transported solutes. Consequently, changes in membrane potential will alter the direction and/or rate of solute flow. These properties are highlighted in the chapter by Professor Lorin J. Mullins, in which he discusses the influence of membrane potential on sodium-calcium exchange, and its relevance to cardiac muscle function. This topic serves as a bridge between the theoretical and the experimental aspects of electrogenic transport and forms an appropriate introduction to the third section of this book. The remainder of the third section addresses the physiological role of electrogenic transport in nerve, cardiac muscle, and smooth muscle.

Electrogenic transport is certainly not limited to the classical excitable tissues, nerve, and muscle. The fourth section of this volume deals with the fundamental properties and physiologic significance of electrogenic transport systems in other cell types: red blood cells, pancreatic β cells, neurospora, *E. coli*, and higher plants.

Data show that electrogenic transport systems present in these various cells are simply variations on a common theme. Clearly, there is something to be learned from an eclectic approach.

The final section focuses on the newer investigational methods for studying electrogenic transport. Included in this section is a chapter describing the use of noise measurements to obtain new information about some of the intermediate steps in carrier-mediated transport processes. Another novel approach described here is one in which voltage-clamp currents and tracer ion fluxes are measured simultaneously to obtain direct information about the stoichiometry of Na-K (sodium pump) and Na-Ca exchange transport. Also included are chapters describing recent work on the relationship of the electrogenic sodium pump current to acidification in neurons, and on the determination of pump conductance and reversal in developing embryos.

This volume attempts to provide the reader with an up-to-date, critical review of the status of primary and secondary electrogenic transport systems. We hope that, by incorporating both theoretical and experimental information, this volume will serve as a long-lasting reference for investigators and students.

MORDECAI P. BLAUSTEIN, M.D.
MELVYN LIEBERMAN, Ph.D.

Acknowledgments

We thank the Director and Staff of the Marine Biological Laboratory, Woods Hole, Massachusetts, for their contribution to the smooth running of the Symposium. We also thank Dr. Paul De Weer and Ms. Kate Eldred for help with the local arrangements, and Mrs. Arlene Renninger for secretarial assistance in preparing this volume. We are indebted to Drs. W. Jonathan Lederer and Roger C. Thomas for organizing the very timely workshop, and are especially grateful to all of the invited speakers and symposium participants. Their enthusiastic presentations and lively exchange of ideas helped to make this a most worthwhile meeting, and we trust that much of the excitement has been maintained in this book.

The Society of General Physiologists' 1982 Symposium was supported in part by grants from the National Institutes of Health (HL-28553), the Upjohn Company, the U.S. Army Research Office, and the U.S. Army Research and Medical Development Command. The views, opinions, and/or findings contained in this report are those of the authors and should not be construed as an official Department of the Army position, policy, or decision, unless so designated by other documentation.

The Editors

Contents

Fundamental Principles of Primary Electrogenic Transport

1 Electrogenic Pumps: Theoretical and Practical Considerations
  P. De Weer

17 Thermodynamics and Kinetics of Electrogenic Pumps
  J. Brian Chapman

33 The Electrogenic Sodium Pump
  I. M. Glynn

Fundamental Principles and Properties of Secondary Electrogenic Transport

49 Electrochemical Driving Forces for Secondary Active Transport: Energetics and Kinetics of Na^+-H^+ Exchange and Na^+-Glucose Cotransport
  Peter S. Aronson

71 Chemiosmotic Coupling and Its Application to the Accumulation of Biological Amines in Secretory Granules
  Robert G. Johnson and Antonio Scarpa

93 Electric Aspects of Co- and Countertransport
  E. Heinz and S. M. Grassl

105 Contributions of Electrogenic Pumps to Resting Membrane Potentials: The Theory of Electrogenic Potentials
  R. A. Sjodin

129 The Energetics and Kinetics of Sodium-Calcium Exchange in Barnacle Muscles, Squid Axons, and Mammalian Heart: The Role of ATP
  Mordecai P. Blaustein

149 The Na^+/Ca^{2+} Exchanger of Heart Sarcolemma is Regulated by a Phosphorylation-Dephosphorylation Process
  Pico Caroni, Luciano Soldati, and Ernesto Carafoli

Electrogenic Transport in Cardiac Muscle, Smooth Muscle, and Nerve

161 An Electrogenic Saga: Consequences of Sodium-Calcium Exchange in Cardiac Muscle
L.J. Mullins

181 Physiologic Criteria for Electrogenic Transport in Tissue-Cultured Heart Cells
Melvyn Lieberman, C. Russell Horres, Ron Jacob, Elizabeth Murphy, David Piwnica-Worms, and David M. Wheeler

193 The Electrogenic Na Pump in Mammalian Cardiac Muscle
D.A. Eisner, W.J. Lederer, and R.D. Vaughan-Jones

215 Influence of the Sodium Pump Current on Electrical Activity of Cardiac Cells
David C. Gadsby

239 Effects of Membrane Potential on Sodium-Dependent Calcium Transport in Cardiac Sarcolemma Vesicles
Robin T. Hungerford, George E. Lindenmayer, William P. Schilling, and Edwin Van Alstyne

253 Functional Significance of Electrogenic Pumps in Neurons
David O. Carpenter and Robert A. Gregg

271 Physiological Role of Electrogenic Pumps in Smooth Muscle
John A. Connor

Electrogenic Transport in Animal and Plant Cells

287 Membrane Electrical Parameters of Normal Human Red Blood Cells
Joseph F. Hoffman and Philip C. Laris

295 The Sodium Pump of Mouse Pancreatic β-Cells: Electrogenic Properties and Activation by Intracellular Sodium
H.P. Meissner and J.C. Henquin

307 Electrical Kinetics of Proton Pumping in *Neurospora*
Clifford L. Slayman and Dale Sanders

323 The *lac* Carrier Protein from *Escherichia coli*
H. Ronald Kaback, Nancy Carrasco, David Foster, Maria Luisa Garcia, Tzipora Goldkorn, Lekha Patel, and Paul Viitanen

331 Electrogenic Ion Transport in Higher Plants
Roger M. Spanswick and Alan B. Bennett

New Investigational Methods for Electrogenic Transport

345 Electrogenic Sodium Pumping in *Xenopus* Blastomeres: Apparent Pump Conductance and Reversal Potential
Luca Turin

353 Electrogenic Sodium Pump Current Associated with Recovery from Intracellular Acidification of Snail Neurons
R. C. Thomas

365 Sodium-Dependent Calcium Efflux and Sodium-Dependent Current in Perfused Barnacle Muscle Single Cells
M. T. Nelson and W. J. Lederer

373 The Effects of Na-Ca Exchange on Membrane Currents in Sheep Cardiac Purkinje Fibres
W. J. Lederer, S.-S. Sheu, R. D. Vaughan-Jones, and D. A. Eisner

381 Photoelectric Properties of the Light-Driven Proton Pump Bacteriorhodopsin
E. Bamberg, A. Fahr, and G. Szabo

395 Subject Index

Contributors

P. S. Aronson
Departments of Medicine and
 Physiology
Yale School of Medicine
New Haven, Connecticut 06510

E. Bamberg
Max Planck Institut für Biophysik
D-6000 Frankfurt 70
Federal Republic of Germany

A. D. Bennett
Section of Plant Biology
Division of Biological Sciences
Cornell University
Ithaca, New York 14853

M. P. Blaustein
Department of Physiology
University of Maryland School of
 Medicine
Baltimore, Maryland 21201

E. Carafoli
Laboratory of Biochemistry
Swiss Federal Institute of
 Technology (ETH)
8092 Zurich, Switzerland

P. Caroni
Department of Biochemistry and
 Biophysics
University of California
San Francisco, California 94143

D. O. Carpenter
Center for Laboratories and
 Research
New York State Department of
 Health
Albany, New York 12201

N. Carrasco
Laboratory of Membrane
 Biochemistry
Roche Institute of Molecular
 Biology
Nutley, New Jersey 07110

J. B. Chapman
Department of Physiology
Monash University
Clayton
Victoria 3168, Australia

J. B. Connor
Bell Laboratories
600 Mountain Avenue
Murray Hill, New Jersey 07974

P. J. De Weer
Department of Physiology and
 Biophysics
Washington University School of
 Medicine
St. Louis, Missouri 63110

D. A. Eisner
Department of Physiology
University College London
Gower Street
London WC1E 6BT, UK

A. Fahr
Universität Konstanz
Fakultät für Biologie
D-755 Konstanz
Federal Republic of Germany

D. Foster
Laboratory of Membrane
 Biochemistry
Roche Institute of Molecular
 Biology
Nutley, New Jersey 07110

D. C. Gadsby
Laboratory of Cardiac Physiology
The Rockefeller University
New York, New York 10021

M. L. Garcia
Laboratory of Membrane
 Biochemistry
Roche Institute of Molecular
 Biology
Nutley, New Jersey 07110

I. M. Glynn
Physiological Laboratory
University of Cambridge
Downing Street
Cambridge CB2 3EG, UK

T. Goldkorn
Laboratory of Membrane
  Biochemistry
Roche Institute of Molecular
  Biology
Nutley, New Jersey 07110

S. M. Grassl
Department of Physiology
Cornell University Medical Center
New York, New York 10021

R. A. Gregg
Center for Laboratories and
  Research
New York State Department of
  Health
Albany, New York 12201

E. Heinz
Department of Physiology
Cornell University Medical Center
New York, New York 10021

J. C. Henquin
1. Physiologisches Institut und
  Medizinische Klinik
Universität des Saarlandes
D-6650 Homburg/Saar
Federal Republic of Germany

J. F. Hoffman
Department of Physiology
Yale University School of Medicine
New Haven, Connecticut 06510

C. R. Horres
Department of Physiology
Duke University Medical Center
Durham, North Carolina 27710

R. T. Hungerford
Department of Pharmacology
Medical University of South
  Carolina
Charleston, South Carolina 29425

R. Jacob
Department of Physiology
Duke University Medical Center
Durham, North Carolina 27710

R. G. Johnson
Department of Biochemistry and
  Biophysics
University of Pennsylvania School
  of Medicine
Philadelphia, Pennsylvania 19104

H. R. Kaback
Laboratory of Membrane
  Biochemistry
Roche Institute of Molecular
  Biology
Nutley, New Jersey 07110

P. C. Laris
Department of Biological Sciences
University of California
Santa Barbara, California 93106

W. J. Lederer
Department of Physiology
University of Maryland School of
  Medicine
Baltimore, Maryland 21201

M. Lieberman
Department of Physiology
Duke University Medical Center
Durham, North Carolina 27710

G. E. Lindenmayer
Department of Pharmacology
Medical University of South
  Carolina
Charleston, South Carolina 29425

H. P. Meissner
1. Physiologisches Institut und
  Medizinische Klinik
Universität des Saarlandes
D-6650 Homburg/Saar
Federal Republic of Germany

L. J. Mullins
Department of Biophysics
University of Maryland School of
  Medicine
Baltimore, Maryland 21201

E. Murphy
Department of Physiology
Duke University Medical Center
Durham, North Carolina 27710

M. T. Nelson
Department of Physiology
University of Maryland School of
  Medicine
Baltimore, Maryland 21201

L. Patel
Laboratory of Membrane
  Biochemistry
Roche Institute of Molecular
  Biology
Nutley, New Jersey 07110

D. Piwnica-Worms
Department of Physiology
Duke University Medical Center
Durham, North Carolina 27710

D. Sanders
Department of Physiology
Yale School of Medicine
New Haven, Connecticut 06510

A. Scarpa
Department of Biochemistry and
  Biophysics
University of Pennsylvania School
  of Medicine
Philadelphia, Pennsylvania 19104

W. P. Schilling
Department of Pharmacology
Medical University of South
  Carolina
Charleston, South Carolina 29425

S.-S. Sheu
Department of Physiology
University of Maryland School of
  Medicine
Baltimore, Maryland 21201

R. A. Sjodin
Department of Biophysics
University of Maryland School of
  Medicine
Baltimore, Maryland 21201

C. L. Slayman
Department of Physiology
Yale School of Medicine
New Haven, Connecticut 06510

L. Soldati
Laboratory of Biochemistry
Swiss Federal Institute of
  Technology (ETH)
8092 Zurich, Switzerland

R. M. Spanswick
Section of Plant Biology
Division of Biological Sciences
Cornell University
Ithaca, New York 14853

G. Szabo
University of Texas Medical Branch
Department of Physiology and
  Biophysics
Galveston, Texas 77550

R. C. Thomas
Department of Physiology
Bristol University
Bristol BS8 1TD, UK

L. Turin
E.R. Biologie du Developement
Station Zoologique
La Darse
F-06230 Villefranche/Mer, France

E. Van Alstyne
Department of Pharmacology
Medical University of South
  Carolina
Charleston, South Carolina 29425

R. D. Vaughan-Jones
Department of Pharmacology
Oxford University
Oxford OX1 3QT, UK

P. Viitanen
Laboratory of Membrane
  Biochemistry
Roche Institute of Molecular
  Biology
Nutley, New Jersey 07110

D. M. Wheeler
Department of Anesthesiology
University of Alabama Hospitals
Birmingham, Alabama 35233

Electrogenic Transport: Fundamental Principles and Physiological Implications, edited by Mordecai P. Blaustein and Melvyn Lieberman. Raven Press, New York © 1984.

Electrogenic Pumps: Theoretical and Practical Considerations

Paul De Weer

Department of Physiology and Biophysics, Washington University School of Medicine, St. Louis, Missouri 63110

This chapter describes some common characteristics, *mutatis mutandis,* of all electrogenic pumps, and examines the extent to which, depending on the particular pump or species, these properties will be apparent or easily accessible to the investigator. Conceptually, although not chronologically, research in the field of electrogenic pumps can roughly be seen as occurring in three stages. (For reviews, see references 9, 22, and 25, and elsewhere in this volume.) The first stage was mainly concerned with establishing, often by ingenious means, the existence of electrogenic pumps in a variety of animal and plant cells, distinguishing them from artifacts or other mechanisms (such as neutral pumps in series with restricted-diffusion spaces, or conductance changes unrelated to pumping), and examining some of their properties. In the second stage, the physiological role of such pumps in the behavior of cells or cell systems was explored: theoretical and experimental assessments of the contribution of electrogenic pumps to the cell membrane potential, to pacemaker tuning, to sensory adaptation, to the modulation of action potentials, to the control of signaling in neuronal networks, etc. A more recent endeavor has been to exploit the electrogenic behavior of a given pump as an experimental *tool* in the further study of the molecular kinetics, stoichiometry, and thermodynamics of the pump itself. It was with this latter approach in mind that the following was written.

DEFINITION

These remarks will be restricted to primary electrogenic pumps, that is, chemically [adenosine triphosphate (ATP)]-driven transmembrane pumps that translocate a greater electric charge in one direction than in the other during their cyclic operation. Not considered will be secondary (e.g., H^+ or Na^+) gradient-driven, current-carrying, carrier-mediated, coupled fluxes, or single-ion facilitated-diffusion carriers, which are current carriers by definition. (The use of the term electrogenic to characterize the latter mechanisms is, in the author's view, inappropriate.) Examples of electrogenic pumps that fit the above definition are Dean's (8) original sodium pump, which was assumed to transport only sodium ions, today's sodium pump, which is commonly supposed to export three sodium ions and import two potassium ions per cycle, and today's ATP-driven proton pumps of algae, molds, and higher plants. An electrogenic pump is thus a molecular device, embedded in the cell membrane, capable of generating an electric current across that membrane, at the expense of a chemical reaction. Like any other electrochemical device, the pump must therefore be characterized by an electromotive force and an internal resistance or conductance.

EQUIVALENT CIRCUIT

The equivalent circuit (1,10,15,24) of an electrogenic pump (i.e., a multitude of pump molecules operating in parallel) embedded in a cell membrane is illustrated in Fig. 1 (top). The pump's equivalent electromotive force (EMF) acts against the cell's membrane potential (V_m), over the pump's internal resistance, R_p. The current pathway is closed over the remainder of the cell membrane,

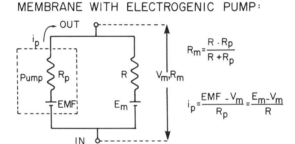

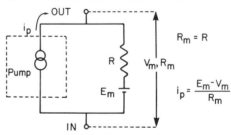

FIG. 1. Equivalent circuits for membranes with electrogenic pumps. **Top:** general case. No assumption is made regarding magnitude of the pump's EMF or resistance. **Bottom:** special case. Rheogenic pump where EMF is large compared to membrane potential, and R_p large and constant, so that reasonable variations in V_m have no detectable effect on pump current, and the pump appears to operate as a constant current generator. [From Abercrombie and De Weer (1), with permission.]

and the passage of this current raises V_m over the diffusion potential (labeled E_m), which would exist in the absence of the electrogenic pump. In other words, the electrogenic current can be thought of as driven either by the potential drop (EMF$-V_m$) over the pump resistance, R_p, or by the difference between the actual membrane potential V_m, and the underlying diffusion potential, E_m, over R, the resistance of the cell membrane exclusive of the pump resistance. Clearly, if membrane potential V_m somehow became equal to EMF, no pump current would flow. And if V_m were artificially raised beyond EMF, the battery would undergo charging, i.e., the pump would be reversed. Thus EMF is operationally equal to the reversal potential for the pump.

An obvious corollary of this arrangement is that the total membrane conductance R_m consists of both the pump conductance and the intrinsic conductance of the rest of the membrane. How much each contributes to the "black box" measured by the experimenter cannot be stated without further independent information. Knowledge of both EMF and pump resistance is prerequisite to any quantitative description of the electrogenic pump under study.

ELECTROMOTIVE FORCE

For a chemically driven pump, EMF is given by the excess of the free energy of the chemical reaction (i.e., ATP hydrolysis) over the reversible osmotic (concentration) work expended in the transfer of stoichiometric amounts of the relevant ions, divided by the net charge translocated across the membrane during the pump's operating cycle:

$$\text{EMF} = \frac{\Delta G \text{ (ATP hydrolysis)} - \text{concentration work}}{\text{Net charge transported}} \quad (1)$$

For example, the "classical" sodium pump presumably transfers 3 Na$^+$ and 2 K$^+$ against their respective concentration gradients per molecule of ATP split, for a net transfer of a single electronic charge.

Published estimates of the free energy of hydrolysis of intracellular ATP

TABLE 1. *Estimates for* $-\Delta G$ *of ATP hydrolysis*

Tissue	kcal/mol	meV/molecule	Authors	Reference
Squid axon	~14	~610	Caldwell & Shirmer (1965)	3
Liver mitochondria	15.4	670	Cockrell et al. (1966)	6
Frog muscle	~11	~480	Kushmerick (1969)	16
Neurospora	~12	~520	Slayman et al. (1973)	23
Rat hepatocytes	~12	~520	Akerboom et al. (1978)	2
Rat heart	12–14	520–610	Nishiki et al. (1978)	20
Rat muscle	~14	~610	Veech et al. (1979)	26
Frog muscle	~13	~560	Dawson et al. (1980)	7
Rat muscle	~15	~650	Meyer et al. (1982)	18

vary among tissues and among authors. Table 1 lists, in chronological order, a number of published values. The most recent estimates tend to be the higher ones, and the reason for this is that estimates for free cytosolic [adenosine diphosphate (ADP)], i.e., not bound to mitochondria or cytoplasmic proteins, have dramatically decreased in recent years. Also, estimates for the cytoplasmic concentration of orthophosphate (P_i) obtained by nuclear magnetic resonance (NMR) techniques have been lower than those obtained by analytical methods, where contamination from breakdown of labile organic phosphate compounds is more likely to occur. At any rate, it seems safe to assume that, in many cells, the negative free energy of ATP hydrolysis is about 13 to 15 kcal/mole ATP, or roughly 600 millielectron volt (meV) per molecule of ATP.

Turning now to the osmotic work to be expended in the reversible translocation of the relevant ions (Table 2), it can be seen that, for a typical animal cell possessing a 3 Na:2 K:1 ATP pump, not more than about $(3 \times \sim60) + (2 \times \sim90) = \sim360$ meV in osmotic work will be performed per molecule of ATP hydrolyzed. From Eq. (1), the EMF of a typical animal sodium pump can be computed:

$$\text{EMF (Na pump)} = \frac{-(\sim600 \text{ meV}) + (\sim360 \text{ meV})}{(1 \text{ charge for a } 3:2 \text{ pump})} = -(\sim250) \text{ mV}$$

In other words, the reversal potential for a typical animal cell 3:2 sodium:potassium pump is expected to be about -250 mV, well beyond normal resting potentials. The ATP hydrolysis thus provides more than ample free energy for the uphill movements of Na and K in animal cells, and the transport reaction is far from equilibrium. In plant cell electrogenic hydrogen pumps, assuming a cytoplasmic pH near that of the extracellular fluid, it is clear that a 1 H^+:1 ATP electrogenic pump should have a reversal potential numerically equal to the free energy of ATP hydrolysis (expressed in meV/molecule); the reversal potential for a 2 H^+:1 ATP pump would be half as large. Finally, if acid-secreting cells do so by means of an electrogenic H^+-ATPase, it follows that no more than a single proton can be transported per molecule of ATP hydrolyzed.

In the reasoning followed above, the expected reversal potential for a given electrogenic pump was computed from a knowledge of the free energy of intracellular ATP hydrolysis and from an assumed ion stoichiometry for the pump under study. Conversely, an experimenter may be able to determine clearly the reversal potential of a given pump and thus deduce from it the pump's

TABLE 2. *Typical concentration gradients for sodium, potassium, and hydrogen*

Ion	Organism	Concentration gradient	Work per ion
Na^+	Animal	$\sim$10-fold	$\sim$60 meV
K^+	Animal	$\sim$30-fold	$\sim$90 meV
H^+	Plant	0.2- to 5-fold	$\pm$30 meV
	Animal (acid-secreting)	10^6- to 10^7-fold	360–420 meV

exact stoichiometry. Unambiguous answers may be difficult to obtain in the case of complicated stoichiometries, since the differences between the reversal potentials for various plausible pump models may not be very large. Relatively simple questions, however, such as "does this putative ATP-driven electrogenic proton pump transport one or two protons at a time?" may be resolved by a clear determination of reversal potential. It should be realized, however, that any stoichiometry deduced from a reversal potential experiment is valid only at the reversal potential. Should the pump possess variable stoichiometry as a function of V_m, for example, this would never be apparent from a reversal potential test. It is not possible to deduce pump stoichiometries from purely electrical measurements at voltages other than the reversal potential.

The foregoing discussion of reversal potentials should not be construed as implying that electrogenic pump reversal potentials can or will be easily measured in normal cells (5). The expected reversal potential for a classical Na:K:ATP pump with 3:2:1 stoichiometry (−250 mV) is beyond dielectric breakdown for animal cells. Only in cells with artificially lowered phosphorylation potentials may it be possible to observe electrogenic sodium pump reversal potentials. Plant cells, however, appear capable of sustaining much larger membrane potentials (for review see ref. 22), and the prospects for observing a *bona fide* proton pump reversal appear brighter there, at least for pump stoichiometries greater than 1 H^+:1 ATP.

The closest approach to electrogenic pump reversal in an intact cell appears to have been achieved by Gradmann et al. (11) in *Neurospora crassa*. These cells could not be hyperpolarized beyond −300 mV without irreversible damage, but from the extrapolated pump reversal potential of −400 mV, the authors concluded that the H^+:ATP stoichiometry of *Neurospora*'s electrogenic proton pump must be 1:1.

PUMP CHORD CONDUCTANCE

The second parameter within the black box of Fig. 1 (top), in addition to the EMF, is the pump's equivalent internal resistance or conductance. This conductance (properly called chord conductance) is defined as the proportionality factor between the current generated by the electrogenic pump, and its driving force, i.e., the potential difference over which the current flows (see Fig. 1, top).

$$I_{pump} = g_{pump} (EMF - V_m)$$

Careful distinction should be made between this chord conductance and what is measured experimentally (i.e., an approximation to the slope conductance) when small voltage steps are imposed on the membrane, and the resulting current excursions noted. As will be discussed, the pump conductance measured experimentally in this fashion may be equal to, or larger or smaller than the chord conductance defined above.

Electrogenic sodium pump current densities in animal cells are usually of the order of a few $\mu A/cm^2$. The sodium pump current is driven by a relatively large EMF (approx -250 mV) against a relatively modest membrane potential (~ -60 mV), which gets raised just a few mV beyond its passive diffusion potential value. It follows from the equivalent circuit in Fig. 1 (top) that the electrogenic sodium pump's conductance is much smaller than that of the remainder of the cell membrane:

$$\frac{g_{\text{Na pump}}}{g_{\text{remainder}}} = \frac{E_m - V_m}{\text{EMF} - V_m} = \frac{\text{a few mV}}{\sim 200 \text{ mV}} \text{ (for animals)} \qquad (2)$$

Electrogenic proton pump current densities in plant cells can be an order of magnitude larger than the sodium pump current densities of animal cells (11,24). Plant membrane diffusional resistances are generally much higher than those of animal cells, and the proton pump-generated hyperpolarizations over and beyond the resting diffusion potential are truly enormous, of the order of ≥ 50 to 150 mV in some cases (15,23,24). Consequently, the values of EMF $- V_m$) and ($E_m - V_m$) in Fig. 1 (top) are of similar magnitude, and the pump conductance, g_p, is a major fraction of the total membrane conductance; it may be as large as or larger than that of the remainder of the cell membrane:

$$\frac{g_{\text{proton pump}}}{g_{\text{remainder}}} = \frac{E_m - V_m}{\text{EMF} - V_m} \simeq 1 \text{ (for plants)} \qquad (3)$$

This comparison between animal sodium pumps and plant proton pumps should not be taken too categorically. Its main purpose is to emphasize that animal sodium pumps appear, in general, to have a rather high internal resistance compared to the rest of the membrane, whereas, thus far, the opposite holds, in general, for some well-studied proton pumps in plant cells. There are, of course, animal cells that exhibit relatively high active sodium-transport rates (certain epithelia, for example) and cell membranes that have relatively high resistances.

The fact that the electrogenic sodium pump of animal cells is a coupled Na/K pump puts an interesting constraint on the magnitude of the chord conductance of this pump for cells in the steady state. Mullins and Noda (19) have derived an equation for the membrane potential, in the steady state, of cells that possess an electrogenic Na/K pump as their only major pump:

$$V_m = \frac{RT}{F} \ln \frac{rP_K[K]_o + P_{Na}[Na]_o}{rP_K[K]_i + P_{Na}[Na]_i} \qquad (4)$$

where r is the Na:K pump coupling ratio, P_K and P_{Na} are permeabilities, and the other symbols have their usual meaning. Ascher (see refs. 9 and 25) has deduced from Eq. 4 that the maximal contribution ($V_m - E_m$) to the membrane potential that a 3:2 electrogenic pump could make in the steady state is about 10 mV. Clearly then, with EMFs of the order of -250 mV, and assuming typical membrane potentials, the electrogenic sodium pump's chord conductance can never be more than a small fraction of the apparent cell membrane conductance in the steady state. Transient states are conceivable, however (e.g., a high-resistance cell recovering rapidly from a large sodium load), where the pump conductance may not be negligible compared to the cell's total conductance. It should be noted, also, that the electrogenic Na/K pump of epithelial cell membranes is not subject to the above steady-state constraint. Here, the pump may be embedded in a high-resistance basolateral membrane, yet be supplied with large quantities of sodium ions entering passively through the apical cell membrane. Consequently, the electrogenic pump located in the basolateral membrane may contribute a sizeable hyperpolarization, and a sizeable conductance, to that membrane in the steady state.

Although pump chord conductance, as formally described in this section, cannot predict exactly what contribution a given electrogenic pump will make to the apparent, experimentally determined, total membrane conductance, it can suggest an order of magnitude of what to expect. Evidence to date suggests that, for most part, animal cell electrogenic sodium pumps will contribute little to overall membrane conductance, whereas plant cell electrogenic proton pumps may constitute a major fraction of it. Plant cell equivalent circuits, therefore, should continue to keep the complete form of Fig. 1 (top). Accurate experimental measurement of pump current in these cells is not straightforward. It does not suffice to divide $V_m - E_m$ (the potential drop on stopping the pump) by the experimentally measured membrane conductance, since the latter contains an appreciable contribution from the pump itself. Animal cell pumps, on the other hand, with their low conductance, allow the equivalent circuit to be simplified to the form shown in Fig. 1 (bottom); namely, as a constant-current source in parallel with the usual passive permeability circuits of the remainder of the membrane. The term rheogenic pump should probably be reserved exclusively for this limiting case. Here, no appreciable error is made by equating R_m, the experimentally measured total membrane resistance, with R, the intrinsic resistance of the nonpump part of the membrane.

Unfortunately, the above distinction between electrogenic pump (general case) and rheogenic pump (limiting case of high-impedance pump) is not completely satisfactory, since the label rheogenic implies not only high impedance, but constant (voltage-independent) impedance as well. These two properties are not necessarily linked, as will be discussed in the next section.

CURRENT–VOLTAGE RELATIONSHIP FOR ELECTROGENIC PUMPS

Thermodynamics demands that if a pump contributes to the cell's resting potential by producing a net current, it in turn is affected by membrane potential. Both Finkelstein (10) and Rapoport (21) have pointed out that a voltage-sensitive electrogenic pump will behave operationally as an apparent membrane conductance. In Finkelstein's model (10), voltage-sensitivity came about for diffusion-kinetic reasons; in Rapoport's model (21), a nonequilibrium thermodynamic approach was used in which it was assumed that the pumping rate depends linearly on how far the reaction is removed from equilibrium. Neither simplification is satisfactory: the limits imposed on a reaction by its thermodynamics cannot be ignored; yet enzyme reactions are, in general, not driven at rates proportional to their driving force, especially when far from equilibrium. The first attempt to incorporate both kinetic and thermodynamic elements into a model for the voltage-dependence of an electrogenic pump was made by Chapman and Johnson (4). A systematic analysis of current-voltage relationships in pump models with carrier-kinetic and thermodynamic constraints has been initiated by Hansen et al. (14).

The actual contribution that an electrogenic pump makes to the conductance of the membrane in which it is embedded is determined by the pump's slope conductance, i.e., the voltage derivative of the pump's current-voltage (I–V) diagram. The task at hand is to construct the curve that describes the magnitude of the electrogenic current as a function of the imposed membrane potential. Experimentally, this task appears to be more likely to be completed soon for plant cells than for animal cells. The reason for this is twofold. Not only is the range of accessible voltages much wider in plant cells, but the pump's conductance is likely to be a sizeable fraction, if not the bulk, of the total membrane conductance. Consequently, it may be possible to construct the plant cell pump's I–V diagram by difference, i.e., from the total membrane I–V diagram before and after pump inhibition. Animal physiologists are not so lucky. First, the reversal potential for the electrogenic sodium pump is almost certainly beyond reach (unless the free energy of ATP hydrolysis is artificially lowered) and, second, the contribution of the pump to the membrane's total conductance is not expected to exceed a few percent. Construction of the electrogenic sodium pump's I–V diagram will thus be a laborious affair, with the actual pump current at each potential being measured specifically, e.g., by means of cardiotonic steroid inhibition. What scant literature there is on the subject (for review see reference 9) suggests that the electrogenic sodium pump of animal cells is little affected by membrane potential in the physiological range.

Future experiments on voltage-sensitive pumps may benefit from a prior analysis of the I–V characteristics of various hypothetical models. Analysis of experimental data in terms of preconceived models can be useful inasmuch as certain classes of models may sometimes be shown to be incompatible with the findings. Unfortunately, the discussion that follows may prove disheartening for those experimenters who hope to make fine distinctions among various classes of electrogenic pump models solely on the basis of their observed I–V characteristics. It appears that even the most reductionist of models allows for a wide variety of I–V characteristics, depending on the values of a few adjustable parameters.

In principle, if a complete kinetic description for a given electrogenic pump were available, with proper rate coefficients for each intermediate step in the reaction cycle, and known voltage sensitivity of these rate coefficients where appropriate, it would be possible to compute the pump's $I-V$ characteristic in a straightforward manner. Even the best-studied pumps, however, are too far from being understood at the required level of refinement even to attempt realistic modeling.

Consider the bare outline for an electrogenic 3:2 sodium:potassium pump model shown in Fig. 2. At least 12 rate coefficients are required to describe the model. It may be reasonable to assume that only the translocating (horizontal) steps will be affected by the transmembrane potential. But, depending on what net charge one chooses to assign to the empty carrier, the net charges translocated in the individual steps could be $+3$ outward and -2 inward, $+2$ outward and $+1$ inward, and $+1$ outward and 0 inward, 0 outward and -1 inward, etc., all of which will have distinct consequences on the voltage sensitivity of the overall reaction. Needless to say, no information on the charge of the empty carrier is available at present. The simplest approximation, then, is to assume that one of the translocating steps (e.g., the sodium transport step) carries a single charge, and the other none. In this scheme, only a single step, translocating a single positive charge, is voltage sensitive, and all others are neutral and voltage-insensitive. But this is probably a gross simplification. For purposes of investigating the effect of membrane potential on the overall pump rate, all voltage-insensitive steps and their rate coefficients will be lumped into a single step with a single pair of rate coefficients. The model of Fig. 2 is thus pared down to its most irreducible form: a two-state model with four rate coefficients, two of them voltage-dependent. (Because of the thermodynamic constraint that the pump flux must vanish at the reversal potential, only three of the four parameters are adjustable.)

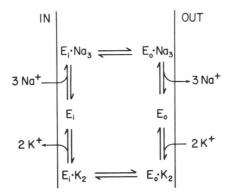

FIG. 2. Outline for a carrier model of the electrogenic Na:K pump. It is assumed that 3 sodium ions are carried from the cell interior to the cell exterior, followed by 2 potassium ions in the opposite direction. Subscripts i and o denote inward- and outward-facing forms of the carrier enzyme E. Not shown are reactions involving phosphorylation (by ATP) and dephosphorylation of the carrier.

The precise physical basis of the voltage sensitivity of a given translocation step is, again, a matter of speculation at this point. Among the possibilities are: Nernst-Planck diffusion regime behavior (10), and migration over an Eyring-type activation energy barrier (17), which may (17) or may not (13) be symmetrically located within the membrane. The various assumptions again predict different shapes for the resulting $I-V$ diagrams, but nothing at this stage justifies assuming anything more complicated (mathematically) than the simplest of models, namely, a single activation-energy barrier that the translocating species must overcome, located halfway through the membrane and additive to the (presumed linear) transmembrane electric field (17).

The next question in the analysis of the two-state, single-symmetric barrier model concerns the identity of the rate-limiting step(s). Somewhere along the reaction cycle illustrated in Fig. 2, the energy of ATP hydrolysis is dissipated. For simplicity one can envision two extreme cases: (a) one in which the free energy is dissipated in the voltage-insensitive reactions, and (b) one in which the free energy is dissipated in the voltage-sensitive translocation. These two extremes are represented schematically in Fig. 3.

In the first case (Fig. 3A), the forward and backward rate coefficients of the voltage-sensitive step are slow (high-energy barrier) and equal (in the absence of a membrane potential, Fig. 3A, left), since no energy is dissipated by hypothesis. Imposition of a membrane field (Fig. 3A, middle) will alter the activation-energy barriers in such a way that the forward rate slows down and the reverse rate is enhanced, causing a reduction of net electrogenic current. Figure 3A, therefore, predicts strong voltage sensitivity in the neighborhood of zero membrane potential. The other extreme is represented in Fig. 3B. Here the slow, rate-limiting step is assumed to be the voltage-insensitive one. If, by hypothesis, energy dissipation occurs in the voltage-sensitive translocation step, the barrier should be low in the forward direction and high in the reverse direction. The reaction rate being limited by the voltage-insensitive step, the overall velocity will be quite insensitive to modest membrane potentials. It will take considerable hyperpolarization before the voltage-sensitive step is slowed down sufficiently to become rate-limiting. The sodium pump is best approximated by this second class. Under physiological circumstances, the sodium-translocating step (presumed here to be charge translocating) is irreversible, as evidenced by the low level of pump-mediated Na:Na exchange.

Typical $I-V$ curves generated by the two extreme models discussed above are illustrated in Fig. 4. (The curves were constrained to have a reversal potential at -200 mV.) The curve representing the model of Fig. 3A shows strong voltage-sensitivity in the neighborhood of zero membrane potential. It also illustrates an experimental fallacy. Although this is a pump with large EMF and large internal chord resistance, it does *not* behave as a constant current generator, as simplistic reasoning might have suggested. (The explanation is that, for kinetic reasons, the pump's internal resistance, although large, is voltage-sensitive.) Curve A also illustrates an experimental nightmare. An investigator trying to

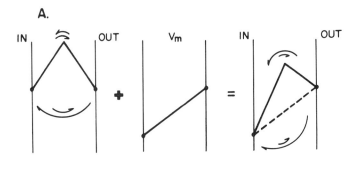

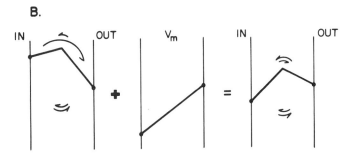

FIG. 3. Two extreme two-state models for the voltage sensitivity of a hypothetical sodium pump. **A:** Model with rate-limiting, voltage-sensitive step and energy dissipation in the voltage-insensitive step. **B:** Model with rate-limiting, voltage-insensitive step and energy dissipation in the voltage-sensitive step. The voltage-sensitive steps are depicted as transitions over a triangular, symmetrically located Eyring-type activation energy barrier, the shape of which is modified by linear addition of the (assumed linear) membrane field. Further description in the text.

determine the pump's reversal potential (and, hence, thermodynamic driving force, stoichiometry, etc.) might easily be misled into believing that the extrapolated reversal potential is much less negative than it really is. Experimental determination of reversal potential in a real pump displaying this type of I–V behavior would be very difficult.

Curve B in Fig. 4 illustrates how, at the other extreme, a pump can remain essentially unaffected by membrane potential except in the immediate neighborhood of reversal potential. Such pumps, over a wide voltage range, can be adequately represented as constant-current sources (Fig. 1B). Curve C represents the irreversible-thermodynamics model (21), where the pumping rate is assumed to be proportional to the distance from reversal potential. It is highly unlikely that any mechanistic model will display such linearity over such wide ranges of membrane potential.

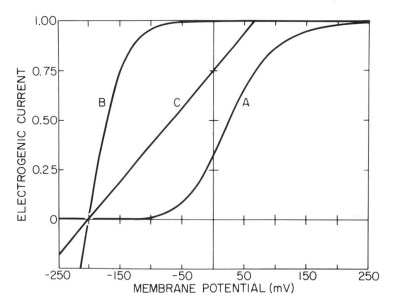

FIG. 4. Current–voltage diagrams for three models of electrogenic pump. Current is in arbitrary units, and all three models are constrained to reverse at −200 mV. **A:** Model of Fig. 3A, where the voltage-sensitive step is rate limiting. **B:** Model of Fig. 3B, where the voltage-insensitive step is rate limiting. **C:** Nonequilibrium thermodynamic model, where current is proportional to driving force.

A final question to be addressed is, can the I–V diagram of an electrogenic pump have domains of negative conductance where, as V_m approaches EMF, the electrogenic current *increases* contrary to simple thermodynamic expectations? Perhaps the simplest model in which such a behavior can be realized is a combination of the 2-state model of Fig. 3, with a voltage-sensitive gate. The gate may be literal, in the sense of a voltage-sensitive access path to the pump, or simply result from the voltage-dependent partitioning of pump molecules between an active and an inactive state.

Figure 5 illustrates schematically the behavior of a hypothetical gated electrogenic pump. The kinetics of the pump itself are assumed to be of the kind illustrated in Fig. 3B and in Fig. 4, curve B; the gate is assumed to close with depolarization. Their combination leads to a region of negative conductance in the I–V diagram of the pump. Other, more complicated models include (12) pump cycles with more than one charge-translocating (hence, voltage-sensitive) step.

CONCLUSION

Both the sodium pump of animal cells and the proton pump of plant cells have very negative reversal potentials; in the case of the sodium pump, this

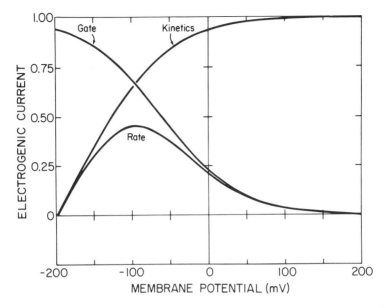

FIG. 5. Current–voltage diagram (labeled Rate) for a voltage-gated electrogenic pump. Pump kinetics are assumed to be of the kind described in Fig. 3B, i.e., where the rate-limiting step is the voltage-insensitive one. Access to the pump, or availability of the pump, is assumed to be controlled by membrane potential, in a fashion described by the curve labeled Gate, i.e., access or availability is higher at more negative potentials.

reversal potential is almost certainly beyond the dielectric breakdown potential of animal cells. Determination of reversal potential, where possible, should in principle allow calculation of the pump's stoichiometry, provided the free energy of ATP hydrolysis is known; however, this stoichiometry may only be valid at reversal and not necessarily beyond.

Sodium pump conductances, unlike plant proton pump conductances, are generally very small compared to total membrane conductance. Thus, it may be technically feasible to construct a plant cell proton pump's I–V diagram simply by difference (i.e., with and without the pump operating), but that approach is unlikely to succeed for the sodium pump. In any case, the I–V diagrams of certain electrogenic pumps may be characterized by a shallow dependence on membrane potential as it approaches reversal potential (curve B in Fig. 4) making it, in practice, extremely difficult or impossible to determine EMF experimentally. *A priori* calculation of a given pump's I–V diagram will not be feasible in the near future, since entirely too little is known about the kinetics and thermodynamics of the manifold intermediate steps of pump cycles. Nor, unfortunately, will it be possible to extract much detailed kinetic information about these steps from experimentally established I–V diagrams (if the only measured variable is net current). Even the most stripped-down of models can generate I–V curves that run the gamut from those that remain essentially unaffected by membrane potential (true rheogenic pumps), to very voltage-sensitive ones that plunge down rapidly to very low rates and then hover near the baseline over very wide ranges of membrane potential, to those that display negative-conductance domains in seeming violation of thermodynamic expectations. Because the sodium pump allows many more kinetically relevant measurements (i.e., one-way label fluxes of Na^+ and K^+, as well as net charge transfer), a thorough investigation of the effect of membrane potential on the behavior of this pump is more likely to yield useful kinetic information. At any rate, it is clear that further progress in the understanding of electrogenic pumps will be made at the laboratory bench rather than at the computer terminal.

ACKNOWLEDGMENT

The author's work was supported by NIH grant No. NS 11223.

REFERENCES

1. Abercrombie, R. F., and De Weer, P. (1978): Electric current generated by squid giant axon sodium pump: External K and internal ADP effects. *Am. J. Physiol.,* 235:C63–C68.
2. Akerboom, T. P. M., Bookelman, H., Zuurendonk, P. F., van der Meer, R., and Tager, J. M. (1978): Intramitochondrial and extramitochondrial concentrations of adenine nucleotides and inorganic phosphate in isolated hepatocytes from fasted rats. *Eur. J. Biochem.,* 84:413–420.
3. Caldwell, P. C., and Schirmer, H. (1965): The free energy available to the sodium pump of squid giant axons and changes in the sodium efflux on removal of the extracellular potassium. *J. Physiol. (Lond.),* 181:25P.
4. Chapman, J. B., and Johnson, E. A. (1976): Current-voltage relationships for theoretical electrogenic sodium pump models. *Proc. Austr. Physiol. Pharmacol. Soc.,* 7:69P.
5. Chapman, J. B., and Johnson, E. A. (1978): The reversal potential for an electrogenic sodium pump. A method for determining the free energy of ATP breakdown? *J. Gen. Physiol.,* 72:403–408.
6. Cockrell, R. S., Harris, E. J., and Pressman, B. C. (1966): Energetics of potassium transport in mitochondria induced by valinomycin. *Biochemistry,* 5:2326–2335.
7. Dawson, J. J., Gadian, D. G., and Wilkie, D. R. (1980): Mechanical relaxation rate and metabolism studied in fatiguing muscle by phosphorus nuclear magnetic resonance. *J. Physiol. (Lond),* 299:465–484.
8. Dean, R. B. (1941): Theories of electrolyte equilibrium in muscle. *Biol. Symp.,* 3:331–348.
9. De Weer, P. (1975): Aspects of the recovery processes in nerve. In: *Neurophysiology, vol. 3,* edited by C. C. Hunt, pp. 231–278. Butterworths, London.
10. Finkelstein, A. (1964): Carrier model for active transport of ions across a mosaic membrane. *Biophys. J.,* 4:421–440.
11. Gradmann, D., Hansen, U.-P., Long, W. S., Slayman, C. L., and Warncke, J. (1978): Current-voltage relationships for the plasma membrane and its principal electrogenic pump in *Neurospora crassa:* I. Steady-state conditions. *J. Membr. Biol.,* 39:333–367.
12. Gradmann, D., Hansen, U.-P., and Slayman, C. L. (1982): Reaction-kinetic analysis of current-voltage relationships for electrogenic pumps in *Neurospora* and *Acetabularia.* In: *Electrogenic Ion Pumps, Current Topics in Membranes and Transport, vol. 16,* edited by C. L. Slayman, pp. 257–276. Academic Press, New York.
13. Hall, J. E., Mead, C. A., and Szabo, G. (1973): A barrier model for current flow in lipid bilayer membranes. *J. Membr. Biol.,* 11:75–97.
14. Hansen, U.-P., Gradmann, D., Sanders, D., and Slayman, C. L. (1981): Interpretation of current-voltage relationships for "active" ion transport systems: I. Steady-state reaction-kinetic analysis of class-I mechanisms. *J. Membr. Biol.,* 63:165–190.

15. Keifer, D. W., and Spanswick, R. M. (1978): Activity of the electrogenic pump in *Chara corallina* as inferred from measurements of the membrane potential, conductance, and potassium permeability. *Plant Physiol.,* 62:653–661.
16. Kushmerick, M. J. (1969): Free energy and enthalpy of ATP hydrolysis in the sarcoplasm. *Proc. R. Soc. Lond. [Biol.],* 174:348–353.
17. Laüger, P., and Stark, G. (1970): Kinetics of carrier-mediated ion transport across lipid bilayer membranes. *Biochim. Biophys. Acta,* 211:458–466.
18. Meyer, R. A., Kushmerick, M. J., and Brown, T. R. (1982): Application of ^{31}P-NMR spectroscopy to the study of striated muscle metabolism. *Am. J. Physiol.,* 242:C1–C11.
19. Mullins, L. J., and Noda, K. (1963): The influence of sodium-free solutions on the membrane potential of frog muscle fibers. *J. Gen. Physiol.,* 47:117–139.
20. Nishiki, K., Erecinska, M., and Wilson, D. F. (1978): Energy relationships between cytosolic metabolism and mitochondrial respiration in rat heart. *Am. J. Physiol.,* 234:C73–C81.
21. Rapoport, S. I. (1970): The sodium-potassium exchange pump: Relation of metabolism to electrical properties of the cell. I. Theory. *Biophys. J.,* 10:246–259.
22. Slayman, C. L. (1982): *Electrogenic Ion Pumps.* (*Current Topics in Membranes and Transport,* vol. 16.) Academic Press, New York.
23. Slayman, C. L., Lang, W. S., and Lu, C. Y.-H. (1973): The relationship between ATP and an electrogenic pump in the plasma membrane of *Neurospora crassa. J. Membr. Biol.,* 14:305–338.
24. Spanswick, R. M. (1972): Evidence for an electrogenic ion pump in *Nitella translucens.* I. The effects of pH, K^+, Na^+, light and temperature on the membrane potential and resistance. *Biochim. Biophys. Acta,* 288:73–89.
25. Thomas, R. C. (1972): Electrogenic sodium pump in nerve and muscle cells. *Physiol. Rev.* 52:563–594.
26. Veech, R. L., Lawson, J. W. R., Cornell, N. W., and Krebs, H. A. (1979): Cytosolic phosphorylation potential. *J. Biol. Chem.,* 254:6538–6547.

Electrogenic Transport: Fundamental Principles and Physiological Implications,
edited by Mordecai P. Blaustein and Melvyn Lieberman. Raven Press, New York © 1984.

Thermodynamics and Kinetics of Electrogenic Pumps

J. Brian Chapman

Department of Physiology, Monash University, Clayton, Victoria 3168, Australia

Electrogenic pumps are membrane-located enzymes that catalyze vectorial metabolic reactions in which some of the chemical reactants differ from their respective products only with regard to their location and electrochemical potential, not with regard to their chemical nature or bonding. For example, the electrogenic sodium pump (40) is the plasmalemmal Na,K-ATPase (36) that couples the extrusion of three intracellular sodium ions (Na^+) and uptake of two extracellular potassium ions (K^+) to the splitting of adenosine triphosphate (ATP) according to the following chemiosmotic reaction:

$$ATP + 3\,Na^+_{in} + 2\,K^+_{out} \rightleftharpoons ADP + P_i + 3\,Na^+_{out} + 2\,K^+_{in} \qquad (1)$$

where Na^+_{in}, K^+_{in}, Na^+_{out}, and K^+_{out} are the intracellular and extracellular forms of Na and K ions, respectively, and ADP and P_i are adenosine diphosphate and inorganic phosphate, respectively (5).

The electrogenic property of reaction 1 derives from the fact that the number of Na ions extruded exceeds the number of K ions taken up. Thus, the stoichiometry of reaction 1 determines that an outward transmembrane electric current is an obligatory component of metabolism. As a consequence, the transmembrane voltage difference, or membrane potential, is an important thermodynamic variable influencing the kinetics of the pump reaction. In particular, the equilibrium constant of reaction 1 depends on the work of extruding net charge against the membrane potential as well as on the standard free energy ATP splitting.

It is generally believed, however, that reaction 1 proceeds according to a complex reaction mechanism consisting of several elementary steps. Of the various schemes proposed the mechanism summarized by Guidotti (13) for the plasmalemmal Na,K-ATPase will be used here to illustrate some general features of thermodynamic and kinetic properties of electrogenic pumps. The choice of the Na,K-ATPase relates to its immediate applicability to the electrophysiology of continuously active tissue such as cardiac muscle (7,22,27), whereas the guiding chemiosmotic principles of vectorial metabolism were developed by Mitchell (31–33) in relation to the electrogenic proton pumps of mitochondria and chloroplasts.

THE MODEL

The sequential reaction mechanism shown in Fig. 1 is a six-step vectorial adaptation of the scheme published by Guidotti (13). Because the enzyme is normally surrounded under physiological conditions by sufficient reactants and products to reduce the quantity of free enzyme to negligible proportions, the scheme is taken to begin by reaction between Na_{in}^+ and energized enzyme (step 1), followed by release of ADP (step 2). The Na^+ extrusion is completed by a conformational change (translocation) and dissociation of the resulting complex (step 3). The K_{out}^+ then dephosphorylates the enzyme (step 4), after which a further translocation occurs in which the enzyme is reenergized (step 5). The K^+ uptake is completed by dissociation of the resulting complex (step 6), which restores the energized form of the enzyme for further reaction with Na_{in}^+ at step 1. Because the overall reaction is voltage dependent, it follows that at least one of the six elementary steps must have a voltage-dependent equilibrium constant, and it seems natural to suppose that the voltage dependence is most likely to occur on one or both of the translocational steps, since it is there that movement of charge may occur.

However, the effect of membrane potential on a voltage-dependent elementary step is to multiply the unidirectional rate coefficients by exponential functions of voltage according to the Butler-Volmer equation governing charge transfer (1). This means that the arbitrary choice of which step or steps are voltage-dependent can be deferred until reasonable chemical reactivity coefficients have been obtained to simulate the nonvectorial behavior of the isolated enzyme where voltage does not arise as a kinetic or thermodynamic factor. Table 1 lists the values for the unidirectional rate coefficients of each step in the scheme of Fig. 1. These coefficients were arrived at in collaborative studies with E. A.

$$E\ ATP + 3Na_{in} \underset{b_1}{\overset{f_1}{\rightleftharpoons}} Na_3E\ ATP$$

$$Na_3E\ ATP \underset{b_2}{\overset{f_2}{\rightleftharpoons}} Na_3E\cdot P + ADP$$

$$Na_3E\cdot P \underset{b_3}{\overset{f_3}{\rightleftharpoons}} E'P + 3Na_{out}$$

$$E'P + 2K_{out} \underset{b_4}{\overset{f_4}{\rightleftharpoons}} K_2E' + P_i$$

$$K_2E' + ATP \underset{b_5}{\overset{f_5}{\rightleftharpoons}} K_2E\ ATP$$

$$K_2E\ ATP \underset{b_6}{\overset{f_6}{\rightleftharpoons}} E\ ATP + 2K_{in}$$

FIG. 1. Reaction scheme for the vectorial Na,K-ATPase of the electrogenic sodium pump adapted from the mechanism summarized by Guidotti (13). E and E' refer to conformationally (translocationally) distinct forms of the enzyme, P_i = inorganic phosphate, ATP and ADP are adenosine tri- and diphosphate, respectively; Na_{in}, Na_{out}, K_{in}, K_{out} refer to intracellular and extracellular Na^+ and K^+, respectively; f_i, b_i are the forward and reverse rate coefficients for step i as listed in Table 1.

TABLE 1. *Reactivity coefficients for the six steps of the scheme of Fig. 1*

Coefficient	Value	Units
f_1	2.5×10^{11}	liter^3 mol^{-3}sec^{-1}
b_1	10^5	sec^{-1}
f_2	10^4	sec^{-1}
b_2	10^5	liter mol^{-1}sec^{-1}
f_3	172	sec^{-1}
b_3	1.72×10^4	liter^3 mol^{-3}sec^{-1}
f_4	1.5×10^7	liter^2 mol^{-2}sec^{-1}
b_4	2×10^5	liter mol^{-1}sec^{-1}
f_5	2×10^6	liter mol^{-1}sec^{-1}
b_5	30	sec^{-1}
f_6	1.15×10^4	sec^{-1}
b_6	6×10^8	liter^2 mol^{-2}sec^{-1}

Johnson and J. M. Kootsey (*unpublished observations;* see also reference 6) in which the following criteria were met:

(a) The ionic activation curves for nonvectorial ATPase activity had to follow the same general shape as those reported by Skou (36,37) for the enzyme extracted from crab nerve.

(b) The maximum nonvectorial molecular turnover rate had to be no more than 167 sec^{-1} (23).

(c) The pseudo-first-order rate coefficients had to be no more than 10^4 sec^{-1} for translocational steps (protein conformational changes) or steps involving group transfer by covalent bonds, whereas the pseudo-second-order rate coefficients for diffusion-limited binding steps were not to exceed 1×10^8 liters mol^{-1} sec^{-1} (15; J. A. Reynolds and C. Tanford, *personal communication*).

(d) The standard free energy of ATP splitting was taken as -31.9 kJ/mol (17,42), resulting in an equilibrium constant (at zero voltage) for the overall reaction of 238,722.0 mol/liter.

Because the original development of this model was directed toward exploring the role of electrogenic transport in cardiac muscle (6; see also references 7,22,27), the choice of rate coefficients was further constrained by the need to satisfy two criteria applying to the vectorial situation where the enzyme is located in the cardiac sarcolemma. First, the resting electrogenic pump current for a quiescent cardiac cell had to be between 0.1 and 0.2 μA/cm^2 with a capability of rising to values around 1 μA/cm^2 following repetitive electrical activity (22). Second, the slope conductance of the pump current–voltage (I–V) relation had to be small in the physiological range (5,7).

METHODS

Simple mass action rate equations were used for each forward and reverse reaction of the six elementary steps of Fig. 1. The control concentrations of the enzyme (9,30), ions, and adenine nucleotides (17) were chosen as appropriate for mammalian cardiac muscle and are listed in Table 2. For the dimensions of the rate coefficients listed in Table 1, the reaction rates have the units of mol cm^{-2} sec^{-1}. Given that each mole of overall reaction results in the extrusion of one Faraday of charge, transport rates in the present work are all expressed as electrogenic pump current per unit area of membrane; 1 μA/cm^2 of pump current corresponds to 82.9 molecular turnovers per second for the ATPase activity.

In the case of voltage-dependent steps, the respective rate coefficients were multiplied by the appropriate exponential functions of membrane potential as given by the Butler-Volmer equation using a symmetry factor of 0.5 (1,10).

Steady-state solutions for the concentrations of enzyme intermediates were found by gaussian elimination and substituted into the appropriate unidirectional rate equations for each step in the reaction sequence. All equations were solved on a digital computer with results displayed graphically.

KINETIC INFLUENCES OF VOLTAGE

Location of the Voltage-Dependent Step

If all of the thermodynamically required voltage dependence is assigned to the translocational step 3, this implies that step 3 is a univalent outward charge movement requiring simultaneously either symport of two anions or antiport of two cations (e.g., protons) to balance the three Na$^+$ extruded. Symport of anions might be achieved by translocation of negatively charged binding sites on the enzyme across the voltage gradient. This further requires that the two balancing charges be returned electroneutrally at step 5 to balance the two K$^+$ taken up. The corresponding I–V relation is shown in Fig. 2 (curve A).

If all of the voltage dependence is assigned to the translocational step 5, then this implies that step 5 involves a univalent outward charge movement requiring simultaneously either symport of three anions or antiport of three

TABLE 2. *Chemical boundary conditions for simulation of electrogenic sodium pump in cardiac muscle, including adenine nucleotide levels (17) and total Na,K-ATPase concentration per unit area of membrane (9,30)*

Parameter	Value	Units
Total ATPase	1.25 × 10^{-13}	mol/cm^2
[Na]$_{in}$	9.6	mmol/liter
[Na]$_{out}$	140	mmol/liter
[K]$_{in}$	150.4	mmol/liter
[K]$_{out}$	5.4	mmol/liter
[ATP]	4.99	mmol/liter
[P$_i$]	4.95	mmol/liter
[ADP]	0.06	mmol/liter
Temperature	310	°K

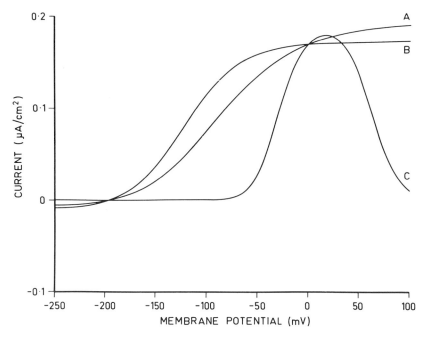

FIG. 2. Pump current–voltage relations for the scheme of Fig. 1 for different locations of the voltage-dependent step(s). **A:** Voltage dependence on step 3; $f_3 = 172 \exp(FV/2RT)$ sec^{-1}, $b_3 = 1.72 \times 10^4 \exp(-FV/2RT)$ liter^3 mol^{-3}sec^{-1}. **B:** Voltage dependence on step 5; $f_5 = 2 \times 10^6 \exp(FV/2RT)$ liter mol^{-1}sec^{-1}, $b_5 = 724 \exp(-FV/2RT)$ sec^{-1}. **C:** Voltage dependence on both steps 3 and 5; $f_3 = 172 \exp(3FV/2RT)$ sec^{-1}, $b_3 = 1.72 \times 10^4 \exp(-3FV/2RT)$ liter^3 mol^{-3} sec^{-1}, $f_5 = 2.10^6 \exp(FV/RT)$ liter mol^{-1}sec^{-1}, $b_5 = 724 \exp(-FV/RT)$ sec^{-1}. All other rate coefficients were as specified in Table 1. All chemical boundary conditions were as listed in Table 2.

cations to balance the two K$^+$ taken up. This also requires that the three balancing charges be moved in the opposite direction at step 3 to balance the three Na$^+$ extruded at that translocational step. The resulting I–V relation is shown as curve B in Fig. 2.

The third possibility considered here is that the three Na$^+$ translocated at step 3 and the two K$^+$ translocated at step 5 are totally unbalanced at each step, requiring the outward movement of three charges at step 3 and the inward movement of two charges at step 5. The resulting I–V relation is shown as curve C in Fig. 2.

It is clear from these results that voltage is a very strong inhibitor of active transport on either side of zero membrane potential if the voltage dependence is assigned fully to all of the sequential ionic translocations. Since electrogenic pumps, including the Na,K-ATPase, are generally thought to behave as constant current sources over most of the physiological range of membrane potentials (7,19,20,38), it seems more likely that voltage dependence is restricted to a single step according to curves A or B. Although in other collaborative studies

(6) we have chosen arbitrarily to assign all of the voltage dependence to step 5 (curve B of Fig. 2), whichever choice was made for a single voltage-dependent step would necessarily imply that all of the membrane potential appears across the reaction distance associated with that step. Therefore, the choice of the electrical symmetry factor of 0.5 in the Butler-Volmer equation (see The Model) is appropriate (1,10) and need not be justified on the basis of the supposed spatial location of the voltage-dependent step in relation to some assumed continuum distribution of the electric field within the membrane, as is commonly done for carrier-mediated diffusion (14,21,28).

The necessary implication that all of the membrane potential appears across the reaction distance of a single step is similar to the concepts of proton wells and potential wells in Mitchell's chemiosmotic mechanisms for the generation and consumption of proticity across mitochondrial and other biological membranes (31–33).

Influence of Voltage-Dependent Step on Ionic Activation

The choice of voltage-dependent step or steps has marked effect on the ionic activation curves for electrogenic transport in the presence of voltage. Figure 3 shows plots of electrogenic pump current activated by Na^+_{in} for voltage dependence at step 3 (A), step 5 (B), and both steps 3 and 5 (C) at a membrane potential of -85 mV. The broken line shows the ionic activation curve obtained at zero membrane potential and is, of course, identical for all three choices of voltage-dependent steps. Curves A and B are similar, showing a sharply rising activation at low $[Na^+]_{in}$ values, becoming more linear between 10 and 20 mM (11) and saturating at higher values (26,35). However, with voltage dependence assigned to both steps 3 (trivalent) and 5 (divalent), the pump is essentially incapable of being activated by Na^+_{in} at a membrane potential of -85 mV.

Influence of Voltage-Dependent Step on Hyperpolarizing Capability

The curves shown in Fig. 3 apply to ionic activation at fixed values of the membrane potential. In the physiological situation any stimulation of electrogenic transport by raised intracellular $[Na^+]$ results in a hyperpolarization of the membrane potential (40,41). An accurate prediction of the physiological hyperpolarizations resulting from raised $[Na^+]_{in}$ for the present model would require an explicit electrochemical model for the entire array of electrodiffusive processes occurring in the membrane (7). Nevertheless, an approximate set of predictions can be made by using the Gauss-Seidel method of successive displacements to obtain numerical solutions. Accordingly, the effect was estimated by allowing the return of the outward electrogenic pump current as an inward passive current through a membrane resistance (R) appropriate for cardiac muscle of 20 k$\Omega \cdot$cm^2 (29,34). Solutions were obtained by an iterative process

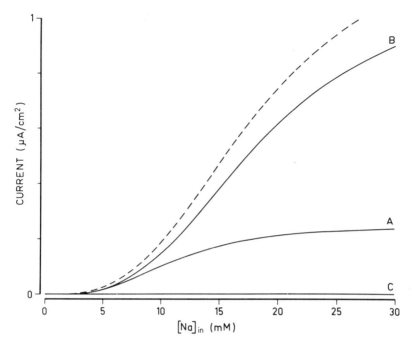

FIG. 3. Activation of vectorial ATPase activity (electrogenic pump current) by intracellular Na^+ at membrane potentials of zero (*broken line*) and −85 mV (*solid lines*) for voltage dependence on step 3 **(A)**, step 5 **(B)**, and both steps 3 and 5 **(C)** of the scheme of Fig. 1. All rate coefficients were as specified in Fig. 2 and Table 1. All other chemical boundary conditions were as listed in Table 2.

in which the membrane potential was first set at −85 mV and the electrogenic pump current (I μA/cm^2) was calculated for a particular value of $[Na^+]_{in}$. The voltage was then set at $-(85 + I \cdot R)$ mV, and a new solution for pump current was calculated. The process was reiterated until successive estimates of the hyperpolarization at a given value of $[Na^+]_{in}$ agreed within 1 μV.

The resulting hyperpolarizations are illustrated in Fig. 4 for the three possible voltage dependences considered. Note that curve A shows less hyperpolarizing capability than curve B because of the correspondingly greater inhibition by voltage in the physiological range (cf. Figs. 2A and 3A), whereas curve C (trivalent Na^+ with divalent K^+ movement) reflects inability to produce any of the significant hyperpolarizations commonly observed under physiological conditions in response to intracellular sodium loads. It is important to note that any attempt to adjust the rate coefficients of Table 1 to allow the mechanism of curve C to produce realistic hyperpolarizing responses to raised $[Na^+]_{in}$ would result in unacceptably high maximal turnover rates for the enzyme in the nonvectorial situation.

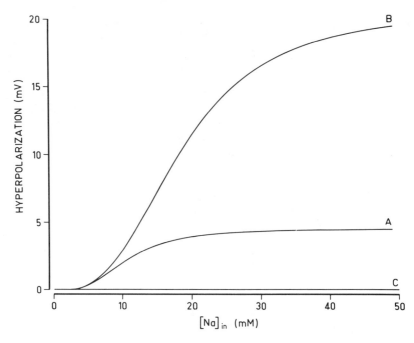

FIG. 4. Hyperpolarization produced by the reaction scheme of Fig. 1 in response to activation by raised intracellular [Na$^+$] (accompanied by equimolar depletion of intracellular [K$^+$]) for voltage dependence on step 3 **(A)**, step 5 **(B)**, and both steps 3 and 5 **(C)**, assuming a passive membrane resistance of 20 k$\Omega \cdot$cm^2. All rate coefficients were as specified in Fig. 2 and Table 1. All other chemical boundary conditions were as listed in Table 2.

Voltage-Dependent Reversal of Electrogenic Transport

It has been shown previously (5) that there must exist a membrane potential that, for a given set of chemical boundary conditions, corresponds to the thermodynamic equilibrium or reversal potential beyond which reaction 1 will proceed backward, coupling the synthesis of ATP to the extrusion of K$^+$ and uptake of Na$^+$. However, the ease of experimental demonstrability of voltage-dependent reversal will be limited by two factors. First, the thermodynamic reversal potential may lie beyond the experimentally feasible range of hyperpolarizations; the chemical boundary conditions specified in Table 2 determine a reversal potential of -198 mV. Second, the relative magnitude of inward electrogenic pump current beyond the reversal potential might be quite small relative to the outward currents observed in the physiological range.

Slayman and co-workers (12,16) have explored the shapes of I–V relations for a range of models and have reported behavior in some cases similar to that illustrated in Fig. 2A and B. Variation of the rate constants for the six steps of the scheme of Fig. 1 has yielded results in general agreement with

Slayman's findings in that there is an inverse relation between the relative ease of reversibility of the pump beyond the reversal potential and the degree of slope conductance observed in the physiological range of membrane potentials.

The Maximum Slope Conductance of the Electrogenic Sodium Pump

Figure 5 shows a family of steady-state I-V relations for the scheme of Fig. 1 obtained at various fixed values of $[Na^+]_{in}$. The thick line is the control curve appropriate for 9.6 mM $[Na^+]_{in}$ and shows very little slope conductance in the physiological range of membrane potentials. However, raising the intracellular $[Na^+]$ clearly increases the relative slope conductance in this range and, at 20 mM $[Na^+]_{in}$, the I-V relation has a slope conductance > 5 μS/cm², corresponding to a resistance of about 190 kΩ·cm² at around -100 mV. Given that the constraint of maximum turnover number for the enzyme is 167 sec^{-1} (see The Model), the maximum possible membrane current capable of being generated by the present model is 2.01 μA/cm². This places an upper limit of about 13 μS/cm² on the possible contribution of the electrogenic sodium pump to the diastolic slope conductance of 50 μS/cm² in cardiac muscle. It is hard to imagine from the present simulations, however, under what conditions that might occur;

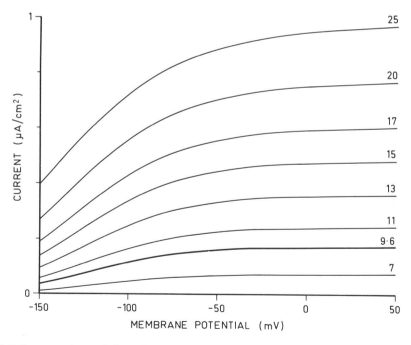

FIG. 5. Current–voltage relations of the scheme of Fig. 1 at different intracellular Na$^+$ concentrations for voltage dependence on step 5. All rate coefficients and other chemical boundary conditions were as specified in Fig. 2B and Tables 1 and 2.

and indeed, under normal physiological conditions, the contribution will be much less, amounting to less than 0.5 $\mu S/cm^2$ or 1% of the diastolic slope conductance of cardiac muscle.

IONIC FLUXES, TRANSPORT RATES, AND THERMODYNAMIC BOUNDARY CONDITIONS

A kinetic expression of the Second Law of Thermodynamics relates the free energy dissipated per mole of reaction (ΔG_{dis}) to the ratios of the forward and reverse unidirectional reaction rates under isothermal conditions:

$$\Delta G_{dis} = -RT \times \ln(r_f/r_b) \tag{2}$$

where r_f and r_b are the forward and reverse rates of reaction, respectively, and R and T have their usual meanings (2,8).

Equation 2 is straightforward if the reaction is an elementary process but becomes experimentally difficult to apply to complex reactions such as the six-step scheme of Fig. 1. The difficulty derives from defining an experimentally measurable unidirectional rate of overall reaction (25). Boudart (2) has shown that Eq. 2 is generally true for single, unbranched chemical reactions (such as the scheme of Fig. 1) in the steady state. However, inspection of the scheme of Fig. 1 reveals that there is no flux of ions or intermediates that would serve as an experimental measure of the unidirectional rates of overall reaction which, following Boudart's definition (2), would be given by:

$$r_f = rf_1 \cdot rf_2 \cdot rf_3 \cdot rf_4 \cdot rf_5 \cdot rf_6/D$$
$$r_b = rb_1 \cdot rb_2 \cdot rb_3 \cdot rb_4 \cdot rb_5 \cdot rb_6/D$$

where

$$D = rf_2 \cdot rf_3 \cdot rf_4 \cdot rf_5 \cdot rf_6 + rb_1 \cdot rf_3 \cdot rf_4 \cdot rf_5 \cdot rf_6$$
$$+ rb_1 \cdot rb_2 \cdot rf_4 \cdot rf_5 \cdot rf_6 + rb_1 \cdot rb_2 \cdot rb_3 \cdot rf_5 \cdot rf_6$$
$$+ rb_1 \cdot rb_2 \cdot rb_3 \cdot rb_4 \cdot rf_6 + rb_1 \cdot rb_2 \cdot rb_3 \cdot rb_4 \cdot rb_5$$

and the rf_i, rb_i are the unidirectional mass action rates of the ith step of the scheme of Fig. 1.

According to these definitions r_f would correspond to the experimentally measurable exit of tracer as an exclusive product of step 6, having entered as an exclusive reactant at step 1. An entry or exit of the same tracer at any intermediate steps other than the first and last, respectively, would nullify the tracer's ability to serve as a measure of the overall unidirectional rates of reaction. Clearly, no such tracer can be introduced into the scheme of Fig. 1, because the Na ions traverse only the first three steps, whereas the K ions traverse only the last three steps. The enzyme itself cannot serve as a tracer because all intermediate forms of the enzyme are capable of being generated by two pathways (forward and reverse).

Consequently, unidirectional ion fluxes cannot be expected to serve as measures of unidirectional reaction rates related to thermodynamic boundary conditions as had been suggested previously (3). Moreover, unidirectional ion fluxes can even be apparently uncorrelated with net ion transport as shown by the results of voltage-dependent variation of Na^+ and K^+ fluxes for the scheme of Fig. 1 illustrated in Fig. 6. Although there is a good correlation between pump-mediated unidirectional Na^+ efflux and net Na^+ extrusion, both the pump-mediated unidirectional K^+ efflux and K^+ influx show a continuous decline over a range of voltages where net K^+ uptake is actually increasing.

The net extrusion of Na^+ shown in Fig. 6 is always 1.5 times the net uptake of K^+ as determined by the stoichiometry of reaction 1. On the other hand, there is no particular meaning to the ratio of pump-mediated unidirectional Na^+ efflux to unidirectional K^+ influx. Therefore, it would be incorrect to attribute a variation in the stoichiometric coupling ratio for net Na^+/K^+ antiport to any observed variation in unidirectional flux ratios (4).

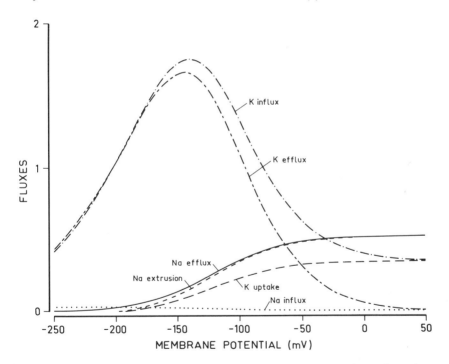

FIG. 6. Unidirectional (isotopic) and net fluxes of Na^+ and K^+ as functions of membrane potential. All rate coefficients and chemical boundary conditions were as specified in Fig. 2B and Tables 1 and 2. Fluxes were calculated by using the method of Boudart (2) to define unidirectional rates of reaction across the first three steps of the scheme of Fig. 1 for Na^+ movements and across the last three steps for K^+ movements (see ref. 4 for fuller details of this method). Only first-order interactions between tracer and enzyme were included in the calculations. Fluxes expressed in electrical units of $\mu A/cm^2$.

THERMODYNAMIC FORCES AND REACTION FLUXES

The modern discipline of irreversible thermodynamics has been frequently applied to biological transport problems. In such treatments it is customary to postulate linear relations between flows of matter and the thermodynamic forces producing such flows, the linearity being thought to apply provided the processes are not occurring far from equilibrium (e.g., 18,24). Although, in this approach, thermodynamic force is usually defined in terms of free energy difference divided by the absolute temperature, it is clear that for an isothermal process the linear relation must also be presumed to hold between rate of reaction and the molar free-energy dissipation. This approach is held to be useful in cases where little is known about the reaction mechanism; the assumptions and predictions are generally thought to be independent of mechanism.

For a reaction mechanism as well defined stoichiometrically and kinetically as the scheme of Fig. 1, the approach of irreversible thermodynamics to establish phenomenological coefficients linking ionic movements to the thermodynamic force producing hydrolysis of ATP would serve no obvious purpose. Indeed, as will become apparent, the approach would be inherently unsound because reaction (1) and the scheme of Fig. 1 both define a single chemiosmotic reaction with a readily calculable thermodynamic force driving net reaction. The coupling of Na^+ and K^+ movements to ATP splitting is kinetically and stoichiometrically fixed and cannot be described in terms of the looser concepts of thermodynamic coupling (25).

Nevertheless, it is of interest to explore the extent to which the linear assumptions of irreversible thermodynamics apply to the relation between the net rate of reaction (1) and the isothermal thermodynamic force driving net reaction. Figure 7 illustrates a number of steady-state relations between net electrogenic pump current and the molar free-energy dissipation of the overall process. For each curve, the free energy dissipated is the difference between the free energy available from ATP splitting and the free energy conserved in the electrochemical work of ion translocation. This quantity is identical to the sum of the steady-state free-energy dissipations of each of the six steps of the scheme of Fig. 1 given by Eq. 2. The curves shown in Fig. 7 were obtained by varying the thermodynamic force in a variety of ways described in the figure legend, i.e., by varying the membrane potential or the concentrations of the ions or adenine nucleotides.

It is clear that all of these curves are highly nonlinear and that most of them have their quasilinear portions (points of inflection in the enzymatic working range) far from equilibrium (zero dissipation). Even the slopes (phenomenological coefficients) of the different curves near to equilibrium are different (see Fig. 7, inset), depending on how the affinity is varied. Thus, there is no such thing, either near to or far from equilibrium, as a unique phenomenological coefficient acting as a proportionality constant between flow of reaction and the thermodynamic force. This observation, although illustrated for the particu-

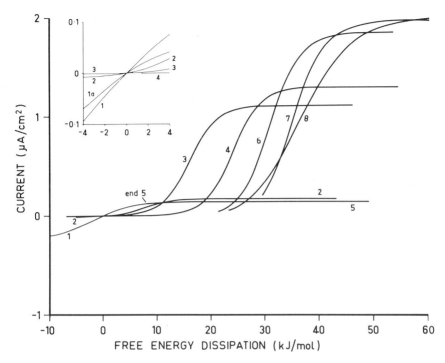

FIG. 7. Relations between electrogenic pump current (net vectorial ATPase reaction) and thermodynamic force (molar free-energy dissipation) varied in several ways. **1:** [ATP] varied from 0.01 to 5.00 mM at −85 mV membrane potential with [ADP] + [ATP] constrained to sum to 5.05 mM and [ATP] + [P_i] constrained to sum to 9.94 mM. **1a:** As for curve 1 but at zero membrane potential. **2:** Membrane potential varied from −250 to +250 mV. **3:** [Na^+]$_{in}$ varied from 1 to 150 mM at −85 mV membrane potential with [Na^+]$_{in}$ + [K^+]$_{in}$ constrained to sum to 160 mM. **4:** As for curve 3 but at zero membrane potential. **5:** [Na^+]$_{out}$ varied from 1 to 200 mM at −85 mV membrane potential. (Note this curve lies partly on top of the thinner curve 1). **6:** [ATP] varied from 0.01 to 5.00 mM at −85 mV with constraints as for curve 1 but with [Na^+]$_{in}$ = [Na^+]$_{out}$ = 100 mM and [K^+]$_{in}$ = [K^+]$_{out}$ = 50 mM. **7:** As for curve 6 but at zero membrane potential. **8:** Membrane potential varied from −350 to +100 mV with [Na^+]$_{in}$ = [Na^+]$_{out}$ = 100 mM and [K^+]$_{in}$ = [K^+]$_{out}$ = 50 mM. All rate coefficients were as specified in Fig. 2B and Table 1. Unless specified otherwise for each curve, all chemical boundary conditions were as listed in Table 2. Note that all curves, if projected sufficiently, intersect at the origin of coordinates (thermodynamic equilibrium), although some curves also intersect at a point corresponding to the diastolic steady state of cardiac muscle appropriate for the rate coefficients and boundary conditions of Tables 1 and 2. *Inset:* Detail of the equilibrium intersection of some of the curves showing the different slopes of intersection.

lar mechanism of Fig. 1 and the chosen reactivity coefficients of Table 1, can hardly be assumed to be unique to the present model. Similar observations are also implicit in the more general results of Slayman and co-workers (12,16), and they would seem to raise some doubt as to the general validity or usefulness of irreversible thermodynamics as a tool for understanding chemiosmotic transport processes.

The results illustrated in Fig. 7 also have some bearing on the question raised by Tanford (39) as to why the Na,K-ATPase operates physiologically so far from equilibrium. The nonequilibrium intersection point of curves 1, 2, 3, and 5 in Fig. 7 corresponds to the diastolic steady state in cardiac muscle for the boundary conditions listed in Table 2 and is displaced about 11 kJ/mole (or 113 mV of membrane potential) away from thermodynamic equilibrium. This is the free energy dissipated by the model simply to balance the passive diastolic leaks of Na^+ and K^+ in the quiescent steady state (7) and corresponds to a diastolic thermodynamic efficiency of about 81% for the electrogenic sodium pump in cardiac muscle. Examination of the shapes of curves 2 and 3 to the right of the diastolic intersection point shows that the transport rate is relatively insensitive to voltage but quite sensitive to raised intracellular $[Na^+]$ (see also Fig. 3).

Thus, the kinetic properties of the enzyme might seem to be optimized in such a way that the transport rate is relatively constant over the large changes in membrane potential occurring in a typical cardiac cycle, yet readily stimulated by increased $[Na^+]_{in}$ as would occur following an increase in heart rate. The resulting increase in outward electrogenic pump current would thereby shorten the duration of the cardiac action potential as already demonstrated for a simplified cardiac membrane model (7,22) and suggested by Isenberg and Trautwein (19). On the other hand, a pump that operated too close to its thermodynamic equilibrium position might exhibit too much slope conductance in its current–voltage relation and so become inhibited by the electrophysiological consequences of any ionic activation by $[Na^+]_{in}$ (cf. curves A and B, of Figs. 3 and 4).

CONCLUSIONS

(a) Relations between thermodynamic boundary conditions and reaction kinetics for electrogenic pumps are readily determined by considering the pumps as catalyzing stoichiometrically defined chemiosmotic reactions in which transmembrane electric current is an obligatory component of metabolism.

(b) A pump transferring several ions per cycle, such as the plasmalemmal Na,K-ATPase, appears likely to operate so as to minimize the number of charges transferred at the various steps in the reaction sequence. It appears unlikely that the electrovalence of any single step will exceed that of the overall reaction.

(c) If an electrogenic pump displays too much slope conductance in its current–voltage relation over the physiological range of membrane potentials, then its hyperpolarizing capability in response to altered ionic concentrations will be restricted.

(d) Unidirectional isotopic fluxes are unreliable as measures of either the stoichiometric coupling or the overall unidirectional rate of reaction of complex active transport mechanisms.

(e) The linear assumptions of irreversible thermodynamics, when applied to a realistic model of the electrogenic Na,K-ATPase, are totally unworkable and misleading.

(f) The electrogenic character of ionmotive pumps appears to add an extra dimension of biological versatility to vectorial metabolic enzymes, allowing them to influence membrane potential in response to altered ionic conditions. This property would seem to be optimized by having the pumps operating physiologically far from thermodynamic equilibrium, so as to behave approximately as constant current sources.

ACKNOWLEDGMENTS

The author is indebted to B. A. W. Coller of the Department of Chemistry, Monash University, and to J. A. Reynolds and C. Tanford of the Department of Physiology, Duke University Medical Center, for valuable discussions of various aspects of the present work.

This work has been supported by a National Heart Foundation of Australia Grant-in-Aid (G1490).

REFERENCES

1. Bockris, J. O'M., and Reddy, A. K. N. (1970): *Modern Electrochemistry: An Introduction to an Interdisciplinary Area.* Plenum, New York.
2. Boudart, M. (1976): Consistency between kinetics and thermodynamics. *J. Phys. Chem.,* 80:2869–2870.
3. Chapman, J. B. (1973): On the reversibility of the sodium pump in dialyzed squid axons. A method for determining the free energy of ATP breakdown? *J. Gen. Physiol.,* 62:643–646.
4. Chapman, J. B. (1982): A kinetic interpretation of "variable" stoichiometry for an electrogenic sodium pump obeying chemiosmotic principles. *J. Theor. Biol.,* 95:665–678.
5. Chapman, J. B., and Johnson, E. A. (1978): The reversal potential for an electrogenic sodium pump: a method for determining the free energy of ATP breakdown? *J. Gen. Physiol.,* 72:403–408.
6. Chapman, J. B., Johnson, E. A., and Kootsey, J. M. (1982): On the use of metabolic poisons to determine the current-voltage relation of an electrogenic active transport mechanism. *Proc. Austr. Physiol. Pharmacol. Soc.,* 13:120P.
7. Chapman, J. B., Kootsey, J. M., and Johnson, E. A. (1979): A kinetic model for determining the consequences of electrogenic active transport in cardiac muscle. *J. Theor. Biol.,* 80:405–424.
8. Chapman, J. B., and McKinnon, I. R. (1978): Consistency between thermodynamics and kinetic models of ion transport processes. *Proc. Austr. Soc. Biophys.,* 2:3–7.
9. Daut, J., and Rudel, R. (1981): Cardiac glycoside binding to the Na/K-ATPase in the intact myocardial cell: Electrophysiological measurement of chemical kinetics. *J. Mol. Cell. Cardiol.,* 13:777–782.
10. Fried, I. (1973): *The Chemistry of Electrode Processes.* Academic Press, London.
11. Gadsby, D. C., and Cranefield, P. F. (1979): Direct measurement of changes in sodium pump current in canine cardiac Purkinje fibers. *Proc. Natl. Acad. Sci. USA,* 76:1783–1787.
12. Gradmann, D., Hansen, U.-P., and Slayman, C. L. (1981): Reaction kinetic analysis of current-voltage relationships for electrogenic pumps in *Neurospora* and *Acetabularia. Curr. Top. Membranes & Transport,* 16:257–276.
13. Guidotti, G. (1979): Coupling of ion transport to enzyme activity. In: *The Neurosciences, Fourth Study Program,* edited by I. O. Schmitt and F. G. Worden, pp. 831–840. MIT Press, Boston.
14. Hall, J. E., Mead, C. A., and Szabo, G. (1973): A barrier model for current flow in lipid bilayer membranes. *J. Membr. Biol.,* 11:75–97.
15. Hammes, G. G., and Schimmel, P. R. (1970): Rapid reactions and transient states. In: *The Enzymes,* 3rd Ed, Vol. II, edited by P. D. Boyer, pp. 67–114. Academic Press, New York.

16. Hansen, U.-P., Gradmann, D., and Slayman, C. L. (1981): Interpretation of current-voltage relationships for "active" ion transport systems. I. Steady-state reaction-kinetic analysis of class-I mechanisms. *J. Membr. Biol.*, 63:165–190.
17. Hassinen, I. E., and Hiltunen, K. (1975): Respiratory control in isolated perfused rat heart. Role of the equilibrium relations between the mitochondrial electron carriers and adenylate system. *Biochim. Biophys. Acta*, 408:319–330.
18. Hill, T. L. (1968): *Thermodynamics for Chemists and Biologists.* Addison-Wesley, Reading, Massachusetts.
19. Isenberg, G., and Trautwein, W. (1974): The effect of dihydro-ouabain and lithium ions on the outward current in cardiac Purkinje fibers. *Pfluegers Arch.*, 350:41–54.
20. Isenberg, G., and Trautwein, W. (1975): Temperature sensitivity of outward current in cardiac Purkinje fibers. Evidence for electrogenicity of active transport. *Pfluegers Arch.*, 358:225–234.
21. Jack, J. J. B., Noble, D., and Tsien, R. W. (1975): *Electric Current Flow in Excitable Cells.* Clarendon Press, Oxford.
22. Johnson, E. A., Chapman, J. B., and Kootsey, J. M. (1980): Some electrophysiological consequences of electrogenic sodium and potassium transport in cardiac muscle: A theoretical study. *J. Theor. Biol.*, 87:737–756.
23. Jorgensen, P. L. (1980). Sodium and potassium ion pump in kidney tubules. *Physiol. Rev.*, 60:864–917.
24. Katchalsky, A., and Curran, P. F. (1965): *Nonequilibrium Thermodynamics in Biophysics.* Harvard University Press, Cambridge, Massachusetts.
25. Keizer, J. (1975): Thermodynamic coupling in chemical reactions. *J. Theor. Biol.*, 49:323–335.
26. Keynes, R. D., and Swan, R. C. (1959): The effect of external sodium concentration on the sodium fluxes in frog skeletal muscle. *J. Physiol. (Lond.)*, 147:591–625.
27. Kootsey, J. M., Johnson, E. A., and Chapman, J. B. (1981): Electrochemical inhomogeneity in ungulate Purkinje fibers: Model of electrogenic transport and electrodiffusion in clefts. *Adv. Physiol. Sci.*, 8:83–92.
28. Lauger, P., and Stark, G. (1970): Kinetics of carrier-mediated ion transport across lipid bilayer membranes. *Biochim. Biophys. Acta*, 211:458–466.
29. Lieberman, M., Sawanobori, T., Kootsey, J. M., and Johnson, E. A. (1975): A synthetic strand of cardiac muscle. Its passive properties. *J. Gen. Physiol.*, 65:527–550.
30. Michael, L. H., Schwartz, A., and Wallick, E. T. (1979): Nature of the transport adenosine triphosphatase-digitalis complex. XIV. Inotropy and cardiac glycoside interaction with cat ventricular muscle. *Mol. Pharmacol.*, 16:135–146.
31. Mitchell, P. (1961): Coupling of phosphorylation to electron and hydrogen transfer by a chemiosmotic type of mechanism. *Nature*, 191:144–148.
32. Mitchell, P. (1968): *Chemiosmotic Coupling and Energy Transduction.* Glynn Research, Bodmin.
33. Mitchell, P. (1979): Compartmentation and communication in living systems. Ligand conduction: A general catalytic principle in chemical, osmotic and chemiosmotic reaction systems. *Eur. J. Biochem.*, 96:1–20.
34. Mobley, B. A., and Page, E. (1972): The surface area of sheep cardiac Purkinje fibers. *J. Physiol. (Lond.)*, 220:547–563.
35. Mullins, L. J., and Frumento, A. S. (1963): The concentration dependence of sodium efflux from muscle. *J. Gen. Physiol.*, 46:629–654.
36. Skou, J. C. (1957): The influence of some cations on an adenosine-triphosphatase from peripheral nerves. *Biochim. Biophys. Acta*, 23:394–401.
37. Skou, J. C. (1975): The ($Na^+ + K^+$) activated enzyme system and its relationship to transport of sodium and potassium. *Q. Rev. Biophys.*, 7:301–434.
38. Slayman, C. L., and Gradmann, D. (1975): Electrogenic proton transport in the plasma membrane of *Neurospora*. *Biophys. J.*, 15:968–971.
39. Tanford, C. (1981): Equilibrium state of ATP-driven ion pumps in relation to physiological ion concentration gradients. *J. Gen. Physiol.*, 77:223–229.
40. Thomas, R. C. (1972): Electrogenic sodium pump in nerve and muscle cells. *Physiol. Rev.*, 52:563–594.
41. Vassalle, M. (1970): Electrogenic suppression of automaticity in sheep and dog Purkinje fibers. *Circ. Res.*, 27:361–377.
42. Veech, R. L., Lawson, J. W. R., Cornell, N. W., and Krebs, H. A. (1979): Cytosolic phosphorylation potential. *J. Biol. Chem.*, 254:6538–6547.

*Electrogenic Transport: Fundamental Principles and Physiological Implications*, edited by Mordecai P. Blaustein and Melvyn Lieberman. Raven Press, New York © 1984.

The Electrogenic Sodium Pump

I. M. Glynn

Physiological Laboratory, University of Cambridge, Cambridge CB2 3EG, England

This chapter will try to answer three questions: (a) Why do we believe the sodium pump to be electrogenic? (b) What fraction of the work done by the pump is electrical work rather than osmotic work? (c) How does the pump generate an outward current? These questions will be dealt with in order, though the answers will, unfortunately, become progressively more vague.

EVIDENCE FOR ELECTROGENICITY

Why do we believe the sodium pump to be electrogenic? There is, of course, much evidence from a variety of tissues, starting with Kernan's (44) demonstration that the active expulsion of sodium from sodium-loaded frog muscle was accompanied by a hyperpolarization beyond the potassium-equilibrium potential. A full historical account of this evidence was assembled by Thomas (70) in his 1972 review, by De Weer (17) in his 1975 review, and, most recently and in a wider context, by Slayman (67) in the interesting historical introduction to the volume reporting the meeting on electrogenic ion pumps, held at Yale last year. In this chapter I will review only four series of experiments—three of them old and one recent—chosen as much for their elegance as for their persuasiveness.

The first experiments to be discussed are those conducted by Rang and Ritchie (62) to investigate the cause of posttetanic hyperpolarization in the rabbit vagus nerve. Posttetanic hyperpolarization—the increase in resting membrane potential that develops during, and persists for a time after, the repeated activity of excitable cells—was described by Hering (35) in 1884. Seventy-five years later, Connelly (13) noted that it could be the result of increased activity of the sodium pump, coping with the extra sodium accumulated during a large number of action potentials, if one assumed that the pump was electrogenic. There were, however, two other hypotheses. Two years before Connelly's paper, Ritchie and Straub (63), suggested that the hyperpolarization came about because increased activity of the sodium pump lowered the potassium concentration in the space between the axon and the neurilemma and so made the potassium equilibrium potential more negative. In 1964, Gage and Hubbard (24) suggested that posttetanic hyperpolarization in the rat phrenic nerve resulted from increased potassium permeability, and therefore a closer approach to the potassium equilibrium potential—a hypothesis that had been proposed earlier by Coombs et al. (14) to explain afterpolarization in motoneurons.

Of these three hypotheses, only the first two involved the sodium pump; and, since Ritchie and Straub had already shown that the posttetanic hyperpolarization in the rabbit vagus was abolished by ouabain, the choice lay between these two. Did increased pump activity hyperpolarize because the pump was electrogenic or because it depleted the extraaxonal space of potassium? To decide Rang and Ritchie did a rather simple experiment. First, to make the posttetanic hyperpolarization easier to measure, they worked with solutions in which isethionate replaced chloride. Because isethionate does not penetrate the axon membrane, short-circuiting by chloride ions was avoided and the hyperpolarization was increased by 2 to 4 mV to about 20 mV. They then showed that removing potassium from the bathing solution slightly reduced the magnitude and greatly reduced the duration of the posttetanic hyperpolarization. Finally, they showed that replacing potassium in the bathing solution just after the abbreviated hyperpolarization had ended caused an immediate hyperpolarization which then decayed in the normal manner. Since adding potassium to the bathing solution could hardly have caused depletion of potassium in the extraaxonal space they concluded that the pump must be electrogenic.

The second set of experiments is that of Nakajima and Takahashi (55) who investigated the after potentials of the crayfish stretch-receptor neuron. This neuron shows after potentials of two kinds. After a single impulse, or a small number of impulses, the membrane is hyperpolarized by a few millivolts, returning to a normal resting potential with a half-time of 50 to 80 msec. This short after potential is distinct from the posttetanic hyperpolarization, which is also shown by the stretch-receptor neuron but which occurs only after a greater number of impulses and decays with a half-time of 2 to 6 secs. The posttetanic hyperpolarization could be abolished by ouabain, and again the question was: Was it caused by the electrogenic effect of the sodium pump or by potassium depletion in the extracellular fluid? First, Nakajima and Takahashi (55) showed that applying hyperpolarizing current to the neuron diminished and eventually reversed the sign of the short after potential. Furthermore, the reversal potential varied with the potassium concentration in the medium in a way that could be explained by supposing that the short after potential was caused by maintained potassium permeability, and that the reversal potential was the potassium equilibrium potential. The voltage at the trough of the short after potential, or the magnitude of the reversal potential, could, therefore, be used to monitor changes in the potassium equilibrium potential. If, during posttetanic hyperpolarization, extracellular potassium depletion made the potassium equilibrium potential more negative, a stimulus given during the period of posttetanic hyperpolarization should generate an impulse with a short after potential, the trough of which was also more negative. In the event, the after potential of an impulse generated during the period of posttetanic hyperpolarization reached a trough only about 1 mV more negative than that of an impulse generated before the tetanus, and separate experiments showed that even that small change could be accounted for by the change in starting potential. Again, the conclusion was that posttetanic hyperpolarization must be attributed to the electrogenic effect of the sodium pump.

The third set of experiments to be discussed is that of Thomas (69), on the expulsion of sodium from snail ganglion neurons. If Rang and Ritchie's (62) experiments are impressive for their simplicity, and Nakajima and Takahashi's for their ingenuity, Thomas's are impressive for the technical tour de force that made them possible. The neurons that Thomas worked with are about 200 μm in diameter and into them he poked five microelectrodes. Of these, one was to measure membrane potential, one was connected to a slowly responding voltage clamp (slowly responding so that it did not interfere with the spontaneous firing of the neuron), one was a sodium-sensitive electrode, and the other two were filled with sodium acetate and potassium acetate, respectively, so that by driving current between them, graded amounts of sodium and acetate ions could be injected into the cell. With the voltage clamp turned off, Thomas injected sodium ions into the cell and showed that this led to a hyperpolarization that decayed with a half-time of a few minutes. This hyperpolarization must have been brought about by increased activity of the sodium pump, since it did not occur in the presence of ouabain, or in a potassium-free bathing medium, but did occur as soon as potassium was restored to the potassium-free medium. When the voltage-clamp was switched on, injection of sodium did not, of course, cause any change in potential, but the clamp current increased linearly during the injection and then decayed exponentially with a half-time of 4 to 5 min. Simultaneous recording from the sodium-sensitive electrode showed that the clamp current and the sodium activity ran parallel courses. These experiments provide strong evidence for the electrogenicity of the sodium pump, but perhaps the most interesting finding was that the total charge delivered by the clamp was only one-third to one-quarter of the total charge on the injected sodium ions. The suggested explanation was that for every 3 (or 4) sodium ions expelled by the pump, 2 (or 3) potassium ions were pumped into the cell.

The three sets of experiments described thus far were all done many years ago, but the fourth set, by Dixon and Hokin (21), were reported only in 1980. By that time, a number of different workers (4,34,36,68) had succeeded in incorporating more or less purified preparations of Na,K-ATPase into artificial lipid vesicles and showing that the addition of Mg^{2+} and adenosine triphosphate (ATP) to the bathing medium led to an uptake of sodium ions and a loss of potassium ions. The explanation for these net movements of the ions was, of course, that although the enzyme was almost certainly inserted randomly into the vesicle membranes, ATP was present only outside the vesicles, so that only enzyme molecules sitting in the membrane with their originally intracellular face facing outward would have been able to hydrolyze it. Dixon and Hokin (21) worked with Na,K-ATPase prepared from the outer medulla of dog kidney and incorporated into artificial lipid vesicles prepared from soya bean lipids. With lipid vesicles, it is not possible to measure membrane potentials directly, but they were able to estimate the potential by observing the distribution of the lipid-penetrating anion ^{14}C-thiocyanate. They found that when Mg^{2+} and ATP were added to the vesicle suspension, the uptake of sodium ions and the expulsion of potassium ions was accompanied by the generation of a membrane potential of 14 mV, positive on the inside. This itself did not prove that the pump was electrogenic, since the potential might have been a diffusion potential resulting from the potassium gradient set up by the pumping. They therefore added nigericin to the suspension, which collapsed the sodium and potassium gradients but left a membrane potential of 9 mV, positive on the inside. Their interpretation was that the 9 mV potential was generated directly by the pump, and the extra 5 mV represented a diffusion potential. This interpretation was supported by experiments showing that extravesicular vanadate (which would be expected to stop transport abruptly) caused a rapid loss of the potential observed in the presence of nigericin, but only a slow loss of the diffusion potential in experiments without nigericin. Intravesicular ouabain also abolished potentials, both in the absence and presence of nigericin, but because it was necessary to incorporate ouabain into the vesicles before the experiment, the rate of development of its effects could not be determined.

Dixon and Hokin (21) lay some stress on the resemblance between the 9-mV electrogenic potential observed in their experiments and the 9-mV electrogenic potential observed in the *Amphiuma* red cell membrane by Hoffman et al. (38). The resemblance must be fortuitous, however, since the potential generated in the artificial lipid vesicles must be a function of the arbitrary density of Na,K-ATPase molecules in the vesicle membranes as well as of the conductance of those membranes.

ELECTRICAL WORK IN PUMPING: SENSITIVITY OF THE PUMP TO MEMBRANE POTENTIAL

Now for the question, what fraction of the work done by the pump is electrical work? If, for the moment, we accept the conventional wisdom that under physiological conditions each cycle of the pump expels three sodium ions and takes in two potassium ions, the total work done by the pump in each cycle is given by the expression

$$3RT \ln \frac{[Na^+]_o}{[Na^+]_i} + 2 RT \ln \frac{[K^+]_i}{[K^+]_o} + EF,$$

where the last term represents the electrical work. For human red cells, where the potential is only about 10 mV, the fraction of electrical work is only about 2%; for mammalian muscle or nerve, with a resting potential of, for example,

90 mV, the fraction works out at about 19%. These figures refer, of course, to the net amount of electrical work done by the pump on its surroundings in one complete cycle. How much electrical work is done by or on the pump in each part of the cycle cannot be determined unless we know how the pump works. From the figures for the entire cycle, however, it is clear that the pump must be sensitive to the electrical potential across the membrane, although the effects of potential may not be easily detectable in conditions that are convenient for the experimenter. To take an extreme case, if the electrical potential is made so great that the energy required to expel three sodium ions and take up two potassium ions exceeds the energy available from the hydrolysis of one molecule of ATP, the pump must either run backwards or change its stoichiometry. Alternatively, if the levels of ATP, adenosine diphosphate (ADP) and orthophosphate, and the concentration gradients of sodium and potassium, are adjusted so that the energy available is just equal to the energy required for osmotic work, the application of relatively small potentials must (unless there is a change in stoichiometry) make the pump run forward or backward depending on the polarity.

In experiments on the H^+-transporting ATPase of *Neurospora*, which is much more strongly electrogenic than the Na,K-ATPase, Warncke and Slayman (71) have shown that changes in membrane potential can lead to large rate changes; the stoichiometry also varies depending on the energy available. Interestingly, however, until Rakowski and De Weer (61) reported their results, the sodium pump seemed remarkably insensitive to changes in membrane potential under most of the conditions that had been investigated.

The first attempt to look for effects of membrane potential on sodium transport through the pump was made by Hodgkin and Keynes (37) in cuttle fish axons. These investigators were primarily interested in determining whether the drop in sodium efflux caused by removing potassium from the bathing medium indicated a coupling between the movements of the two ions or was merely a result of the hyperpolarization that followed the removal of potassium. By applying hyperpolarizing currents to a short length of axon they were able to show that hyperpolarizations of up to 28 mV had no significant effect on the sodium efflux.

A more thorough investigation of the effects of potential on sodium and potassium fluxes through the pump was made by Brinley and Mullins (12,53), using their capillary dialysis technique and altering the membrane potential either by passing hyperpolarizing currents or by changing $[K^+]_i$ or $[K^+]_o$. The upshot of their experiments was that, except in one set of conditions, varying the membrane potential in the range 10 to 90 mV (inside negative) had no consistent effect on the ATP-dependent sodium efflux or potassium influx. The exception was an experiment in which an axon, immersed in artificial sea water in which all of the sodium (but not the potassium) had been replaced by lithium, was perfused with a solution containing 5 mM ADP instead of ATP. The ATP/ADP ratio was not measured but was thought to be about 1. When this axon was hyperpolarized by 50 mV, the sodium efflux, which had already been reduced by the removal of sodium from the bathing medium, was further reduced by about 23%. At first sight, it is tempting to suggest that the low ATP/ADP ratio made the sodium efflux sensitive to potential by bringing the reversal potential for the system closer to the physiological range. Such a thermodynamic explanation is not tenable, however, in view of the absence of sodium in the bathing medium. Brinley and Mullins did not report the effects of hyperpolarization when axons immersed in artificial sea water containing normal concentrations of sodium were perfused with ADP, but Abercrombie and De Weer (1) were unable to detect any effect of hyperpolarizations up to 40 mV on the ouabain-sensitive efflux of sodium from axons in which the ADP level had been greatly increased by the injection of arginine. It is also uncertain whether the effect of potential observed by Brinley and Mullins was an effect on the sodium pump or on the Na^+–Ca^{2+} exchange mechanism which, it is now clear, is affected by ATP (6,18) and by membrane potential (19,54).

In frog muscle too, evidence for an effect of membrane potential on the sodium pump is doubtful. The earlier work of Horowicz and Gerber (40) suggesting that stimulation of the sodium pump by a raised extracellular potassium concentration or by azide was a consequence of the fall in membrane potential has not been supported by the more recent work of Beaugé et al. (8,9). Horowicz and Gerber used isolated fibers of semitendinosus muscle and Beaugé et al. (9) used sartorius muscles, but it seems unlikely that this can account for the disagreement. The subject has been reviewed recently by Sjodin (66).

Zade-Oppen et al. (75) looked for effects of membrane potential on the sodium pump in high-potassium sheep red cells, varying the membrane potential between -10 and $+60$ mV by suspending the cells in media containing different concentrations of chloride. Because the cell membranes were much more permeable to chloride than to any of the other ions present, the ratio $[Cl^-]_i/[Cl^-]_o$ determined the membrane potential; however, because of rapid Cl^-–OH^- exchange through the anion exchange mechanism, the chloride ratio also determined the pH drop across the membrane. Zade-Oppen et al. (75), therefore, did the experiment in two ways, keeping either the internal or the external pH constant as they varied the potential. It turned out that, provided the internal pH was kept constant, sodium efflux through the pump was not affected by changes in potential. The earlier results of Cotterell and Whittam (15) on human red cells, showing that hyperpolarization appeared to cause a small reduction in efflux, can almost certainly be explained by the fall in internal pH that must have occurred in their experiments.

In addition to experiments investigating possible effects of the membrane potential on the pump fluxes measured directly, there have been a number of experiments in which pump current has been measured as a function of membrane potential. Nakajima and Takahashi (55), working with the crayfish stretch-receptor neuron, Meunier and Tauc (51), Marmor (50), and Thomas (70) working with mollusc ganglion neurons, and Isenberg and Trautwein (41) working with cardiac Purkinje fibers, all found that the pump current was unaffected by changes in membrane potential—in other words, the pump behaved as a constant-current generator. The apparent voltage-sensitivity of pump current in snail ganglion neurons described by Kostyuk et al. (48) was later shown by Konenenko and Kostyuk (47) to be the result of changes in potassium conductance.

THE MECHANISM OF CURRENT GENERATION

Let us now turn to the third question: How does the sodium pump generate an outward current? An obvious answer is that the pump pumps three sodium ions out for every two potassium ions that it pumps in, and that the outward current represents the charge deficit. Before accepting this explanation, however, we must consider whether the conventional view about stoichiometry is soundly based, and whether there are any associated movements of hydrogen ions or anions that should be entered in the balance sheet.

It must be admitted that the statement that, under physiological conditions, the ratio of sodium ions pumped out to potassium ions pumped in is 3:2 is sometimes made in a manner more appropriate to the recitation of a creed than to the recounting of an experimental finding. There are, in fact, substantial difficulties in determining this ratio. The sodium pump is never the sole pathway for sodium and potassium ions through the membrane; it follows that, even if conditions are chosen that minimize fluxes through other pathways, one ultimately has to have some criterion—ATP-dependence or ouabain-sensitivity—for distinguishing fluxes through the pump. And even where it is safe to assume that, for example, ouabain inhibits the pump completely and affects no other pathway, there is still the problem that, except in very brief experiments, changes in potassium and especially in sodium concentration in the ouabain-poisoned cells will ensure that the fluxes in those cells will not be identical with the nonpump fluxes in the unpoisoned cells. Experiments can generally be made briefer by the use of tracers to measure unidirectional fluxes, but the picture is then complicated by the possible contributions of K^+-K^+ exchange and Na^+-Na^+ exchange, which can be ignored in experiments on net fluxes. Whichever method is adopted, one ends up taking the ratio of two estimates each of which represents a difference between two measurements, with the summation of errors that that involves. There is then a tendency to assume that the nearest simple numerical ratio must be the true ratio, despite the fact that we know that the pump can operate in different modes so that true ratio over a period containing many cycles need not be simple. Finally, there is no escaping the awkward fact that once there is a conventional view of what the stoichiometry is, investigators tend to be less critical of their results if those results conform to that view.

Despite these difficulties, the conventional view is probably justified.

In human red cells—cells in which the 3:2 ratio was first described (60)—a great deal of subsequent work has yielded results that support a 3:2 ratio more or less closely (23,28,58,64,72,73; cf. 65).

As has been mentioned, in recent years several groups of workers have demonstrated ATP-dependent sodium and potassium fluxes in artificial lipid vesicles with partially purified Na,K-ATPase incorporated into their membranes (4, 34,36,42,68). The results of all of these groups are compatible with the 3:2 ratio, although not all of them provide convincing evidence for it. In order to determine the true coupling ratio, it is not sufficient merely to determine the uptake of sodium into and the loss of potassium from the vesicles in a known time, because once the equilibrium is disturbed net passive fluxes will occur. It is therefore necessary either to compare initial flux rates or to measure, in addition, the passive permeabilities to the two ions. The former procedure was adopted by Karlish and Pick (42) in experiments using Na,K-ATPase from pig kidney outer medulla, the latter by Sweadner and Goldin (68) using enzyme prepared from dog brain gray matter and by Goldin (34) using enzyme from dog kidney outer medulla. All three series of experiments gave ratios close to 3:2 and not significantly different from it.

In squid axons (1,5,11,12,53,61) the position is much less clear, and the interpretation of some of the experimental findings is complicated by the existence of a second ATP-dependent pathway—the Na^+-Ca^{2+} exchange mechanism—by which sodium ions can cross the membrane. At present evidence is not sufficient to establish a definite coupling ratio, and it is possible that the ratio may vary with the conditions. However, comparisons between measured fluxes and also between measurements of fluxes and of pump current make it likely that under physiological conditions more sodium ions are pumped out than potassium ions are pumped in. In muscle there is also some disagreement between estimates of the coupling ratio and some uncertainty about whether that ratio is fixed (2,16,52,66).

Epithelia, with their multiplicity of membranes and channels, might seem unpromising material for investigating the coupling ratio of the sodium pump, but there have recently been two ingenious approaches by Nielsen (56,57), using frog skin. According to the two-membrane hypothesis of Koefoed-Johnson and Ussing (46), sodium ions are pumped out of the cell across the inward-facing membrane in exchange for potassium ions, which then leak out again through the same membrane. The short-circuit current is therefore equivalent to the flux of sodium through the pump. If the potassium channels in the inward-facing membrane could be blocked, and a pathway for potassium ions could be created in the outward-facing membrane, the pump would transport potassium ions outwards and the short-circuit current would then be equivalent to the sodium flux through the pump minus the rate of loss of potassium from the solution bathing the inner face. By measuring this rate of loss of potassium, and the short-circuit current, it would, therefore, be possible to calculate both the sodium and the potassium fluxes and, hence, the coupling ratio. Nielsen (56) did this experiment using barium ions to block the potassium channels in the inward-facing membrane and the polyene antibiotic, filipin, to create a nonspecific channel in the outward-facing membrane. The ratio usually came out rather greater than 3:2, but this is probably because backward leakage of potassium across the inward-facing membrane, despite the presence of barium ions, made the estimate of potassium influx too low.

Nielsen's second approach (57) was even simpler. If the coupling ratio is 3:2, then if the potassium channels in the inward-facing membrane are suddenly blocked by the addition of barium ions, the pump should transfer only one charge per cycle instead of three, and the short-circuit current should immediately drop to one-third of its value. This is what happened. Of course, the accumulation of potassium ions should then hyperpolarize the cell, reducing the entry of sodium through the outward-facing membrane, thus gradually slowing the pump and reducing the current still further. In fact, the current remained at the one-third level for several minutes and then gradually rose again. The explanation of the rise is probably that the effect of barium ions on the membrane resistance was transient, but the events subsequent to the initial drop in short-circuit current have not been fully explained.

An Na^+/K^+ ratio greater than 1 will make the pump electrogenic only if the charge deficit for the movements of the alkali metal ions is not balanced by the associated transport of other ions. Indeed, in interpreting the work just discussed, we have implicitly assumed the absence of such associated transport. There are two classes of ions that must be considered: anions and hydrogen ions.

Until recently, although there seems to have been no systematic investigation of the effects of anions on the sodium pump, there was no reason to think that the movement of anions was ever linked directly, i.e., not by electrical coupling, to pump activity. Wieth (73) found that when bicarbonate replaced most of the intra- and extracellular chloride in a suspension of red cells, active sodium efflux and potassium influx were both increased, but the effect could be wholly explained by a rise in intracellular sodium concentration secondary to increased passive sodium permeability. Unless chloride and bicarbonate had equivalent effects on the pump, it seemed to follow that anions were not involved. The recent work of Dissing and Hoffman (20) has made it necessary to qualify this statement. They used lipid-penetrating dyes to determine the membrane potential in human red cells. When the cells were pumping normally in media containing potassium, the pump, as expected, was electrogenic. When the cells were incubated in media containing neither sodium nor potassium, the pump continued to expel sodium at a slower rate—also as expected (see reference 27)—but, surprisingly, the efflux was not electrogenic and yet was accompanied by the efflux of anions. The explanation of this wholly unexpected effect remains a mystery, although it may be relevant that in human red cells there is recent evidence suggesting that the Na,K-ATPase may be closely associated with the anion transport protein (band 3) (22). Despite this association, however, there is no reason to think that the normal working of the sodium pump is accompanied by anion transport.

For hydrogen ions, the position is more complicated, for the pump is certainly sensitive to pH—running more slowly when intracellular pH falls (45)—and cells must have an outwardly directed hydrogen ion pump if they are to maintain their intracellular pH within normal limits in the face of the inward driving force corresponding to the resting potential. Observations that the muscles of potassium-depleted or ouabain-treated rats had a low intracellular pH led to the hypothesis that hydrogen ions could be moved outwards by the sodium pump (see references 10,25,74). Such a transport, if it occurred, would, of course, increase the electrogenicity of the pump. In fact, it is now clear from the work of Aickin and Thomas (3) that ouabain has no immediate effect on the outward transport of hydrogen ions from mouse muscle, and that the effects described in the literature come about because both ouabain and potassium-depletion cause a rise in intracellular sodium concentration. This rise decreases the rate of hydrogen ion expulsion by the H^+-Na^+ exchange mechanism, which seems to be largely responsible for maintaining the intracellular pH of mammalian muscle.

The electrogenic effect of the sodium pump therefore seems to be the straightforward result of the discrepancy between the number of sodium ions pumped out and the number of potassium ions pumped in. Why should those numbers be three and two? The stoichiometry obviously fits well with the experimental finding that the magnitudes of potassium influx and sodium efflux through the pump vary with the cation concentrations on the two sides of the membrane in a way that suggests that sodium ions activate at three internal sites and potassium ions activate at two external sites [see (30) for references]. It is there-

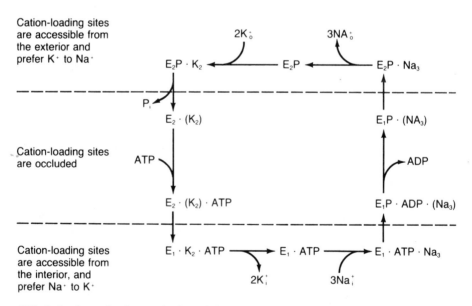

FIG. 1. A scheme for the mechanism of the sodium pump; modified from Karlish et al. (43) (with permission) by (a) the omission of reactions not immediately relevant to the present discussion, and (b) an assumption of 3:2 stoichiometry. The symbols Na and K within brackets denote occluded ions. E_1 and E_2, and E_1P and E_2P, refer, respectively, to different conformations of the unphosphorylated and phosphorylated enzyme.

fore tempting (Fig. 1) to suppose that (a) phosphorylation of the enzyme by ATP can occur only if three intracellular sites are occupied by sodium—these sodium ions becoming occluded within the phosphoenzyme and then escaping to the exterior after a conformational change, and (b) rapid hydrolysis of the phosphoenzyme can occur only if two extracellular sites are occupied by potassium—these potassium ions becoming occluded within the dephosphoenzyme and then escaping to the interior after a conformational change.

This hypothesis would be supported if we could show that the newly phosphorylated enzyme contained three sodium ions and that the newly dephosphorylated enzyme contained two potassium ions. Unfortunately, although there is indirect evidence (29,43) and some preliminary direct evidence (33) for the occlusion of sodium ions within the newly formed phosphoenzyme, there is no information about the number of sodium ions occluded. In the potassium half of the cycle there is now very good evidence that the hydrolysis leads to the occlusion of potassium ions within the newly formed dephosphoenzyme (7,32,59), but there is some doubt about whether two or three ions are occluded.

Another test of the hypothesis would be to see whether the rates of phosphorylation and of dephosphorylation varied appropriately with the concentrations of sodium and potassium ions. The dependence of phosphorylation rate on sodium concentration has been investigated by Mårdh and Post (49) and showed no sign of the expected sigmoidicity, but the results were puzzling in that the sodium concentration also had a large effect on the steady-state level of phosphorylation, which was not easy to explain. There is no evidence about the form of the relation between potassium concentration and the rate of phosphoenzyme hydrolysis.

I will not venture further into details of the possible mechanism of the sodium pump at this stage. I should like, instead, to return to the question of the apparent insensitivity of the pump to changes in membrane potential, recognizing that this insensitivity may need to be reassessed in the light of the new findings of Rakowski and De Weer (61).

Insensitivity implies that none of the steps that are rate limiting in the conditions of the experiment are sensitive to the applied potential. That can only be because the rate-limiting steps do not involve large net movements of charge with or against the electric field—either because charged groups are not moved, or because their movement occurs perpendicular to the direction of the field or in a region of membrane with no field, or conceivably because the effects of the field on the movements of different charged groups cancel out. Since the pump is electrogenic, however, at least one of the steps in the cycle must be voltage sensitive, and it would tell us something about the mechanism if we knew which steps were. In principle, this could be determined by altering conditions to make each step rate limiting in turn, and testing for voltage sensitivity in each case. That is hardly a practicable proposition, but it is worth noting that most of the experiments showing insensitivity to changes in membrane potential were done under conditions in which the low intracellular sodium concentration limited the rate of pumping. It is, therefore, unlikely that the phosphorylation step catalyzed by intracellular sodium ions is voltage sensitive. A similar argument would tend to exclude the hydrolysis step catalyzed by extracellular potassium ions, but less convincingly, because in most cases the external potassium concentration was not very low compared with the concentration required to saturate the pump.

Two obvious candidates for the voltage-sensitive step (or steps) are (a) the conformational change $[E_2 \cdot (K_2) \rightarrow E_1 \cdot K_2]$ catalyzed by ATP, which precedes the release of potassium ions from the occluded K form to the cell interior (see references 31,32,43) and (b) the conformational change $[E_1P \cdot (Na_3) \rightarrow E_2P \cdot Na_3]$ that is thought to precede the release of sodium ions from the occluded Na form to the cell exterior (see references 29,33,43). Whether the conformational change $[E_2 \cdot (K_2) \rightarrow E_1 \cdot K_2]$ that precedes the release of occluded potassium is voltage-sensitive might be investigated by looking at the effects of changing the membrane potential on the efflux of sodium from Na-rich, ATP- (and ADP)-poor cells incubated in a medium containing enough potassium to saturate the pump. To look for effects of membrane potential on the rate of conversion of $E_1P \cdot (Na_3)$ to $E_2P \cdot Na_3$ is more difficult, because without using inhibitors there is no known way of ensuring that that step is the major rate-limiting step. During $Na^+ - Na^+$ exchange in bathing media of low sodium concentration, however, the rate of the reverse change $[E_2P \cdot Na_3 \rightarrow E_1P \cdot (Na_3)]$ is, presumably, rate limiting. It might therefore be worth looking for effects of membrane potential on sodium efflux from ATP-poor, ADP-rich cells incubated in media containing no potassium (or potassium cogeners) and a low concentration of sodium. If changing the membrane potential turned out to have no effect, the interconversion of the two forms of phosphoenzyme could not be voltage sensitive. A positive answer would be more ambiguous. It would imply either that the interconversion is voltage sensitive, or that the relation between the sodium concentration is the immediate vicinity of the sodium loading sites and in the external bulk phase is voltage sensitive, as it could be if the loading sites are situated at some depth in the membrane. It might be thought perverse to look for effects of membrane potential on an exchange which would not be expected to lead to a transfer of charge (26) and which does not appear to do so (1); but, of course, the absence of net charge transfer merely implies that any change in potential must change the inward and outward fluxes in the same proportion. In principle, both of the experiments suggested could be done either with squid axons or with red cells, although they do not look easy. Whether they are worth doing remains to be seen, but if there are any snags it is difficult to think of anybody more likely to be aware of them than readers of this volume.

Having just become aware of the recent experiments of Rakowski and De Weer (61) I should like to add one comment. An intriguing feature of those experiments is that not only did hyperpolarization, contrary to previous findings, slow the sodium pump when it was running forwards, but, paradoxically, hyperpolarization also slowed the pump when it was running backward and, presumably, generating an inward current. Earlier in this volume, De Weer pointed out that there was nothing thermodynamically improper in this, and that a voltage-sensitive gate in series with the pump would be one way to explain the observed behavior. That is correct, but I suspect that the true explanation is that the rate-limiting steps were different in the two parts of the experiment. If both the forward- and the backward-running pump involved (different) rate-limiting steps in which there was a net movement of positive charge against the field imposed by the membrane potential, the paradox would be resolved. We might imagine, for example, that the conversion of $E_1P \cdot (Na_3)$ to $E_2P \cdot Na_3$ was rate-limiting under the conditions in which the pump ran forward, that the conversion of $E_1 \cdot K_2$ to $E_2 \cdot (K_2)$ was rate-limiting under the conditions in which the pump ran backward, and that both conversions moved positive charge against the field. The detailed explanation may be wrong but I suspect that it is the right kind of explanation.

REFERENCES

1. Abercrombie, R. F., and De Weer, P. (1978): Electric current generated by squid giant axon sodium pump: External K and internal ATP effects. *Am. J. Physiol.*, 235:C63–C68.
2. Adrian, R. H., and Slayman, C. L. (1966): Membrane potential and conductance during transport of sodium, potassium and rubidium in frog muscle. *J. Physiol. (Lond.)*, 184:970–1014.
3. Aickin, C. C., and Thomas, R. C. (1977): An investigation of the ionic mechanism of intracellular pH regulation in mouse soleus muscle fibres. *J. Physiol. (Lond.)*, 273:295–316.
4. Anner, B. M., Lane, L. K., Schwartz, A., and Pitts, B. J. R. (1977): A reconstituted $Na^+ + K^+$ pump in liposomes containing purified $(Na^+ + K^+)$-ATPase from kidney medulla. *Biochim. Biophys. Acta,* 467:340–345.
5. Baker, P. F., Blaustein, M. P., Keynes, R. D., Manil, J., Shaw, T. I., and Steinhardt, R. A. (1969): The ouabain-sensitive fluxes of sodium and potassium in squid axons. *J. Physiol. (Lond.)*, 200:459–496.
6. Baker, P. F., and Glitsch, H. G. (1973): Does metabolic energy participate directly in the Na^+-dependent extrusion of Ca^{2+} ions from squid giant axons. *J. Physiol. (Lond.)*, 233:44P–46P.
7. Beaugé, L. A., and Glynn, I. M. (1979): Occlusion of K ions in the unphosphorylated sodium pump. *Nature,* 280:510–512.
8. Beaugé, L. A., and Sjodin, R. A. (1976): An analysis of the influence of membrane potential and metabolic poisoning with azide on the sodium pump in skeletal muscle. *J. Physiol. (Lond.)*, 263:383–403.
9. Beaugé, L. A., Sjodin, R. A., and Ortiz, O. (1975): The independence of membrane potential and potassium activation of the sodium pump in muscle. *Biochim. Biophys. Acta,* 389:189–193.
10. Bondani, A., and Withrow, C. D. (1965): Effects of ouabain on intracellular pH. *Fed. Proc.,* 24:487.
11. Brinley, F. J., and Mullins, L. J. (1968): Sodium fluxes in internally dialyzed squid axons. *J. Gen. Physiol.,* 52:181–211.
12. Brinley, F. J., and Mullins, L. J. (1974): Effects of membrane potential on sodium and potassium fluxes in squid axons. *Ann. NY Acad. Sci.,* 242:406–433.
13. Connelly, C. M. (1959): Recovery processes and metabolism in nerve. *Rev. Mod. Phys.,* 31:475–484.
14. Coombs, J. S., Eccles, J. C., and Fatt, P. (1955): The electrical properties of the motoneurone membrane. *J. Physiol. (Lond.)*, 130:291–325.
15. Cotterell, D., and Whittam, R. (1971): The influence of the chloride gradient across red cell membranes on sodium and potassium movements. *J. Physiol. (Lond.)*, 214:509–536.
16. Cross, S. R., Keynes, R. D., and Rybová, R. (1965): The coupling of sodium efflux and potassium influx in frog muscle. *J. Physiol. (Lond.)*, 181:865–880.
17. De Weer, P. (1975): Aspects of the recovery processes in nerve. In: *Neurophysiology, Vol. 3.,* edited by C. C. Hunt, pp. 232–278. Butterworths, London.

18. DiPolo, R. (1974): Effect of ATP on the calcium efflux in dialyzed squid giant axons. *J. Gen. Physiol.*, 64:503–517.
19. DiPolo, R., Rojas, H., and Beaugé, L. A. (1982): Ca entry at rest and during prolonged depolarization in dialyzed squid axons. *Cell Calcium,* 3:19–41.
20. Dissing, S., and Hoffman, J. F. (1983): Anion-coupled Na efflux mediated by the Na:K pump in human blood cells. *Curr. Top. Membranes & Transport,* 19: (*in press*).
21. Dixon, J. F., and Hokin, L. E. (1980): The reconstituted (Na,K)-ATPase is electrogenic. *J. Biol. Chem.*, 255:10681–10686.
22. Fossel, E. T., and Solomon, A. K. (1983): Relation between red cell membrane (Na,K)-ATPase and band 3. *Curr. Top. Membranes & Transport,* 19: (*in press*).
23. Funder, J., and Wieth, J. O. (1967): Effect of ouabain on glucose metabolism and on fluxes of sodium and potassium of human blood cells. *Acta Physiol. Scand.*, 71:113–124.
24. Gage, P. W., and Hubbard, J. I. (1964): Ionic changes responsible for post-tetanic hyperpolarization. *Nature,* 203:653–654.
25. Gardner, L. I., MacLachlan, E. A., and Berman, H. (1952): Effect of potassium deficiency on carbon dioxide, cation, and phosphate content of muscle. *J. Gen. Physiol.*, 36:153–159.
26. Garrahan, P. J., and Glynn, I. M. (1967): The behaviour of the sodium pump in red cells in the absence of external potassium. *J. Physiol. (Lond.),* 192:159–174.
27. Garrahan, P. J., and Glynn, I. M. (1967): The sensitivity of the sodium pump to external sodium. *J. Physiol. (Lond.),* 192:175–188.
28. Garrahan, P. J., and Glynn, I. M. (1967): The stoichiometry of the sodium pump. *J. Physiol. (Lond.),* 192:217–235.
29. Glynn, I. M., and Hoffman, J. F. (1971): Nucleotide requirements for sodium-sodium exchange catalysed by the sodium pump in human red cells. *J. Physiol. (Lond.),* 218:239–256.
30. Glynn, I. M., and Karlish, S. J. D. (1975): The sodium pump. *Annu. Rev. Physiol.,* 37:13–55.
31. Glynn, I. M., and Karlish, S. J. D. (1982): Conformational changes associated with K^+ transport by the Na^+/K^+-ATPase. In: *Membranes and Transport, Vol. 1,* edited by A. N. Martonosi, pp. 529–536. New York, Plenum.
32. Glynn, I. M., and Richards, D. E. (1982): Occlusion of rubidium ions by the sodium-potassium pump: Its implication for the mechanism of potassium transport. *J. Physiol. (Lond.),* 330:17–43.
33. Glynn, I. M., and Richards, D. E. (1983): The existence and role of occluded ion forms of Na^+,K^+-ATPase. *Curr. Top. Membranes & Transport,* 19: (*in press*).
34. Goldin, S. M. (1977): Active transport of sodium and potassium ions by the sodium and potassium ion-activated adenosine triphosphatase from renal medulla. Reconstitution of the purified enzyme into a well defined *in vitro* transport system. *J. Biol. Chem.,* 252:5630–5642.
35. Hering, E. (1884): Ueber positive Nachswankung des Nervenstroms nach elektrischer Reizung. Beiträge zur allgemein Nerven-und Muskelphysiologie. 15 Mitteilung, *Schriftenr. Balkankommn, Akad. Wiss., Wien,* 89:137–158.
36. Hilden, S., and Hokin, L. E. (1975): Active potassium transport coupled to active sodium transport in vesicles reconstituted from purified sodium and potassium ion-activated adenosine triphosphatase from the rectal gland of *Squalus acanthias*. *J. Biol. Chem.,* 250:6296–6303.
37. Hodgkin, A. L., and Keynes, R. D. (1955): Active transport of cations in giant axons from Sepia and Loligo. *J. Physiol. (Lond.),* 128:28–60.
38. Hoffman, J. F., Kaplan, J. H., and Callan, T. J. (1979): The Na:K pump in red cells is electrogenic. *Fed. Proc.,* 38:2440–2441.
39. Horowicz, P., and Gerber, C. J. (1965): Effects of external potassium and strophanthidin on sodium fluxes in frog striated muscle. *J. Gen. Physiol.,* 48:489–514.
40. Horowicz, P., and Gerber, C. J. (1965): Effects of sodium azide on sodium fluxes in frog striated muscle. *J. Gen. Physiol.,* 48:515–528.
41. Isenberg, G., and Trautwein, W. (1974): The effect of dihydro-ouabain and lithium-ions on the outward current in cardiac Purkinje fibers. *Pfluegers Arch.,* 350:41–54.
42. Karlish, S. J. D., and Pick, U. (1981): Sidedness of the effects of sodium and potassium ions on the conformational state of the sodium-potassium pump. *J. Physiol. (Lond.),* 312:505–529.
43. Karlish, S. J. D., Yates, D. W., and Glynn, I. M. (1978): Conformational transitions between Na^+-bound and K^+-bound forms of $(Na^+ + K^+)$—ATPase, studied with formycin nucleotides. *Biochim. Biophys. Acta,* 527:252–264.

44. Kernan, R. P. (1962): Membrane potential changes during sodium transport in frog sartorius muscle. *Nature*, 193:986–987.
45. Keynes, R. D. (1965): Some further observations on the sodium efflux in frog muscle. *J. Physiol. (Lond.)*, 178:305–325.
46. Koefoed-Johnson, V., and Ussing, H. H. (1958): The nature of the frog skin potential. *Acta Physiol. Scand.*, 42:298–308.
47. Kononenko, N. I., and Kostyuk, P. G. (1976): Further studies of the potential-dependence of the sodium-induced membrane current in snail neurones. *J. Physiol. (Lond.)*, 256:601–615.
48. Kostyuk, P. G., Krishtal, O. A., and Pidoplichko, V. I. (1972): Potential-dependent membrane current during the active transport of ions in snail neurones. *J. Physiol. (Lond.)*, 226:373–392.
49. Mårdh, S., and Post, R. L. (1977): Phosphorylation from adenosine triphosphate of sodium- and potassium-activated adenosine triphosphatase. Comparison of enzyme-ligand complexes as precursors to the phosphoenzyme. *J. Biol. Chem.*, 252:633–638.
50. Marmor, M. F. (1971): The independence of electrogenic sodium transport and membrane potential in a molluscan neurone. *J. Physiol. (Lond.)*, 218:599–608.
51. Meunier, J. M., and Tauc, L. (1970): Participation d'une pompe métabolique au potentiel de repos de neurones d'aplysie. *J. Physiol. (Paris.)*, 62:192C–193C.
52. Mullins, L. J., and Awad, M. Z. (1965): The control of the membrane potential of muscle fibers by the sodium pump. *J. Gen. Physiol.*, 48:761–775.
53. Mullins, L. J., and Brinley, F. J. (1969): Potassium fluxes in dialyzed squid axons. *J. Gen. Physiol.*, 53:704–740.
54. Mullins, L. J., and Requena, J. (1981): The "late" Ca channel in squid axons. *J. Gen. Physiol.*, 78:683–700.
55. Nakajima, S., and Takahashi, K. (1966): Post-tetanic hyperpolarization and electrogenic Na pump in stretch receptor neurone of crayfish. *J. Physiol. (Lond.)*, 187:105–127.
56. Nielsen, R. (1979): Coupled transepithelial sodium and potassium transport across isolated frog skin: Effect of ouabain, amiloride and the polyene antibiotic filipin. *J. Membr. Biol.*, 51:161–184.
57. Nielsen, R. (1979): A 3 to 2 coupling of the Na-K pump responsible for the transepithelial Na transport in frog skin disclosed by the effect of Ba. *Acta Physiol. Scand.*, 107:189–191.
58. Post, R. L., Albright, C. D., and Dayani, K. (1967): Resolution of pump and leak components of sodium and potassium ion transport in human erythrocytes. *J. Gen. Physiol.*, 50:1201–1220.
59. Post, R. L., Hegyvary, C., and Kume, S. (1972): Activation by adenosine triphosphate in the phosphorylation kinetics of sodium and potassium ion transport adenosine triphosphatase. *J. Biol. Chem.*, 247:6530–6540.
60. Post, R. L., and Jolly, P. C. (1957): The linkage of sodium, potassium, and ammonium active transport across the human erythrocyte membrane. *Biochim. Biophys. Acta.*, 25:118–128.
61. Rakowski, R. F., and De Weer, P. (1983): Forward and reverse electrogenic pumping by the Na/K pump of squid giant axon. *J. Gen. Physiol. (in press)*.
62. Rang, H. P., and Ritchie, J. M. (1968): On the electrogenic sodium pump in mammalian non-myelinated nerve fibres and its activation by various external cations. *J. Physiol. (Lond.)*, 196:183–221.
63. Ritchie, J. M., and Straub, R. W. (1957): The hyperpolarization which follows activity in mammalian non-medullated fibres. *J. Physiol. (Lond.)*, 136:80–97.
64. Sachs, J. R. (1972): Recoupling of the Na-K pump. *J. Clin. Invest.*, 51:3244–3247.
65. Sachs, J. R. (1977): Kinetic evaluation of the Na-K pump reaction mechanism. *J. Physiol. (Lond.)*, 273:489–514.
66. Sjodin, R. A. (1982): *Transport in Skeletal Muscle.* Wiley, New York.
67. Slayman, C. L. (1982): Historical introduction to proceedings of symposium on electrogenic ion pumps, at Yale in 1981. In: *Current Topics in Membranes and Transport*, 16:xxxi–xxxvii.
68. Sweadner, K. J., and Goldin, S. M. (1975): Reconstitution of active ion transport by the sodium and potassium ion-stimulated adenosine triphosphatase from canine brain. *J. Biol. Chem.*, 250:4022–4024.
69. Thomas, R. C. (1969): Membrane current and intracellular sodium changes in a snail neurone during extrusion of injected sodium. *J. Physiol. (Lond.)*, 201:495–514.
70. Thomas, R. C. (1972): Electrogenic sodium pump in nerve and muscle cells. *Physiol. Rev.*, 52:563–594.

71. Warncke, J., and Slayman, C. L. (1980): Metabolic modulation of stoichiometry in a proton pump. *Biochim. Biophys. Acta,* 591:224–233.
72. Whittam, R., and Ager, M. E. (1965): The connection between active ion transport and metabolism in erythrocytes. *Biochem. J.,* 97:214–227.
73. Wieth, J. O. (1969): Effects of bicarbonate and thiocyanate on fluxes of Na and K, and on glucose metabolism of actively transporting human red cells. *Acta Physiol. Scand.,* 75:313–329.
74. Williams, J. A., Withrow, C. D., and Woodbury, D. M. (1971): Effects of ouabain and diphenylhydantoin on transmembrane potentials, intracellular electrolytes, and cell pH of rat muscle and liver in vivo. *J. Physiol. (Lond.),* 212:101–115.
75. Zade-Oppen, A. M. M., Schooler, J. M., Cook, P., and Tosteson, D. C. (1979): Effect of membrane potential and internal pH on active sodium-potassium transport and on ATP content in high-potassium sheep erythrocytes. *Biochim. Biophys. Acta,* 555:285–298.

Electrogenic Transport: Fundamental Principles and Physiological Implications, edited by Mordecai P. Blaustein and Melvyn Lieberman. Raven Press, New York © 1984.

Electrochemical Driving Forces for Secondary Active Transport: Energetics and Kinetics of Na^+-H^+ Exchange and Na^+-Glucose Cotransport

Peter S. Aronson

Departments of Medicine and Physiology, Yale University School of Medicine, New Haven, Connecticut 06510

Secondary active transport may be defined as a process whereby the net flux of a solute up its electrochemical potential difference across a cell membrane is directly coupled to, and thus energized by, the net flow of one or more other solutes down their electrochemical potential differences across the same membrane (3,38). Such coupling of solute fluxes arises from specific transport systems that mediate the cotransport (symport) and/or exchange (antiport) of solutes. This mechanism contrasts with primary active transport, whereby the uphill flux of a solute is directly coupled to, and thus energized by, the flow of an exergonic chemical reaction such as adenosine triphosphate (ATP) hydrolysis (3,38). The Na^+-coupled transport systems for a variety of solutes including sugars, amino acids, organic acids, phosphate, sulfate, Cl^-, and H^+ have been identified in the plasma membranes of animal cells, supporting the concept that energization of secondary active transport in animal cells is generally achieved by coupling solute flux to the inwardly directed electrochemical Na^+ gradient that exists across the cell membrane due to the primary active extrusion of Na^+ from the cell (15,21,32,35,37). However, in some cases, ion gradients generated by Na^+-coupled transport can themselves serve to energize secondary active transport via transport pathways not directly coupled to Na^+. For example, the outwardly directed electrochemical gradient of OH^- or HCO_3^- generated by plasma membrane Na^+-H^+ exchange provides a possible driving force for the uphill transport of Cl^- or organic anions into the cell via anion exchange (4,13,30).

This chapter will describe how rates of net solute flux through secondary active transport systems depend on the electrochemical ion gradients that are their driving forces. Toward this end, the general principles governing the energetics and kinetics of secondary active transport systems will be outlined. Various properties of two secondary active transport systems located in the luminal (brush border, microvillus) membrane of renal tubular cells, namely, the Na^+-

H⁺ exchanger and the Na⁺-glucose cotransporter, will be described to illustrate these principles.

ENERGETICS OF SECONDARY ACTIVE TRANSPORT

Energetic considerations allow us to establish a series of inequalities that define the conditions under which net solute transport in a given direction may proceed through a particular transport pathway. For any transport process, net solute flux in a given direction may spontaneously occur only when the total free-energy change associated with the transport event is negative ($\Delta G_{total} < 0$). In essence, then, the negative free-energy change for a transport process is the driving force for that process. The source of the negative free energy required to drive solute flux in a given direction depends on the nature of the underlying transport process, as illustrated in Fig. 1.

Let us first briefly consider the case of passive or noncoupled diffusion, whereby the flow of a solute is not coupled to the flows of other solutes, the progress of a chemical reaction, or the flow of solvent. For n_A moles of solute A moving from the outside to the inside compartment (net $J_A^{o \rightarrow i} > 0$), the associated free-energy change is given by:

$$\Delta G_{total} = -n_A \, \Delta \tilde{\mu}_A^{o\text{-}i} \tag{1}$$

where $\Delta \tilde{\mu}_A^{o\text{-}i}$ is the inwardly directed electrochemical potential difference for A. The inwardly directed electrochemical potential difference for A is defined by:

$$\Delta \tilde{\mu}_A^{o\text{-}i} = RT \ln \frac{(A)_o}{(A)_i} + Z_A F(\Psi_o - \Psi_i) + \bar{S}_A(T_o - T_i) + \bar{V}_A (P_o - P_i) \tag{2}$$

where (A), Ψ, T, and P represent A activity, electrical potential, temperature, and pressure, respectively, in the indicated compartment, and $\bar{S}_A$, $\bar{V}_A$, and Z_A signify the partial molar entropy, the partial molar volume, and the charge, respectively, of A. For substances other than water, the hydrostatic pressure

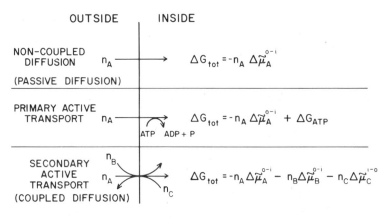

FIG. 1. Energetics of transport processes. [From Aronson (3), with permission.]

term is generally negligible compared to the other terms in Eq. 2. Hence, in the absence of a temperature gradient, the inwardly directed electrochemical potential difference for solute A may be approximated by:

$$\Delta \tilde{\mu}_A^{o\text{-}i} = RT \ln \frac{(A)_o}{(A)_i} + Z_A F (\Psi_o - \Psi_i) \qquad (3)$$

For net flux of A to occur from the outside to the inside compartment (net $J_A^{o \to i} > 0$), ΔG_{total} must be <0 and, thus, there must exist an inwardly directed electrochemical potential difference for A ($\Delta \tilde{\mu}_A^{o\text{-}i} > 0$). In the case of passive or noncoupled diffusion of solute A, the only driving force is the electrochemical potential difference for A across the membrane. This is true regardless of the microscopic mechanism by which passive A transport takes place (e.g., simple diffusion, carrier-mediated transport).

Active transport may be most simply defined (34) as a process that can effect the net transport of a solute against its electrochemical potential difference (i.e., net $J_A^{o \to i} > 0$ despite $\Delta \tilde{\mu}_A^{o\text{-}i} < 0$). In the case of primary active A transport, the additional driving force necessary to allow the net transfer of A up its electrochemical potential difference is provided by coupling A transport to the progress of a chemical reaction, such as ATP hydrolysis, for which there is a large negative ΔG.

In the case of secondary active A transport, the supplemental driving force to allow net transport of A up its electrochemical potential difference is provided by coupling A diffusion to the flow of one or more other solutes down their electrochemical potential differences. For example, Fig. 1 includes a scheme in which n_B moles of B are cotransported per n_A moles of A, and/or n_C moles of C are exchanged for n_A moles of A. For the net inward transfer of each n_A moles of A, the free-energy change is:

$$\Delta G_{total} = - n_A \Delta \tilde{\mu}_A^{o\text{-}i} - n_B \Delta \tilde{\mu}_B^{o\text{-}i} - n_C \Delta \tilde{\mu}_C^{i\text{-}o} \qquad (4)$$

Thus, ΔG_{total} can be brought <0, thereby allowing net inward movement of A to proceed against its own electrochemical potential difference ($\Delta \tilde{\mu}_A^{o\text{-}i} < 0$), if the inwardly directed electrochemical potential difference for B or the outwardly directed electrochemical potential difference for C is sufficiently large. That is, net active transport of A in the inward direction can occur as long as:

$$-\Delta \tilde{\mu}_A^{o\text{-}i} < \frac{n_B}{n_A} \Delta \tilde{\mu}_B^{o\text{-}i} + \frac{n_C}{n_A} \Delta \tilde{\mu}_C^{i\text{-}o} \qquad (5)$$

The greater the magnitude of the inwardly directed $\Delta \tilde{\mu}_B$ or the outwardly directed $\Delta \tilde{\mu}_C$, and the larger the stoichiometries of coupling (n_B/n_A, n_C/n_A), the greater is the outwardly directed $\Delta \tilde{\mu}_A$ against which net inward J_A can proceed. The condition for equilibrium of this coupled diffusion system (i.e., $\Delta G_{total} = 0$, net $J_A^{o \to i} = 0$), will be when $-\Delta \tilde{\mu}_A^{o\text{-}i}$ is exactly balanced by the sum of ($n_B/$

n_A) $\Delta\tilde{\mu}_B^{o-i}$ and (n_C/n_A) $\Delta\tilde{\mu}_C^{i-o}$. Accordingly, the limiting A activity ratio is given by:

$$\frac{(A)_i}{(A)_o} \leq \left[\frac{(B)_o}{(B)_i}\right]^{n_B/n_A} \left[\frac{(C)_i}{(C)_o}\right]^{n_C/n_A}$$

$$\cdot \exp\left[\frac{F}{RT}(\Psi_o - \Psi_i)\left(Z_A + \frac{n_B}{n_A}Z_B - \frac{n_C}{n_A}Z_C\right)\right] \quad (6)$$

where the inequality represents the condition for net inward A flux and the equality represents the condition for equilibrium (net $J_A^{o \to i} = 0$ via this pathway).

Applying Eq. (6) to the case of 1:1 Na⁺-H⁺ exchange allows us to calculate the limiting H⁺ gradient that can be generated by an inward Na⁺ gradient:

$$\frac{(H^+)_o}{(H^+)_i} \leq \frac{(Na^+)_o}{(Na^+)_i} \quad (7)$$

or the limiting Na⁺ gradient that can be generated by an outward H⁺ gradient:

$$\frac{(Na^+)_i}{(Na^+)_o} \leq \frac{(H^+)_i}{(H^+)_o} \quad (8)$$

Likewise, applying Eq. (6) to the case of 1:1 Na⁺-glucose cotransport allows us to calculate the limiting glucose gradient that can be generated by an inwardly directed electrochemical Na⁺ gradient:

$$\frac{(G)_i}{(G)_o} \leq \frac{(Na^+)_o}{(Na^+)_i} \exp\frac{F}{RT}(\Psi_o - \Psi_i) \quad (9)$$

Equations (8) and (9) predict that an outward H⁺ gradient should drive uphill Na⁺ accumulation and that an inward Na⁺ gradient should drive uphill glucose uptake. Both of these phenomena have been demonstrated in luminal membrane vesicles isolated from the proximal tubule and small intestine (8,23,25,26,33), as illustrated in Fig. 2 for rabbit renal microvillus membrane vesicles. These membrane vesicles are oriented right-side-out (5,20), contain an osmotically active internal space (10,23,25), do not metabolize substrates such as glucose (7,25), and thus have no apparent energy source to drive uphill transport other than that which may be imposed in the form of electrochemical ion gradients. Uptake of radiolabeled solutes by these vesicles is easily assayed by rapid-filtration techniques (8,23). The left panel of Fig. 2 demonstrates that an inside-acid pH gradient (pH$_i$ 5.9, pH$_o$ 7.2) stimulates the rate of Na⁺ uptake, and induces the transient accumulation of Na⁺ above its eventual equilibrium level of uptake. The right panel demonstrates that an inward Na⁺ gradient (Na$_o^+$ 100 mM, Na$_i^+$ 0 mM) stimulates the rate of D-glucose uptake and induces the transient accumulation of glucose above its equilibrium level. The uphill accumulation of solutes via secondary active transport is transient and has an overshoot pattern in these membrane preparations because the initially imposed ion gradients dissipate relatively rapidly with time.

Another prediction from Eqs. 8 and 9 is that the transmembrane electrical

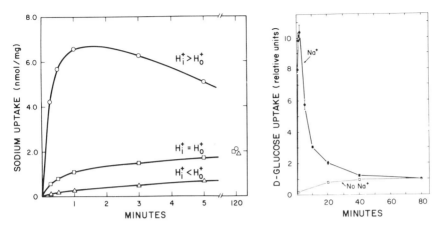

FIG. 2. Secondary active solute accumulation in renal microvillus membrane vesicles. Uptake of 1 mM Na^+ **(left)** was assayed at pH 7.2, using vesicles preloaded with pH 5.9 buffer (○); at pH 7.4, using vesicles preloaded with pH 7.4 buffer (□); or at pH 6.1, using vesicles preloaded with pH 7.4 buffer (△). Uptake of 50 μM D-glucose **(right)** in the presence (●) or absence (○) of 100 mM NaCl was assayed using vesicles preloaded with Na^+-free buffer. [From Kinsella and Aronson (26) and Aronson and Sactor (8), with permission.]

potential difference should be a driving force for the accumulation of glucose via Na^+-cotransport but not for the accumulation of Na^+ via Na^+-H^+ exchange. This prediction has also been verified in studies utilizing luminal membrane vesicles isolated from the intestine and kidney (10,26,31,33). In the experiment illustrated in Fig. 3, rabbit renal microvillus membrane vesicles were preloaded with K^+ and then diluted 8-fold so that an outwardly directed K^+ gradient was imposed. Under this circumstance, addition of the K^+ ionophore valinomycin should move the membrane potential in the direction of greater inside negativity. Whereas Na^+-dependent glucose accumulation is markedly enhanced by the inside-negative membrane potential induced by valinomycin, as shown in the left panel of Fig. 3, the uptake of Na^+ measured in the absence of glucose is unaffected by the ionophore, as shown in the right panel. The latter result is consistent with the fact that Na^+ uptake in rabbit renal microvillus membrane vesicles occurs predominantly by electroneutral Na^+-H^+ exchange even in the absence of an outward H^+ gradient (26,28).

The predictions of Eqs. 8 and 9 concerning the conditions under which the Na^+-H^+ exchanger and the Na^+-glucose cotransporter should be at equilibrium have also been evaluated (29,41), as illustrated in Fig. 4. The panel on the left shows the results of an experiment in which rabbit renal microvillus membrane vesicles were preloaded with Na^+ at pH_i 6.5 and then diluted 1:10 into media of varying pH_o after which the 5-sec net flux of Na^+ was assayed. The condition for equilibrium (i.e., no net flux) was pH_o 7.5, when $(H^+)_i/(H^+)_o$ was equal to $(Na^+)_i/(Na^+)_o$ in accord with Eq. (8). In addition, the condition for no net flux was unaltered by the proton ionophore FCCP, which, in the

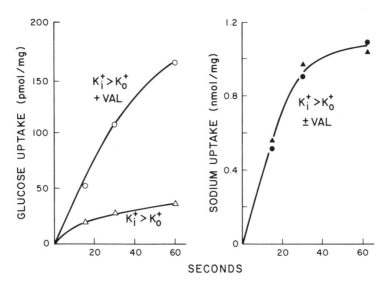

FIG. 3. Effect of membrane potential on solute uptake in renal microvillus membrane vesicles. Uptake of 3.5 μM D-glucose **(left)** in the presence of 132 mM Na$^+$ and 8 mM K$^+$ was assayed using vesicles preloaded with 66 mM K$^+$. Uptake of 1 mM Na$^+$ **(right)** in the presence of 8 mM K$^+$ was assayed using vesicles preloaded with 66 mM K$^+$. These experiments were performed in the absence (△, ▲) or presence (○, ●) of valinomycin (10 μg/mg protein). [From Kinsella and Aronson (26), with permission.]

presence of an inside-acid pH gradient, shifts the membrane potential toward greater inside negativity (10,26). Turner and Moran (41) have similarly evaluated the conditions under which the Na$^+$-glucose cotransporter in rabbit renal outer cortical brush border membranes is at equilibrium, as shown in the right panel of Fig. 4. Here, membrane vesicles were preloaded with glucose at Na$_i$ 20 mM and then diluted 1:6 into media of varying Na$_o^+$ after which the 2-sec net flux of glucose was assayed. In this experiment, the transmembrane electrical potential difference was shunted to 0 mV by adding valinomycin in the presence of $K_i^+ = K_o^+$. The condition for equilibrium was rather precisely determined to be Na$_o^+$ 120 mM, when (Na$^+$)$_o$/(Na$^+$)$_i$ was equal to $(G)_i/(G)_o$ in perfect accord with Eq. 9. Clearly, in the experiment shown on the left any component of noncoupled Na$^+$ transport must have been small compared to the flux of Na$^+$ via 1:1 Na$^+$-H$^+$ exchange, and in the experiment shown on the right, the noncoupled transport of glucose must have been small compared to the flux of glucose via 1:1 Na$^+$-glucose cotransport.

KINETICS OF SECONDARY ACTIVE TRANSPORT

As already discussed, energetic considerations allow us to define the conditions under which a given transport process will be at equilibrium, determine whether there is displacement from equilibrium, and then predict the direction in which

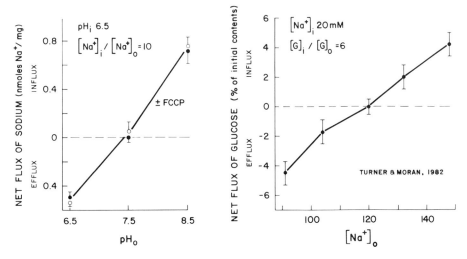

FIG. 4. Determination of equilibrium conditions for coupled transport systems in renal microvillus membrane vesicles. The 5-sec net flux of Na+ **(left)** in the presence of 0.1 mM Na+ and varying external pH was assayed using vesicles preloaded with 1.0 mM Na+ at pH 6.5. This was performed in the absence (●) or presence (○) of FCCP (22 μg/ml). The 2-sec net flux of D-glucose **(right)** in the presence of 83 μM glucose, 100 mM K+, and varying external Na+ was assayed using vesicles preloaded with 500 μM glucose, 100 mM K+ and 20 mM Na+. This was performed in the presence of valinomycin (12.5 μg/mg protein). [Data from Kinsella and Aronson (29) and Turner and Moran (41), with permission.]

net transport must proceed in response to the displacement from equilibrium. Unfortunately, considerations based only on equilibrium thermodynamics do not allow one to predict the rate at which such net transport will proceed. For example, in the case of net Na+ influx across a membrane via 1:1 Na+-H+ exchange, the net thermodynamic driving force is the sum of the inwardly directed electrochemical potential difference for Na+ and the outwardly directed electrochemical potential difference for H+. As long as this sum has a positive value, there will be net influx of outside Na+ in exchange for inside H+. But what is the quantitative relationship between the rate of net Na+ entry and the magnitude of the electrochemical potential differences for Na+ and H+? As will be illustrated, such quantitative kinetic relationships depend uniquely on the microscopic details of the particular secondary active transport process of interest. Such kinetic relationships cannot be predicted on the basis of energetic considerations alone. Indeed, we will see that under certain conditions the rate of solute flux may not vary at all or may even vary inversely with the magnitude of its net thermodynamic driving force. Several factors that may complicate the relationship between transport rate and electrochemical driving force are well illustrated by certain aspects of the kinetics of Na+-H+ exchange and Na+-glucose cotransport in renal microvillus membrane vesicles.

Saturability of Substrate Sites

One obvious reason for the rate of solute transport to be insensitive to alterations in electrochemical driving force is that transport systems function as if they possess saturable binding sites for solutes. For example, the rate of Na^+ influx via Na^+-H^+ exchange in rabbit renal microvillus membranes conforms to simple saturation kinetics with:

$$J_{Na}^{o \to i} = \frac{J_{max}(Na^+)_o}{(Na^+)_o + app\ K_{Na}} \qquad (10)$$

where the apparent (app) K_{Na} has a value of 6 to 13 mM at pH_o 7.5 (9,27,28,42). Clearly, when $(Na^+)_o \gg app\ K_{Na}$, the rate of Na^+ influx approaches its maximal value and becomes insensitive to changes in $(Na^+)_o$. Thus, even though the inwardly directed Na^+ gradient is part of the thermodynamic driving for Na^+ entry, the rate of Na^+ influx is insensitive to changes in the magnitude of the Na^+ gradient when such changes are induced by varying $(Na^+)_o$ in a concentration range well above K_{Na}.

The dependence of the transport rate of a solute on the electrochemical potential difference(s) of the other solute(s) to which its transport is coupled also is characterized by saturability. For example, the outwardly directed H^+ gradient is part of the driving force for Na^+ influx via Na^+-H^+ exchange. In the experiment illustrated in Fig. 5, the initial rate of 0.1 mM Na^+ uptake into vesicles with internal pH 6.0 was measured as a function of external pH (9). The Na^+ influx was inhibited as external pH decreased, the results conforming to a simple titration curve with apparent pK 7.3. That is, the inhibition of Na^+ influx by external H^+ appears to arise from titration of a single site with an apparent pK of 7.3. In studies not illustrated, we have found that external H^+ acts as a simple competitive inhibitor of Na^+ influx (9). Indeed, when inhibition by external H^+ was studied as a function of external Na^+ concentration in these studies, an apparent K_H of 38 nM was determined, in excellent agreement with the apparent pK of 7.3 indicated in Fig. 5. From a kinetic point of view, these results can be easily explained on the basis of a simple model in which external Na^+ and H^+ compete for binding to a single transport site with apparent pK 7.3 to 7.4. It is important to note that although the outwardly directed H^+ gradient is part of the thermodynamic driving force for Na^+ entry, the rate of Na^+ influx has a variable dependence on the magnitude of the transmembrane pH gradient. For example, Fig. 5 indicates that Na^+ influx is insensitive to variation in the transmembrane pH gradient at high values of external pH ($pH_o \gg app\ pK$, $(H^+)_o \ll app\ K_H$) but is strongly dependent on the transmembrane pH gradient at lower values of external pH ($pH_o \leq app\ pK$, $(H^+)_o \geq app\ K_H$). This is in accord with the fact that if external H^+ is a competitive inhibitor of Na^+ influx, then:

$$J_{Na}^{o \to i} = \frac{J_{max}(Na^+)_o}{(Na^+)_o + app\ K_{Na}(1 + (H^+)_o/app\ K_H)} \qquad (11)$$

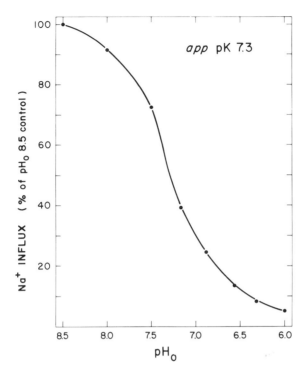

FIG. 5. Effect of external pH on Na$^+$ influx in renal microvillus membrane vesicles. The 2-sec uptake of 0.1 mM Na$^+$ in the presence of varying external pH was assayed using vesicles preloaded with pH 6.0 buffer. [From Aronson et al. (9), with permission.]

Clearly, at $(H^+)_o \ll$ app K_H, the rate of Na$^+$ influx will be insensitive to changes in $(H^+)_o$.

Alternative Transport Modes of Varying Kinetic Favorability

Another factor that can complicate the relationship between transport rate and electrochemical driving force is that transport systems can often function in different modes, which vary in their kinetic favorability. For example, Li$^+$ and NH$_4^+$ each can compete with Na$^+$ and H$^+$ for binding to the external site of the renal microvillus membrane Na$^+$-H$^+$ exchanger (9,27). Indeed, an inwardly directed gradient of Li$^+$ can drive uphill H$^+$ efflux, confirming the presence of a Li$^+$-H$^+$ exchange mode (26,27). (An inwardly directed NH$_4^+$ gradient also generates an inside-alkaline pH gradient (27), although this may reflect entry of NH$_4^+$ via nonionic diffusion of NH$_3$ rather than mediated NH$_4^+$-H$^+$ exchange). In the experiment illustrated in Fig. 6, the effects of external Li$^+$, Na$^+$, and NH$_4^+$ on ^{22}Na efflux were examined (27). Membrane vesicles containing 1 mM ^{22}Na at pH$_i$ 7.5 were diluted 1:20 into ^{22}Na-free medium in the absence or

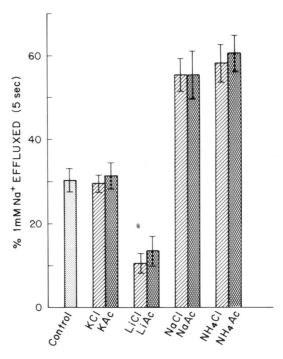

FIG. 6. Effect of external cations on Na$^+$ efflux from renal microvillus membrane vesicles. The 5-sec efflux of ^{22}Na was assayed using vesicles preloaded with 1 mM ^{22}Na at pH 7.5 and then diluted 1:20 into ^{22}Na-free media at pH 7.5 containing no added salts (control) or containing chloride or acetate salts of K$^+$ (10 mM), Li$^+$ (5 mM), Na$^+$ (10 mM), or NH$_4^+$ (10 mM). [From Kinsella and Aronson (27), with permission.]

presence of external K$^+$, Li$^+$, Na$^+$, or NH$_4^+$ at pH$_o$ 7.5, and then the efflux of ^{22}Na was measured over the subsequent 5-sec period. It should be emphasized that Na$^+$ efflux, like influx, is almost entirely sensitive to inhibition by amiloride (6,28), a specific inhibitor of the Na$^+$-H$^+$ exchanger in these membrane vesicles (28). External K$^+$ had no effect on Na$^+$ efflux, consistent with its inability to inhibit Na$^+$ influx or drive H$^+$ efflux in renal microvillus membranes (26). In contrast, external Li$^+$ markedly inhibited the rate of ^{22}Na efflux, whereas external Na$^+$ and NH$_4^+$ caused significant stimulation. The trans effects of these cations were independent of whether Cl or acetate salts were used. Since the effect of external Na$^+$, NH$_4^+$ or Li$^+$ to induce an inside-alkaline pH gradient is prevented by use of acetate as the accompanying anion (27), the effects of external Li$^+$, Na$^+$, and NH$_4^+$ on ^{22}Na efflux in Fig. 6 must have been direct and could not have been secondary to alterations in the transmembrane pH gradient.

The stimulation of ^{22}Na efflux by external Na$^+$ or NH$_4^+$, demonstrating the existence of Na$^+$-Na$^+$ and Na$^+$-NH$_4^+$ exchange modes, in principle, could arise in two ways. First, since the exchanger at pH$_o$ 7.5 is not fully saturated by external H$^+$, additional carriers could be recruited by external Na$^+$ or NH$_4^+$

so that efflux of ^{22}Na in exchange for Na$^+$ or NH$_4^+$ supplements that in exchange for external H$^+$. This does not appear to be a fully adequate explanation, however, because experiments not illustrated have shown that trans stimulation of ^{22}Na efflux by external Na$^+$ at pH$_o$ 7.5 is greater than the stimulation of ^{22}Na efflux that results from reducing external pH to values as low as 6.0 (9). This leaves the second possibility, namely that exchange of internal ^{22}Na for external Na$^+$ or NH$_4^+$ is kinetically favored over exchange for external H$^+$. Such would be the case if the rate limiting step for Na$^+$ or NH$_4^+$ entry (i.e., translocation or debinding at the inside) is faster than that for H$^+$ entry. Thus, displacement of H$^+$ at the external cation binding site by Na$^+$ or NH$_4^+$ would lead to acceleration of ^{22}Na efflux. Conversely, the trans inhibition by Li$^+$ suggests that exchange of internal ^{22}Na for external Li$^+$ is kinetically less favorable than exchange for external H$^+$. This implies that the translocation or debinding of Li$^+$ must be slower than that of H$^+$, thereby explaining why displacement of H$^+$ at the external cation binding site by Li$^+$ leads to inhibition of ^{22}Na efflux. Taken together, these studies suggest that the rate constants for the rate-limiting step in cation influx must be in the order Na$^+ \sim$ NH$_4^+ >$ H$^+ >$ Li$^+$. Consistent with the concept that the rate-limiting step for Li$^+$ entry must be slower than that for Na$^+$ are the studies of Ives et al. (24) showing that the J_{max} for Li$^+$-H$^+$ exchange is only 25% of that for Na$^+$-H$^+$ exchange.

It should be noted that these effects of external cations on ^{22}Na efflux could not have been predicted simply on the basis of the associated changes in electrochemical driving forces. For example, let us compare the driving forces for Na$^+$ efflux in the control case to that in the presence of external Li$^+$. In both cases, the outwardly directed Na$^+$ gradient is the same and there is no transmembrane H$^+$ gradient. Based on energetic considerations, one might have predicted that adding an inwardly directed gradient of a cation such as Li$^+$, which shares the exchanger, could only serve to accelerate Na$^+$ efflux. In fact, just the opposite effect was observed due to the fact that Na$^+$-Li$^+$ exchange is a kinetically less favorable mode than is Na$^+$-H$^+$ exchange.

Transportable Solute Acting as Allosteric Modifier

A third factor that can make the relationship between transport rate and electrochemical driving force less than straightforward is that a solute that is a substrate for transport by a transport system may, in certain cases, also be an allosteric modifier of that same transport system. This is well illustrated by the interaction of internal H$^+$ with the Na$^+$-H$^+$ exchanger in renal microvillus membrane vesicles. In the experiment shown in Fig. 7, the rate of influx of 1 mM Na$^+$ at external pH 7.3 was measured as a function of internal pH (6). As expected, the rate of Na$^+$ influx was progressively stimulated as internal pH was lowered, and this stimulation was abolished by the Na$^+$-H$^+$ exchange inhibitor, amiloride. As discussed earlier, the coupling ratio of the Na$^+$-H$^+$ exchanger appears to be 1.0 and the kinetics of Na$^+$ influx are consistent with

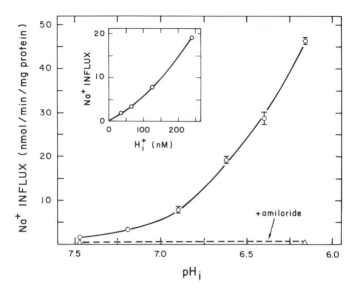

FIG. 7. Effect of internal pH on Na$^+$ influx in renal microvillus membrane vesicles. The 5-sec uptake of 1 mM Na$^+$ in the presence (△) and absence (○) of 1 mM amiloride at external pH 7.3 was assayed using vesicles preloaded with media of varying pH. [Adapted from Aronson et al. (6), with permission.]

the existence of a single transport site for external Na$^+$. Taken together, these findings suggest that the stoichiometry of exchange is 1:1 and thus that there should exist a single transport site for internal H$^+$. However, as indicated in the inset to Fig. 7, a consistent observation was that the Na$^+$ influx rate had a greater than first-order dependence on the internal H$^+$ concentration, implying that in addition to the single transport site for internal H$^+$, there might exist one or more allosteric modifier sites at which internal H$^+$ could bind and activate the exchanger.

Additional experiments were performed to confirm that internal H$^+$, independent of its role as transportable substrate for exchange, could actually activate the Na$^+$-H$^+$ exchanger. In an experiment not illustrated, we examined the effect of internal pH on the rate of ^{22}Na uptake occurring via exchange for internal Na$^+$ (i.e., Na$^+$-Na$^+$ exchange). In the absence of an internal activation site for H$^+$, one would expect that raising internal H$^+$ would, if anything, inhibit the rate of Na$^+$-Na$^+$ exchange, as internal H$^+$ might compete with Na$^+$ at the internal transport site. In fact, we observed that Na$^+$-Na$^+$ exchange was significantly greater at internal pH 6.9 than at internal pH 7.5, supporting the existence of an internal H$^+$ activator site (6).

In the experiment shown in Fig. 8, the effect of internal pH on the rate of amiloride-sensitive ^{22}Na efflux was determined (6). Membrane vesicles were preloaded with 50 mM ^{22}Na at internal pH 7.5 or 6.9, diluted 1:100 into a Na$^+$-

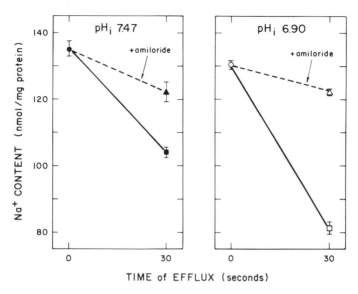

FIG. 8. Effect of internal pH on Na$^+$ efflux from renal microvillus membrane vesicles. The 30-sec efflux of ^{22}Na was assayed using vesicles preloaded with 50 mM ^{22}Na at pH 7.47 **(left)** or pH 6.90 **(right)** and then diluted 1:100 into ^{22}Na-free media at external pH 7.47 in the presence (▲, △) or absence (■, □) of 1 mM amiloride. [From Aronson et al. (6), with permission.]

free medium at external pH 7.5, and then the net efflux of ^{22}Na was assayed over the subsequent 30-sec period. Clearly, Na$^+$ efflux from vesicles with internal pH 6.9 was significantly greater than from vesicles with internal pH 7.5. The internal H$^+$-stimulated component of Na$^+$ efflux was amiloride-sensitive, consistent with its occurring via the Na$^+$-H$^+$ exchanger. Here we have a clear-cut example of transport rate varying inversely with electrochemical driving force. In vesicles with internal pH 7.5, the net thermodynamic driving force for Na$^+$ efflux is the 100:1 outwardly directed Na$^+$ gradient unopposed by any H$^+$ gradient, whereas in vesicles with internal pH 6.9, the same outwardly directed Na$^+$ gradient is opposed by a 4:1 outwardly directed H$^+$ gradient. Thus, efflux was faster under the conditions in which its thermodynamic driving force was smaller. No laws of thermodynamics are violated, of course, since even at internal pH 6.9 the net thermodynamic driving force for Na$^+$ efflux (i.e., $\Delta\bar{\mu}_{Na}^{i-o} - \Delta\bar{\mu}_{H}^{i-o}$) is still positive. As schematically illustrated in Fig. 9, the kinetic effect of internal H$^+$ to allosterically activate the Na$^+$-H$^+$ exchanger is apparently more than sufficient to compensate for the expected effect of internal H$^+$ to compete with Na$^+$ at the internal transport site. In the experiment shown in Fig. 8, a relatively high internal Na$^+$ concentration (50 mM) was employed to minimize H$^+$ inhibition of Na$^+$ binding at the internal transport site. Thus, Na$^+$ efflux was stimulated by internal H$^+$ despite a diminished driving force.

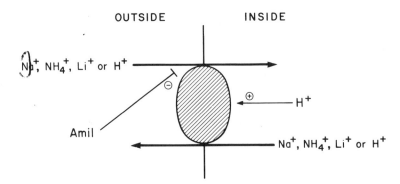

FIG. 9. Interaction of H+ with the activator and transport sites of the renal microvillus membrane Na+–H+ exchanger.

Potential Dependence of Rate-Limiting Step not Corresponding to Net Electrogenicity

A fourth problem that can cloud the relationship between transport rate and thermodynamic driving force is that the electrical potential dependence of the rate-limiting step in transport of a solute may not necessarily correspond to the net electrogenicity of the overall transport process. The rate of solute transport via a truly electrogenic process, under certain conditions, may be rather insensitive to alterations in the membrane potential (i.e., flat current versus voltage curve), whereas the rate of solute transport via an electroneutral process may, at times, be influenced by the membrane potential. This general issue is well illustrated by the potential dependence of D-glucose transport in renal microvillus membrane vesicles.

The Na+-glucose cotransport in renal microvillus membrane vesicles is unquestionably an electrogenic process as evidenced by the observations that in the presence of Na+ an inwardly directed D-glucose gradient generates an inside-positive membrane potential (12) and that an inside-negative membrane potential can drive uphill glucose accumulation in the presence of Na+ but absence of a chemical Na+ gradient [i.e., $(Na^+)_o = (Na^+)_i$] (11). The rate of Na+-dependent glucose transport is potential-dependent in the manner expected for a transport process that is associated with the movement of positive charge (10,31). Maneuvers that move the membrane potential toward greater inside negativity stimulate glucose influx, whereas maneuvers that move the membrane potential toward greater inside positivity causes inhibition. This is one case where changes in transport rate correspond well to alterations in electrochemical driving force.

However, as illustrated in Fig. 10, the rate of D-glucose uptake even in the absence of Na+ is still potential-dependent. In this experiment, conducted by Hilden and Sacktor (22), an outwardly directed K+ gradient was imposed, and then uptakes of D-glucose and L-glucose were measured in the presence and

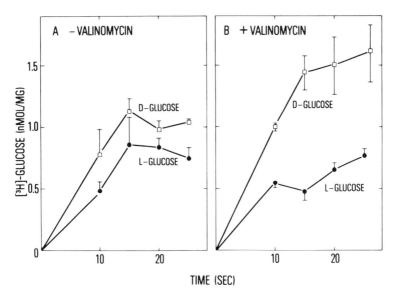

FIG. 10. Effect of membrane potential on Na^+-independent D-glucose transport in renal microvillus membrane vesicles. Uptake of 5 mM D-glucose (□) or L-glucose (●) in the absence of Na^+ and presence of 12.5 mM K^+ was assayed using vesicles preloaded with 200 mM K^+. This was performed in the absence **(A)** or presence **(B)** of valinomycin (100 μg/ml). [From Hilden and Sacktor (22), with permission.]

absence of valinomycin, which, in this circumstance, should serve to move the membrane potential toward greater inside negativity. The L-glucose uptake rate—which in other experiments was shown to be nonsaturable and phlorizin-insensitive (22), consistent with passive diffusion—was unaffected by valinomycin. In contrast, D-glucose uptake was significantly stimulated by the ionophore. It should be noted that in studies not illustrated the potential-sensitive component of D-glucose uptake was found to be saturable, consistent with its representing carrier-mediated transport (22). Indeed, the sensitivity of this potential-dependent component of Na^+-independent D-glucose uptake to inhibition by phlorizin and glucose analogs suggested that it was mediated by the same transport system that performs Na^+-glucose cotransport. It should be emphasized that Na^+-independent D-glucose transport is not electrogenic, as an inwardly directed D-glucose gradient does not generate a membrane potential in the absence of Na (12). Thus, the demonstration that the rate of Na^+-independent D-glucose uptake is potential-dependent underscores the fact that the rate of even an electroneutral process for which the membrane potential is not a thermodynamic driving force can still be potential-sensitive.

To explain how Na^+-independent glucose transport can be potential-dependent, let us first examine the mechanism underlying the potential dependence of Na^+-dependent glucose transport. A general carrier model for Na^+-glucose cotransport is illustrated in Fig. 11. The possible ordered or random nature

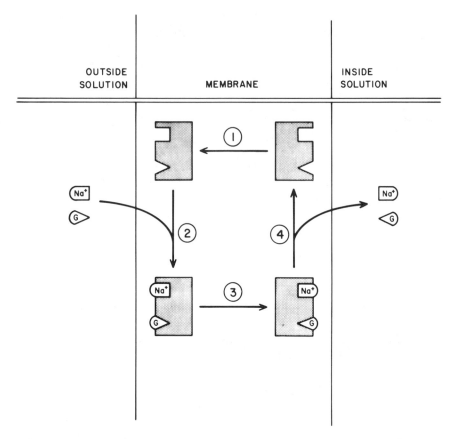

FIG. 11. General carrier model of Na$^+$-glucose cotransport. [From Aronson and Kinsella (5), with permission.]

of Na$^+$ and glucose binding is here ignored. It should be emphasized that the concept of carrier-mediated transport is used to indicate a transport mechanism in which the substrate binding sites functionally alternate between the two surfaces of the membrane. Such functional translocation most likely is accomplished by a conformational change that alternates access to the sites from one side of the membrane to the other without actual motion of the binding sites themselves, rather than by movement of a truly mobile carrier across the entire width of the membrane (39). The most obvious choice for the step at which the membrane potential might influence the rate of glucose transport is step 3, translocation of the Na$^+$-carrier-glucose ternary complex, which would bear a positive charge if the free carrier were uncharged. However, any of the steps in the carrier cycle could be potential-dependent, subject only to the restriction that there must be no net flux when the system is at equilibrium as defined by Eq. (9) for the case of 1:1 cotransport. For example, if there were charged groups in the carrier protein, then the membrane potential might influence its conformation even in the absence of Na^+ binding. Such a system might behave kinetically as if the free carrier were negatively charged. Thus, its cycling to the outer membrane surface (step 1) would be potential-dependent, thereby conferring potential dependence on the overall transport process.

One approach for distinguishing among these possible mechanisms is to investigate the effect of electrical potential on the binding of a substance that is not itself translocated. Phlorizin, a competitive inhibitor of Na^+-dependent glucose transport in renal microvillus membrane vesicles (8), binds to a Na^+-dependent, high-affinity receptor that appears to be identical with the Na^+-glucose cotransport carrier (14,17,19,39). Moreover, phlorizin is not itself transported across the microvillus membrane by the Na^+-glucose cotransport system (2,14).

The effect of valinomycin and K^+ gradients on the rate of Na^+-dependent phlorizin binding to renal microvillus membrane vesicles is shown in the upper panels of Fig. 12 (2). In the presence of an inwardly directed K^+ gradient, when valinomycin should move the membrane potential toward greater inside positivity, the ionophore inhibited the phlorizin uptake rate. In the presence of an outwardly directed K^+ gradient, when valinomycin should enhance inside negativity, the ionophore stimulated the phlorizin uptake rate. Experiments employing additional maneuvers to vary the membrane potential confirmed that inside positivity inhibits and inside negativity stimulates the rate of Na^+-dependent phlorizin binding (2). In contrast, as shown in the lower panels of Fig. 13, valinomycin in the presence of similar K^+ gradients had no effect on the rate of dissociation of previously bound phlorizin. Similar observations have recently been reported using intestinal microvillus membrane vesicles (39).

Can the potential dependence of phlorizin uptake be attributed to potential-dependent translocation of the Na^+-carrier-phlorizin ternary complex to the inside of the membrane (step 3 in Fig. 11)? In that case, the dissociation of previously bound phlorizin would require potential-dependent translocation back to the outer membrane surface. Since the rate of phlorizin dissociation was potential-independent, this hypothesis is unlikely. Another possibility is that the potential dependence of phlorizin uptake reflects the potential dependence of Na^+ binding, as the rate of phlorizin binding is highly Na^+-dependent (2). Such could be the case if the access of Na^+ to its binding site required Na^+ diffusion partially through the membrane. However, external Na^+ markedly inhibits the rate of phlorizin dissociation (2). Nevertheless, the phlorizin dissociation rate was found to be potential-independent even in the presence of external Na^+ (2). Thus, potential-dependent access of Na^+ to its binding site cannot explain the findings in Fig. 12. We are left with the possibility that the membrane potential modulates the availability of binding sites at the membranes surface as if the free carrier were negatively charged (step 1 in Fig. 11), or that the membrane potential modifies carrier activation with respect to sugar binding (step 2). In either case, these studies support the concept that there is potential dependence of some step(s) in the carrier cycle other than the steps of Na^+

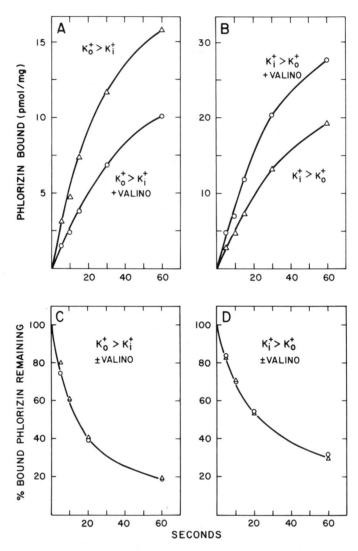

FIG. 12. Effect of membrane potential on phlorizin association and dissociation rates in renal microvillus membrane vesicles. **A:** Uptake of 0.2 μM phlorizin in the presence of 60 mM Na^+ and 60 mM K^+ was assayed using vesicles preloaded with no K^+ in the presence (○) or absence (△) of valinomycin. **B:** Uptake of 0.2 μM phlorizin in the presence of 120 mM Na^+ and 15 mM K^+ was assayed using vesicles preloaded with 75 mM K^+ in the presence (○) or absence (△) of valinomycin. **C:** Release of bound phlorizin was assayed using vesicles that were first pre-equilibrated with 0.2 μM phlorizin in the presence of 120 mM Na^+ with (○) or without (△) valinomycin and then diluted 1:20 into a phlorizin-free medium containing 150 mM K^+. **D:** Release of bound phlorizin was assayed using vesicles that were first pre-equilibrated with 0.2 μM phlorizin in the presence of 60 mM Na^+ and 60 mM K^+ with (○) or without (△) valinomycin and then diluted 1:20 into a phlorizin-free medium containing no K^+. [From Aronson (2), with permission.]

binding or Na⁺ translocation. Potential dependence of carrier recruitment (step 1) or of sugar binding (step 2) could explain the potential dependence of the D-glucose flux that takes place via this transport system in the absence of Na⁺. Finally, it should be noted that the capability of the Na⁺-glucose cotransport system to carry out sugar transport in the absence of Na⁺ does not necessarily contradict the coupling ratio of 1.0 demonstrated in Fig. 4. In the range of Na⁺ concentrations examined in that figure, the noncoupled flux of glucose must have been negligible compared to the Na⁺-coupled flux. Presumably, at very low Na⁺ concentrations the Na⁺:glucose coupling ratio would be <1.0.

SUMMARY AND CONCLUSIONS

It is intuitively appealing to assume that rates of biological processes are linear functions of their driving forces. Indeed, according to the precepts of irreversible (nonequilibrium) thermodynamics, rates of transport processes can be described as linear functions of their driving forces under conditions "near" equilibrium where such linear behavior holds (16,36). However, as illustrated by aspects of the kinetics of Na⁺-H⁺ exchange and Na⁺-glucose cotransport in renal microvillus membrane vesicles, the assumption that transport rates are linear functions of their electrochemical driving forces is not generally valid. In fact, examples were cited in which transport rates actually varied inversely with the magnitudes of the relevant thermodynamic driving forces. To be sure, the transport systems in these cited examples were studied under conditions far from equilibrium. It should be recognized, however, that secondary active transport systems functioning under physiologic conditions are often poised far from equilibrium. For example, mouse soleus muscle cells bathed at extracellular pH 7.4 have an intracellular pH of 7.0 (1), far below the value of approximately 8.4 that would be required to bring a 1:1 Na⁺-H⁺ exchanger to equilibrium in these cells that have an extracellular:intracellular Na⁺ concentration ratio of about 10. Similarly, in isolated, perfused, rabbit proximal tubules, the cell:lumen glucose concentration ratio is approximately 2 (40), far below the value of about 40 that would be required to bring a 1:1 Na⁺-glucose cotransporter to equilibrium if the $\Delta\tilde{\mu}_{Na}$ across the luminal membrane were 95 mV, as found in bullfrog kidney cells (18).

The nonlinear dependence of transport rates on electrochemical driving forces is relevant to assessing the adequacy of strategies commonly used to test whether the transport of a solute occurs by secondary active transport. In general, strategies that evaluate whether the electrochemical potential difference for one solute can act as a driving force to effect the net transport of a second solute against its electrochemical potential difference are more valid for identifying secondary active transport processes than are strategies that merely evaluate whether the electrochemical potential difference for one solute can affect the rate of transport of a second solute. For example, we have seen that under certain conditions imposing an outwardly directed H⁺ gradient can stimulate the rate of Na⁺

efflux from renal microvillus membrane vesicles (Fig. 8). Interpreting such a finding as evidence that the efflux of Na^+ must be coupled to the efflux of H^+ would not be correct. The demonstration that an outwardly directed H^+ gradient can drive net uphill Na^+ accumulation (Fig. 2), an effect not attributable to the generation of an inside-negative membrane potential (Fig. 3), does prove that the influx of Na^+ must be coupled to the efflux of H^+ (i.e., Na^+–H^+ exchange). As another example, we have seen that the membrane potential can modify the rate of Na^+-independent D-glucose transport. Interpreting such a finding as evidence that Na^+-independent glucose transport must be coupled to a transmembrane flow of charge (i.e., be electrogenic) would be incorrect. In contrast, the demonstration that an inside-negative membrane potential can by itself act as a driving force to effect net uphill D-glucose accumulation in the presence of Na^+ but absence of a Na^+ concentration gradient (11) does prove that the Na^+-dependent flux of glucose must be coupled to a transmembrane flow of positive charge.

The rate of a secondary active transport process may have a highly variable dependence on its thermodynamic driving force. The kinetic behavior of a secondary active transport process depends uniquely on the mechanistic details of that transport process, and these must be empirically determined for each case. Surely, much additional experimental work must be performed to define these mechanistic details.

ACKNOWLEDGMENT

Support was provided by USPHS grant No. AM-17433 and an Established Investigator Award from the American Heart Association.

REFERENCES

1. Aickin, C. C., and Thomas, R. C. (1977): An investigation of the ionic mechanism of intracellular pH regulation in mouse soleus muscle fibres. *J. Physiol. (Lond.)*, 273:295–316.
2. Aronson, P. S. (1978): Energy-dependence of phlorizin binding to isolated renal microvillus membranes. Evidence concerning the mechanism of coupling between the electrochemical Na^+ gradient and sugar transport. *J. Membr. Biol.*, 42:81–98.
3. Aronson, P. S. (1981): Identifying secondary active solute transport in epithelia. *Am. J. Physiol.*, 240:F1–F11.
4. Aronson, P. S. (1982): Pathways for solute transport coupled to H^+, OH^-, or HCO_3^- gradients in renal microvillus membrane vesicles. In: *Membranes and Transport, vol. 2*, edited by A. N. Martonosi, pp. 225–230. Raven Press, New York.
5. Aronson, P. S., and Kinsella, J. L. (1981). Use of ionophores to study Na^+ transport pathways in renal microvillus membrane vesicles. *Fed. Proc.*, 40:2213–2217.
6. Aronson, P. S., Nee, J., and Suhm, M. A. (1982): Modifier role of internal H^+ in activating the Na^+-H^+ exchanger in renal microvillus membrane vesicles. *Nature*, 299:161–163.
7. Aronson, P. S., and Sacktor, B. (1974): Transport of D-glucose by brush border membranes isolated from the renal cortex. *Biochim. Biophys. Acta*, 356:231–243.
8. Aronson, P. S., and Sacktor, B. (1975): The Na^+ gradient-dependent transport of D-glucose in renal brush border membranes. *J. Biol. Chem.*, 250:6032–6039.

9. Aronson, P. S., Suhm, M. A., and Nee, J. (1983): Interaction of external H^+ with the Na^+-H^+ exchanger in renal microvillus membrane vesicles. *J. Biol. Chem.*, 258:6767–6771.
10. Beck, J. C., and Sacktor, B. (1975): Energetics of the Na^+-dependent transport of D-glucose in renal brush border membrane vesicles. *J. Biol. Chem.*, 250:8674–8680.
11. Beck, J. C., and Sacktor, B. (1978): The sodium electrochemical potential-mediated uphill transport of D-glucose in renal brush border membrane vesicles. *J. Biol. Chem.*, 253:5531–5535.
12. Beck, J. C., and Sacktor, B. (1978): Membrane potential-sensitive fluorescence changes during Na^+-dependent D-glucose transport in renal brush border membrane vesicles. *J. Biol. Chem.*, 253:7158–7162.
13. Blomstedt, J. W., and Aronson, P. S. (1980): pH gradient-stimulated transport of urate and p-aminohippurate in dog renal microvillus membrane vesicles. *J. Clin. Invest.*, 65:931–934.
14. Chesney, R., Sacktor, B., and Kleinzeller, A. (1974): The binding of phlorizin to the isolated luminal membrane of the renal proximal tubule. *Biochim. Biophys. Acta*, 332:263–277.
15. Crane, R. K. (1977): The gradient hypothesis and other models of carrier-mediated active transport. *Rev. Physiol. Biochem. Pharmacol.*, 78:99–159.
16. Essig, A., and Caplan, S. R. (1979): The use of linear nonequilibrium thermodynamics in the study of renal physiology. *Am. J. Physiol.*, 236:F211–F219.
17. Frasch, W., Frohnart, P. P., Bode, F., Baumann, K., and Kinne, R. (1970): Competitive inhibition of phlorizin binding by D-glucose and the influence of sodium: A study on isolated brush border membrane of rat kidney. *Pfluegers Arch.*, 320:265–284.
18. Fujimoto, M., Naito, K., and Kubota, T. (1980): Electrochemical profile for ion transport across the membrane of proximal tubular cells. *Membr. Biochem.*, 3:67–97.
19. Glossman, H., and Neville, D. M., Jr. (1972): Phlorizin receptors in isolated kidney brush border membranes. *J. Biol. Chem.*, 247:7779–7789.
20. Haase, W., Schafer, A., Murer, H., and Kinne, R. (1978): Studies on the orientation of brush-border membrane vesicles. *Biochem. J.*, 172:57–62.
21. Heinz, E. (Editor) (1972): *Na-Linked Transport of Organic Solutes.* Springer-Verlag, Berlin.
22. Hilden, S., and Sacktor, B. (1982): Potential-dependent D-glucose uptake by renal brush border membrane vesicles in the absence of sodium. *Am. J. Physiol.*, 242:F340–F345.
23. Hopfer, U., Nelson, K., Perrotto, J., and Isselbacher, K. J. (1973): Glucose transport in isolated brush border membranes from rat small intestine. *J. Biol. Chem.*, 248:25–32.
24. Ives, H. E., Yee, V. J., and Warnock, D. G. (1982): Effects of Na^+ and Li^+ on the activity of the Na^+/H^+ antiporter of renal brush border membrane vesicles. *Fed. Proc.*, 41:1262.
25. Kinne, R., Murer, H., Kinne-Saffran, E., Thees, M., and Sachs, G. (1975): Sugar transport by renal plasma membrane vesicles. *J. Membr. Biol.*, 21:375–395.
26. Kinsella, J. L., and Aronson, P. S. (1980): Properties of the Na^+-H^+ exchanger in renal microvillus membrane vesicles. *Am. J. Physiol.*, 238:F461–F469.
27. Kinsella, J. L., and Aronson, P. S. (1981): Interaction of NH_4^+ and Li^+ with the renal microvillus membrane Na^+-H^+ exchanger. *Am. J. Physiol.*, 241:C220–C226.
28. Kinsella, J. L., and Aronson, P. S. (1981): Amiloride inhibition of the Na^+-H^+ exchanger in renal microvillus membrane vesicles. *Am. J. Physiol.*, 241:F374–F379.
29. Kinsella, J. L., and Aronson, P. S. (1982): Determination of the coupling ratio for Na^+-H^+ exchange in renal microvillus membrane vesicles. *Biochim. Biophys. Acta*, 689:161–164.
30. Liedtke, C. M., and Hopfer, U. (1982): Mechanism of Cl^- translocation across small intestinal brush-border membrane. II. Demonstration of Cl^--OH^- exchange and Cl^- conductance. *Am. J. Physiol.*, 242:G272–G280.
31. Murer, H., and Hopfer, U. (1974): Demonstration of electrogenic Na^+-dependent D-glucose transport in intestinal brush border membranes. *Proc. Natl. Acad. Sci, USA*, 71:484–488.
32. Murer, H., and Kinne, R. (1980): The use of isolated membrane vesicles to study epithelial transport processes. *J. Membr. Biol.*, 55:81–95.
33. Murer, H., Hopfer, U., and Kinne, R. (1976): Sodium/proton antiport in brush-border-membrane vesicles isolated from rat small intestine and kidney. *Biochem. J.*, 154:597–604.
34. Rosenberg, T. (1948): On accumulation and active transport in biological systems. I. Thermodynamic considerations. *Acta Chem. Scand.*, 2:14–33.
35. Sacktor, B. (1977). Transport in membrane vesicles isolated from the mammalian kidney and intestine. *Curr. Top. Bioenerg.*, 6:39–81.

36. Schultz, S. G. (1980): *Basic Principles of Membrane Transport,* pp. 15–20. Cambridge University Press, Cambridge.
37. Schultz, S. G., and Curran, P. F. (1970): Coupled transport of sodium and organic solutes. *Physiol. Rev.,* 50:637–718.
38. Stein, W. D. (1967): *The Movement of Molecules Across Cell Membranes,* pp. 207–241. Academic Press, New York.
39. Toggenburger, G., Kessler, M., and Semenza, G. (1982): Phlorizin as a probe of the small-intestine Na^+, D-glucose cotransporter: A model. *Biochim. Biophys. Acta,* 688:557–571.
40. Tune, B. M., and Burg, M. B. (1971): Glucose transport by proximal renal tubules. *Am. J. Physiol.,* 221:580–585.
41. Turner, R. J., and Moran, A. (1982): Stoichiometric studies of the renal outer cortical brush border membrane D-glucose transporter. *J. Membr. Biol.,* 67:73–80.
42. Warnock, D. G., Reenstra, W. W., and Yee, V. J. (1982): Na^+/H^+ antiporter of brush border vesicles: Studies with acridine orange uptake. *Am. J. Physiol.,* 242:F733–F739.

Electrogenic Transport: Fundamental Principles and Physiological Implications,
edited by Mordecai P. Blaustein and Melvyn Lieberman. Raven Press, New York © 1984.

Chemiosmotic Coupling and Its Application to the Accumulation of Biological Amines in Secretory Granules

Robert G. Johnson and Antonio Scarpa

Department of Biochemistry and Biophysics, University of Pennsylvania School of Medicine, Philadelphia, Pennsylvania 19104

The chemiosmotic theory, first proposed by Mitchell (23) in the early 1960s, has revolutionized both understanding and experimental approaches in the area of energy transduction in biological membranes. According to this theory, the metabolic energy provided in the cell through the oxidation of substrates or the absorption of light is coupled to synthesis of adenosine 5'-triphosphate (ATP) or transport of ions or metabolites primarily through the buildup of a proton gradient across the cell or organelle membranes. Accordingly, both those reactions yielding metabolic energy and those requiring it, such as the synthesis of ATP or the transport of ions against concentration gradients, are not linked by high-energy chemical intermediates, but through a higher energy intermediate state, the proton electrochemical potential ($\Delta\bar{\mu}_{H^+}$). Energy transducing membranes are therefore primary electrogenic H^+ pumps and, because of their relative impermeability to H^+, are able to build up a proton electrochemical potential difference across their membranes.

The electrochemical proton gradient is usually expressed in electrical units,

$$\Delta\bar{\mu}_{H^+} = \Delta\psi - (2.3\ RT/F)\Delta pH\ \text{(in mV)},$$

where $\Delta\Psi$ represents the electric component (transmembrane potential), and ΔpH the contribution made by the chemical potential, or pH gradient.

Although a few reservations may still exist on some quantitative or molecular aspects of the chemiosmotic coupling, the basic tenets of the theory have outlasted unparalleled scrutiny over the last 20 years and have facilitated the overall understanding of how ATP is produced in cells and how ion transport is coupled to metabolic energy in mitochondria, chloroplasts, and bacteria (for review, see references 9,19,24).

In recent years the chemiosmotic coupling mechanism has also provided significant insight into the cellular and molecular mechanisms of ion and metabolite transport in a variety of other cells and subcellular components not comprised by the original chemiosmotic formulation. An example is the uptake of biological amines into storage vesicles, which is coupled to ATP hydrolysis of these vesicles through an overall chemiosmotic mechanism. The operation of this mechanism has become evident using bovine chromaffin granules and ghosts (for review, see reference 12), but similar findings have now been documented in 5-hydroxytryptamine containing dense granules of pig platelets (4), mast cells granules (14), and anterior (5) and posterior (32) pituitary granules. The essence of the model is that an H^+-translocating ATPase exists in the chromaffin granule membrane and is responsible for the anisotropic, inwardly directed movement of H^+. This H^+-ATPase, in conjunction with the extremely low permeability of the membrane of these granules to H^+, is responsible for the generation and maintenance of a ΔpH (inside acidic), and $\Delta \psi$ (positive inside). Amine accumulation is thought to proceed by a separate, distinct, carrier-mediated process, inhibitable by reserpine and dependent on one or both components of the electrochemical proton gradient ($\Delta \bar{\mu}_{H^+}$).

STRUCTURE AND FUNCTION OF CHROMAFFIN GRANULES

The chromaffin granule is a subcellular organelle within the chromaffin cell of the adrenal medulla in which most of the catecholamines contained within the adrenal medulla are localized and stored before release. These organelles can be isolated in large amounts and high purity through the use of isotonic gradients of high density, low viscosity supporting materials (3).

The membrane and matrix composition of these granules have been studied in great detail (for review, see reference 34), and are schematically depicted in Fig. 1. Catecholamine content (2,400 nmol/mg protein) and nucleotide content

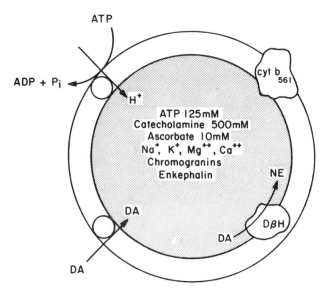

FIG. 1. Diagrammatic representation of the composition of chromaffin granules.

(800 nmoles/mg protein, mostly ATP) constitute 35% of the dry weight of the entire granule. Based on the intravesicular water space, it can be calculated that if catecholamines and ATP measured within the matrix space were free in solution, the apparent concentration would be 0.5 and 0.125 M, respectively. Also found within the matrix space is a high content of ascorbate and chromogranins, a class of large molecular weight acidic proteins, the function of which is virtually unknown. The membrane of the chromaffin granule maintains one of the lowest protein/lipid ratios (0.45 by weight) found within mammalian membranes, and an extremely high percentage of the lipid (10 to 20%) is constituted as lysophosphatidylcholine (7). Major membrane-associated proteins identified thus far are: (a) cytochrome b_{561} and other respiratory chain components, the physiological significance of which is not understood; (b) dopamine beta-hydroxylase, which has been extensively investigated and even purified to homogeneity; (c) an H^+ transporting ATPase.

CATECHOLAMINE ACCUMULATION IN THE GRANULES

Considerable effort has been devoted over the last 20 years to the understanding of the mechanism(s) responsible for the accumulation and storage of the large amount of catecholamines found within the chromaffin granules. Figure 2 sche-

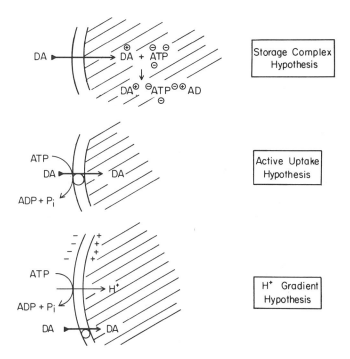

FIG. 2. Present and past models for explaining catecholamine accumulation in chromaffin granules.

matically illustrates the three models proposed in the literature to account for catecholamine accumulation in the granules. In the storage-complex hypothesis, catecholamine uptake is accounted for solely through permeation down a concentration gradient and by strong internal binding (1). According to the active uptake hypothesis, uptake proceeds against a concentration gradient via a carrier-mediated mechanism directly coupled to ATP hydrolysis (2). Both hypotheses are now considered untenable, and, over the last few years, a large body of experimental evidence has resulted in the formulation of the H^+ gradient hypothesis, the basic elements of which are now generally accepted (for review, see reference 18). The model, illustrated at the bottom of Fig. 2, states that a H^+-ATPase exists in the chromaffin granule membrane, and the vectorial movement of H^+ from the cytosol to the chromaffin granule interior results in the generation of a transmembrane electrochemical H^+ gradient. Amine accumulation is thought to proceed by a separate carrier-mediated process, coupled to the electrochemical H^+ gradient.

The essential features of the experimental data from which this model was constructed include: (a) measurement of chromaffin granule membrane permeability and H^+-buffering capacity; (b) measurements of the transmembrane H^+ electrochemical potential; (c) studies on the properties of the H^+-ATPase, which generates and maintains an H^+ gradient across the granule membrane; (d) qualitative and quantitative investigation of catecholamine distribution according to the above H^+ electrochemical potential.

PERMEABILITY AND BUFFERING CAPACITY OF CHROMAFFIN GRANULES

A salient characteristic of the membrane of the chromaffin granule is that it is more impermeable to various divalent and monovalent cations than that of any previously studied subcellular organelle (17) (Fig. 3). For example, the conductance to H^+ is at least one order of magnitude less than that measured in mitochondria. In addition, the intragranular H^+ buffering capacity is very large, approaching 300 μmol H^+/pH unit/g dry wt (17). These two properties may explain the experimental observations that isolated granules can maintain transmembrane H^+ concentration gradients (ΔpH) for several hours (13). At variance with the low permeability to H^+, K^+, Na^+, Ca^{2+}, Mg^{2+}, and ATP, chromaffin granules possess a finite permeability to Cl^-, which may be considered of physiological importance in controlling intravesicular pH and transmembrane potentials (see below).

MEASUREMENTS OF THE H^+ TRANSMEMBRANE GRADIENTS IN CHROMAFFIN GRANULES AND GHOSTS

Since the small size of the granules precludes direct measurement of the intravesicular pH or transmembrane potential ($\Delta\psi$) with microelectrodes, other

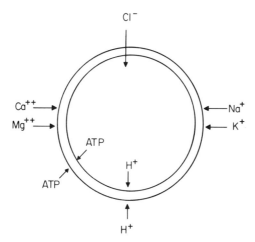

FIG. 3. Schematic representation of relative permeabilities of various ions and ATP across chromaffin granules and ghosts.

less direct techniques widely applied to other systems have been used to measure these two parameters. For measurement of the ΔpH, ^{14}C-methylamine distribution, ^{31}P nuclear magnetic resonance (NMR), and fluorescent dye distribution have been utilized (17,26,31). For measurement of the $\Delta\psi$, both ^{14}C-SCN$^-$ (thiocyanate) and fluorescent dye distributions have been used and have yielded equivalent results (11,31).

Figure 4 shows an example of measurement of intravesicular pH in intact chromaffin granules suspended in a sucrose medium at pH 7.5. Based on the measurement of intravesicular ^3H_2O spaces and ^{14}C-methylamine distribution, a ΔpH across the membrane equivalent to 1.7 pH units was calculated. When the pH of the medium was varied over the range of 4.5 to 7.85, a marked change of ΔpH was observed. Zero ΔpH was observed when the granules were suspended in a medium of pH 5.5. The calculated ΔpH does not arise from the establishment of a Donnan equilibrium distribution, since identical values can be obtained in media of various ionic compositions. These and other data clearly indicate that the intravesicular pH of isolated chromaffin granules is 5.5, and that this acidic pH can be maintained for long periods of time independently of the pH of the suspending medium.

Figure 5 shows the results of an experiment in which a transmembrane potential ($\Delta\psi$) can be measured in intact chromaffin granules. In this experiment the $\Delta\psi$ was measured through the distribution in the granules and in the medium of labeled lipophilic anions or cations. The lipophilic anion ^{14}C-SCN$^-$ permeates biological membranes and accumulates within the intravesicular space when a positive inside potential is present (30). For potentials negative inside, the lipophilic cation ^3H-tetramethylphenylphosphonium (TPMP$^+$) is utilized (29).

When ATP is added to a suspension of chromaffin granules, a time-dependent

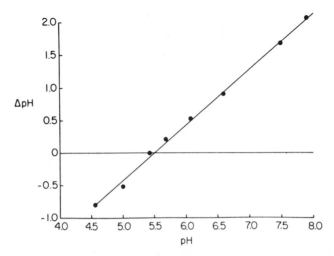

FIG. 4. Measurement of ΔpH between isolated chromaffin granules and media of various pHs. The medium consisted of 0.25 M sucrose buffered with 20 mM each of Tris and 2-[N-morpholino]ethanesulfonate (Mes) at the pH indicated in the figure, 2.1 mg protein/ml of chromaffin granules, and tracer amounts of ^3H_2O and ^{14}C-methylamine. Chromaffin granule isolation and ΔpH measurements were carried out as described previously (3,17). Temperature was 24°C.

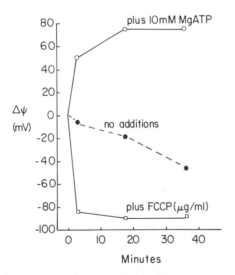

FIG. 5. Measurement of Δψ in isolated chromaffin granules. The medium contained 0.27 M sucrose, 40 mM Tris/maleate (pH 6.85), chromaffin granules (10 mg protein/ml), and either ^{14}C-SCN⁻ and ^3H_2O or ^3H-TPMP⁺ and ^{14}C-polydextran. At 0 time 10 mM MgATP or 1 μg/ml FCCP were added, and at the time indicated 1.2-ml samples were withdrawn and centrifuged in a desk microcentrifuge. Supernatants and pellet were assayed for radioactivity and ΔpH calculated as previously described. Temperature was 37°C. [From Johnson and Scarpa (18), with permission.]

increase in the $\Delta\psi$, which reaches a plateau at 80 mV and can be sustained for over 40 min, is observed (Fig. 5). In the absence of ATP, the $\Delta\psi$ is measured to be near 0 mV at the beginning of the incubation and gradually decreases to more negative values during the incubation. In the presence of carbonyl cyanide p-trifluoromethoxyphenylhydrazone (FCCP), a compound that transports protons electrogenically down their electrochemical gradient, a large and stable potential, negative inside (−90 mV), can be evoked. A similar curve was obtained when ATP was added concomitantly with the FCCP (data not shown). These results suggest that in the presence of ATP an electrogenic event is occurring, which is consistent with the inward movement of H^+. In the presence of FCCP, H^+ can distribute according to its electrochemical potential, and since the diffusion gradient is outward, a $\Delta\psi$ negative inside is measured.

The existence of a ΔpH and a $\Delta\psi$ across the membrane of chromaffin granules leads to a search for a pump that could be responsible for the generation of these gradients. It is now generally accepted that the H^+-translocating ATPase within the chromaffin granule membrane is responsible for the inwardly directed translocation of H^+ and for the development of an electrochemical transmembrane H^+ gradient ($\Delta\bar{\mu}_{H^+}$) across the membrane, composed of interconvertible electrical ($\Delta\psi$) and concentration (ΔpH) components according to the chemiosmotic hypothesis (23):

$$\Delta\bar{\mu}_{H^+} = \Delta\psi - Z\Delta pH$$

where $Z = 2.3\ RT/F$.

The operation of this H^+ translocating ATPase in quantitative terms and its relation to the establishment of the ΔpH, $\Delta\psi$, $\Delta\bar{\mu}_{H^+}$, and catecholamine uptake can be better studied in chromaffin ghosts, which are devoid of the H^+ gradients, ATP, and catecholamines endogenously present in chromaffin granules. Resealed chromaffin ghosts, maintaining internal volumes and permeability characteristics similar to that of chromaffin granules, can be routinely prepared through hypotonic lysis of the granules and resealing in an isotonic medium of choice followed by extensive dialysis (16,27,33). If the ATPase within the membrane is responsible for H^+ accumulation, then the addition of ATP to a suspension of chromaffin ghosts should result in the inwardly directed movement of H^+. In the absence of a permeable anion cotransport, the translocation of H^+ with a positive charge results in the rapid generation of a membrane potential, inside positive, which limits further influx of H^+. The overall result is the generation of a large $\Delta\psi$ and negligible ΔpH, as shown diagrammatically in Fig. 6B. It has been well established that the membrane of the chromaffin granule maintains a selective permeability to anions, with Cl^- possessing a greater endogenous permeability than that of sulfate or isethionate (6,8). Hence, in the presence of Cl^-, the ATPase-dependent establishment of a $\Delta\psi$, positive inside, results in the inward movement of Cl^- down its electrochemical gradient. The cotransport of Cl^- effectively prevents the establishment of a large transmem-

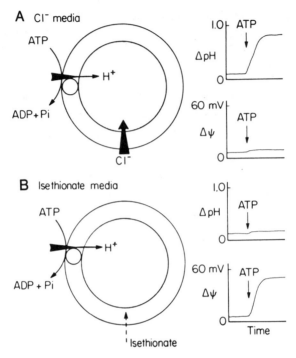

FIG. 6. Schematic illustration of ATP-dependent generation of ΔpH **(A)** or Δψ **(B)** in chromaffin ghosts.

brane potential, positive inside, and further H$^+$ entry can ensue. The overall result is the generation of a large ΔpH and minimal Δψ (Fig. 6A).

A series of experiments has conclusively demonstrated that in chromaffin ghosts the addition of ATP produces ΔpH alone, Δψ alone, or both, depending on the composition of the medium and, in particular, the concentration of [Cl$^-$] (13,15,16,18,25,27). Based on the estimated value of cytosolic pH, cytosolic [Cl$^-$], and [ATP], it is conceivable that the chromaffin granule *in situ* maintains a $\Delta\bar{\mu}_{H^+}$ across its membrane of close to 200 mV, contributed by both ΔpH and Δψ.

H$^+$-ATPASE OF CHROMAFFIN GRANULES

Despite the mechanistic importance of this electrogenic H$^+$ pump in generating and maintaining electrochemical H$^+$ gradients across the chromaffin granule membranes, genuine insight into its basic properties, function, and regulation have lagged behind. Several factors have contributed to the lack of information or disagreement of data on this ATPase in the literature. Among these are: (a) impurity of the preparations used; (b) interference by catecholamines in protein assays; (c) use of intact chromaffin granules containing endogenous ATP and H^+ gradients; (d) technical limitation in simultaneous measurement of ATPase activities, H^+ fluxes, and electrochemical H^+ gradients. This has prompted a reevaluation of the kinetics, regulation, and stoichiometry of this ATPase with state-of-the-art technology and with highly purified chromaffin granules, chromaffin ghosts, or fragmented membranes (12). These properties are summarized in Table 1.

The K_m of the ATPase for ATP (in the presence of a large, fixed Mg^{2+} concentration) was 69 µM. This value is far below the cytosolic ATP concentration, suggesting that ATPase activity *in situ* should not be affected by changes in cytosolic ATP concentrations.

The V_{max} of the ATPase activity was consistently measured within the range of 110 to 130 nmol ATP/mg of protein/min. The variability of previous data in the literature more than likely relates to the varying methods of preparation, the different temperatures under which the experiments were performed, and the inaccurate protein determination due to the unsuitability of the assays used.

The measured pH optimum was 7.40. Indirect evidence is accumulating which indicates that the F_1 portion of the ATPases is spacially located in the cytosolic side of the membrane. Since the presumed cytosolic pH is 7.4, as opposed to the pH 5.5 of the intragranular space, the ATPase would thus be functioning optimally *in vivo*.

Two specific inhibitors of the ATPase were studied and titrated over a wide range of concentrations. The first, dicyclohexylcarbodiimide (DCCD), is known to inhibit the ATPase of mitochondria, chloroplasts, and *Escherichia coli*, and exerts 50% of the inhibition at 310 nmoles/mg of protein. The second, trimethyl tin (TMT), exerts 50% inhibition at 20 nmoles/mg protein, more than one order of magnitude smaller than that required for DCCD.

Two other inhibitors, oligomycin and aurovertin, specifically inhibit other H^+-ATPases. Neither has any effect on the chromaffin granule ATPase. These

TABLE 1. *Properties of ATPase of chromaffin granules*

K_m (for ATP)	69 µM
V_{max} (nmol/mg)	
Granules	10
Granules + FCCP	45
Membranes (25°C)	148 to 250
Specificity[a]	ATP=ITP>GTP>UTP>CTP
Optimum pH	7.3
Temperature dependence on rate (nmol/mg)	16 (5°C) 148 (25°C) 525 (40°C)
Inhibitors (nmol/mg for 50% inhibition)	DCCD, 310
	TMT, 20
	Aurovertin, no effect
	Oligomycin, no effect
	Reserpine, no effect

[a] ATP = adenosine 5'-triphosphate; GTP = guanosine 5'-triphosphate; UTP = uridine 5'-triphosphate; CTP = cytosine 5'-triphosphate. ITP = inosine 5'-triphosphate.

differences in the selectivity of agents, the mechanism of action of which is thought to relate to binding to the F_o portion of the ATPase indicate at least the nonidentity of the mitochondrial and chromaffin granule ATPase. In addition, the absence of inhibition of the chromaffin granule ATPase activity further supports the notion that mitochondrial contamination is negligible.

Reserpine, a specific inhibitor of the catecholamine transporter, has had mixed reports on its ATPase inhibitory ability. The data in Table 1 support the conclusion that reserpine at concentrations that totally inhibit catecholamine accumulation does not alter ATPase activity.

If the ATPase activity within the chromaffin granule membrane is coupled to anisotropic H^+ movement and to the establishment of an electrochemical proton gradient, then it would be predicted that the rate of the ATPase activity may be linked to the magnitude of the $\Delta\bar{\mu}_{H^+}$. In other words, once the maximum proton gradient is established based on the thermodynamic properties of the H^+ pump and the proton leak or backflow through the membrane, any perturbation of the equilibrium state should result in an increased rate of ATP hydrolysis as the enzyme attempts to reestablish the gradient. The relationship between H^+-ATPase activity and $\Delta\bar{\mu}_{H^+}$ in intact chromaffin granules was then investigated by measuring: (a) baseline ATPase activity; (b) ATPase activity after collapse of the ΔpH; and (c) ATPase activity after collapse of the $\Delta\psi$.

When intact chromaffin granules are suspended in a sucrose medium at pH 6.8, an intravesicular pH of 5.5 can be measured, which is independent of the presence of ATP (17). On the other hand, when ATP is added to isolated granules, the membrane potential increases from 0 to 80 mV, positive inside (13,21,31).

In the presence of 5 mM ATP, a linear rate of ATPase activity was observed (Fig. 7). Collapse of the ΔpH from 1.2 pH units to 0.3 pH units with 50 mM NH$_4$Cl resulted in a moderate increase in the ATPase activity (40%). On the other hand, collapse of the membrane potential from 60 to 0 mV with 50 mM SCN$^-$ produced a 120% increase in rates of ATP hydrolysis. Finally, the addition of FCCP, which permits H^+ to equilibrate according to its electrochemical potential, resulted in a stimulation of the ATPase activity of over 300%. The results indicate that the chromaffin granule ATPase is responsive to changes of either component of the electrochemical proton gradient, but is largely modulated by changes in $\Delta\psi$.

Further experiments have also conclusively shown that the H^+ transport across the membrane is stoichiometrically coupled to ATP hydrolysis and that the H^+/ATP ratio is 1.8 (12).

H^+ TRANSMEMBRANE ELECTROCHEMICAL GRADIENTS AND CATECHOLAMINE UPTAKE

The properties of the amine accumulation are consistent with a carrier-mediated active process based on the following empirical evidence: Uptake exhibits

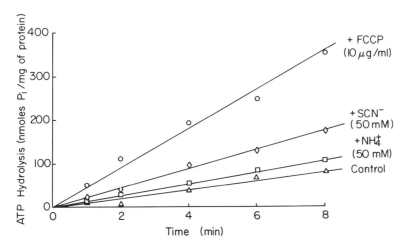

FIG. 7. Effect of FCCP, SCN$^-$, and NH$_4^+$ on ATPase activity of intact chromaffin granules. Chromaffin granules (4.1 mg) were incubated for 5 min in a medium containing 270 mM sucrose, 10 mM Tris-maleate (pH 7.0), and either 50 mM (NH$_4$)$_2$SO$_4$ (□——□), 50 mM KSCN (◇——◇), or 10 μg/ml FCCP (○——○). A control (△——△) was also performed. The reaction was initiated by the addition of 5 mM MgATP, and ATPase activity was measured as described previously (12). Values for ΔpH and Δψ were measured in parallel samples as described in Figs. 4 and 5.

structural specificity and stereospecificity, depends on the presence of Mg and ATP, has a Q_{10} of 4 to 6, and is specifically inhibited by reserpine. It was the coupling of the inward movement of the amine to metabolic energy expenditure that remained an enigma and was largely solved by the finding that these vesicles maintain an electrochemical H$^+$ membrane gradient. If catecholamine permeation of the chromaffin granule membrane occurs via the uncharged species (as does methylamine permeation), then the possibility exists that catecholamines could distribute across the membrane according to the ΔpH. If the intragranular space is maintained at pH 5.5 and the cytosolic pH is in the region of 7.2 to 7.4, then a difference of 1.7 to 1.9 pH units exists. If the catecholamines are capable of distributing according to the ΔpH, then a distribution gradient equivalent to 50 to 80:1 could occur, based purely on the existence of a ΔpH and the passive permeability of the chromaffin granule to uncharged amines.

Experiments in which intact chromaffin granules were used and the internal and external pH was varied revealed that amine uptake was proportional to the ΔpH across their membranes and that the uptake was sensitive to the amine transporter inhibitor reserpine (13,27). Based on these data, a model was generated based on the existence of the H$^+$-translocating ATPase and resultant pH gradient, which explained the empirical data better than previous hypotheses (see Fig. 2).

This simple explanation, however, was not entirely consistent with other experimental data. In the presence of ATP a Δψ is generated across the chromaffin granule membrane without significantly enhancing the already large ΔpH (Fig. 5). Yet amine accumulation increased over that measured in the absence of ATP, when only a ΔpH exists across the granule membrane. Further experiments with chromaffin granules have indicated that accumulation against a net concentration gradient was observed in the presence of a ΔpH alone, or a $\Delta\psi$ alone, with the optimal rate and extent of amine uptake observed in the presence of both a ΔpH and $\Delta\psi$ (18).

The use of the chromaffin ghost system devoid of any endogenous gradients or components, and in which only a ΔpH or a $\Delta\psi$ could be generated, provides a better experimental approach for the investigation of the effect of these two components on catecholamine uptake.

By selection of the experimental conditions (i.e., the presence or the absence of Cl$^-$), ATP generates either a ΔpH alone or a $\Delta\psi$ alone (see Fig. 6). Both ΔpH and $\Delta\psi$ can support biological amine uptake, as shown in Figs. 8 and 9. Fig. 8B indicates that the presence of ATP induces a time-dependent acidification of the chromaffin ghosts. Since permeable Cl$^-$ is present in the incubation medium, and the intravesicular buffering capacity of the ghosts is low, the addition of ATP results in a large ΔpH and virtually no $\Delta\psi$ (not shown). Serotonin accumulation occurred in a time-dependent fashion, reaching a steady state at approximately 24 min. The addition of ammonia collapses the ΔpH to 0 and prompted a time-dependent release of the accumulated serotonin. This experiment and others support the notion that net amine accumulation can occur in the presence of a ΔpH alone.

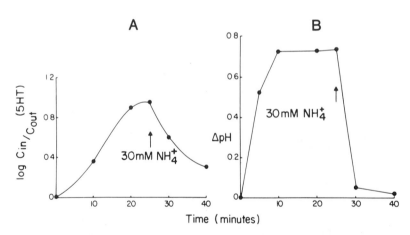

FIG. 8. The measurement of ^{14}C-serotonin accumulation and ΔpH in chromaffin ghosts under identical conditions. The chromaffin ghosts were formed in 185 mM KCl, 1 mM Tris/maleate, pH 6.8. The reaction mixture contained 185 mM KCl, 30 mM Tris/maleate, pH 6.8, chromaffin ghosts (3.4 mg of protein/ml), and 10 mM Mg ATP. The total initial volume was 10 ml. To one chamber ^{14}C-methylamine and ^{3}H$_2$O were added. After 24 min, 30 mM NH$_4$Cl was added to both chambers. The isolation and assay procedures followed were described in the legend to Fig. 4. Temperature was 24°C. Serotonin was used instead of epinephrine or norepinephrine to minimize oxidation and binding.

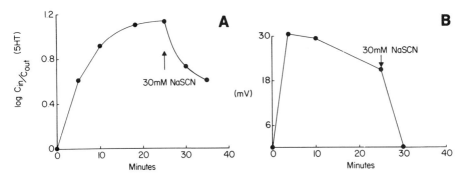

FIG. 9. The measurement of ^{14}C-serotonin accumulation and $\Delta\psi$ under identical conditions. The chromaffin ghosts were formed in 185 mM sodium isethionate, 20 mM Tris/maleate, pH 6.8. The reaction mixtures consisted of 185 mM sodium isethionate, 30 mM Tris/maleate, pH 6.8, chromaffin ghosts (3.3 mg of protein/ml), and 8 mM MgATP. The total initial volume was 10 ml. To one chamber ^{14}C-methylamine and $^{3}H_2O$ were added to measure ΔpH **(A)** and to the other ^{14}C-serotonin and $^{3}H_2O$ **(B)** were added. After 24 min, 30 mM NaSCN was added to both chambers. The isolation and assay procedures followed were described in the legend to Fig. 5. The temperature was 24°C.

On the other hand, chromaffin ghosts that have a high internal buffering capacity and are suspended in a medium containing an impermeable cation or anion (e.g., isethionate) should be expected to establish a large $\Delta\psi$ and small ΔpH when ATP is added. This is due to the fact that the inward movement of H^+ cannot be coupled to outward cation or inward anion movement, and thus a large potential, inside positive, is established, which inhibits further influx of H^+.

The time-dependent influx of ^{14}C-serotonin and simultaneous measurement of the $\Delta\psi$ were monitored under identical conditions (Fig. 9). The addition of 8 mM ATP to a suspension of chromaffin ghosts suspended in 185 mM sodium isethionate, 30 mM Tris/maleate plus 30 mM NH_4^+ (to collapse any ΔpH) resulted in the rapid generation of a membrane potential, inside positive, of 31 mV (Fig. 9B). The ΔpH was zero (data not shown). Commensurate with the membrane potential generation, was a time-dependent influx of ^{14}C-serotonin (Fig. 9A). After 25 min of incubation, 30 mM SCN^- was added. The membrane potential rapidly decreased to zero (Fig. 9B). Correspondingly, a time-dependent efflux of ^{14}C-serotonin ensued (Fig. 9A). This experiment supports the notion that net amine accumulation can occur in the presence of a $\Delta\psi$ alone.

In the previous experiments, net accumulation of amines against concentration gradients, and not simple exchange, is measured. This is documented by the experiment shown in Fig. 10, where *net* uptake and release of epinephrine according to the ΔpH was measured kinetically through two different techniques. The experimental points of Fig. 10, top, indicate high-pressure liquid chromatography (HPLC) determination of concentrations of epinephrine left in the medium after rapid separation of the granules from the medium through Millipore filtration. The trace shown in Fig. 10, bottom, is the continuous readout on a strip-

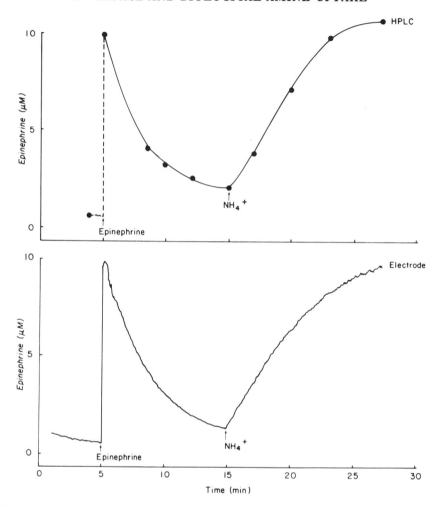

FIG. 10. The kinetic measurement of epinephrine accumulation into chromaffin ghosts by potentiometric techniques and HPLC. The reaction mixture contained 185 mM KCl, 30 mM Hepes (pH 6.80), and chromaffin ghosts (0.34 mg protein/ml). The total volume was 2.1 ml. The catecholamine, reference, and auxiliary electrodes were allowed to equilibrate in the reaction chamber; 2.5 mM MgATP was added to initiate the reaction, and 10 μM epinephrine bitartrate and 20 mM $(NH_4)_2SO_4$ were added at the times indicated. For HPLC determination, 200-μl aliquots of the experimental medium were removed and centrifuged for 7 min in a model 4200 Eppendorf microcentrifuge at the times indicated, and the supernatant was immediately assayed for epinephrine (section 2). Temperature was maintained at 37°C using a water bath. See Hayflick et al. (10).

chart recorder of the free epinephrine as detected by the glassy carbon electrode (10). In the presence of an ATP-driven ΔpH gradient of 1.1 pH, the addition of epinephrine results in a rapid upward deflection of the trace, which is due to the increase in epinephrine in the medium and, consequently, of its oxidation products, which are detected by the glassy carbon electrode (Fig. 10, bottom trace). This is followed by a slower, time-resolvable decrease in free epinephrine present in the medium, which is due to the transport of epinephrine from the medium to the ghost interior. This reaches a plateau after approximately 10 min, when most of the epinephrine added to the medium has been taken up by the ghosts. Addition of NH_4^+, which abolishes the ATP-dependent intravesicular space acidification, prompts a complete release of epinephrine from the ghosts and a consequent increase in free epinephrine in the medium, which is detected by the electrode. The comparison of this measurement with that of Fig. 10, top, where HPLC was used to detect epinephrine in the medium, shows good quantitative and kinetic agreement between the data obtained with the two methods.

In the previous experiments amine accumulation was studied in chromaffin ghosts in which either a ΔpH or $\Delta\psi$ had been induced and maintained by a process coupled to ATP hydrolysis. To investigate amine accumulation under conditions wherein a pH gradient was generated in the absence of ATP, chromaffin ghosts were formed in 185 mM KCl, 1 mM Tris/maleate medium, pH 6.8, and resuspended in 185 mM choline chloride, 30 mM Tris/maleate, pH 6.8. In the absence of nigericin (Fig. 11, left side), no ΔpH was measured, nor was any significant amine uptake noted. However, in the presence of nigericin, a lipophilic carboxylic acid ionophore known to catalyze an electrosilent exchange of K^+ and H^+ across biological membranes, a large ΔpH was generated (Fig. 11, left side). Under this condition, a large accumulation of ^{14}C-epinephrine could be measured (Fig. 11, right side). This experiment, in which the ΔpH was generated at the expense of a K^+ gradient, is further evidence that amine accumulation into chromaffin ghosts is dependent on the existence of a ΔpH across the chromaffin ghost membrane. Coupled with the previous experiments, in which amine accumulation was studied in chromaffin ghosts in which the ΔpH was generated by ATP addition, these observations indicate that amine uptake occurs in accordance with the ΔpH, regardless of the mechanism of its generation.

THE ROLE OF ΔpH AND $\Delta\Psi$ IN CATECHOLAMINE UPTAKE IN CHROMAFFIN GRANULES

The experiments described above and others with isolated chromaffin granules and ghosts support the conclusion that the driving force for amine uptake is the $\Delta\bar{\mu}_{H^+}$. The more challenging question remains: What is the precise quantitative relationship between the $\Delta\bar{\mu}_{H^+}$ and $\Delta\bar{\mu}_A$ (electrochemical gradient for amines)?

Within a chemiosmotic model, the energy for substrate transport systems is derived from the $\Delta\bar{\mu}_{H^+}$ generated from the H^+-pumping ATPase, and the coupling between the transported species (amine and proton, in this case) would approach equilibrium, i.e., net transport would vanish (22), when

$$\Delta\bar{\mu}_A - n\Delta\bar{\mu}_{H^+} = 0$$

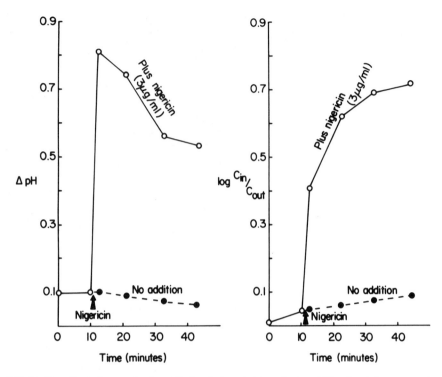

FIG. 11. The effect of a counter-ion diffusion-induced pH gradient on ^{14}C-epinephrine accumulation into chromaffin ghosts. The chromaffin ghosts were formed in 185 mM KCl, 1 mM Tris/maleate, pH 6.8, and then resuspended in 185 mM choline chloride, 30 mM Tris/maleate, including either ^{14}C-methylamine and ^{3}H$_2$O **(left)** or ^{14}C-epinephrine and ^{3}H$_2$O **(right)**. Nigericin (3 μg/ml) was added at the indicated time. An equivalent volume of ethanol was added to those incubations labeled "no addition." The chromaffin ghosts were present at 3.1 mg of protein/ml, and the total volume was 10 ml. The temperature was 24°C. See Johnson et al. (16).

where n is the number of protons translocated in an antiport (opposite) direction to the amine, i.e., n is the stoichiometry of the reaction. The value n determines the relative contribution of the ΔpH and $\Delta \psi$ to the driving force. Because of the ΔpH, inside acidic, and $\Delta \psi$, positive inside, generated by the H$^+$-translocating ATPase, the diffusion gradient for protons is in the outward direction in the chromaffin granule or ghost. Therefore, in the model for amine uptake based on chemiosmotic processes, the influx of amines is coupled to H$^+$ efflux. More precisely, this H$^+$-amine antiport is defined by nonspontaneous obligatory coupling through the existence of a putative reserpine-sensitive translocator.

A highly purified preparation of chromaffin ghosts was utilized in which it was possible to generate and sustain either a large ΔpH or a $\Delta \psi$, or a combination of both, approaching values observed in intact granules. In a series of experiments the $\Delta\bar{\mu}_{H^+}$ and $\Delta\bar{\mu}_A$ were measured simultaneously under a wide range of conditions, in an attempt to differentiate between various mathematically derived models for the driving force and possible H^+/amine stoichiometries. Amine uptake was measured in the presence of (a) ΔpH alone; (b) $\Delta\psi$ alone; (c) $\Delta\bar{\mu}_{H^+}$; and (d) various external pH values.

A reliable method for a titratable decrease in the membrane potential has been the addition of large concentrations of thiocyanate (1 to 60 mM) to the experimental medium (20,21,28). In the first series of experiments (Fig. 12), ghosts were incubated in the presence of ATP and 60 mM thiocyanate in an attempt to completely dissipate any electrical gradients generated, thereby leaving only the ΔpH to drive accumulation.

Conversely, by incubating the ghosts with increasing concentrations of ammonia (1 to 50 mM), the magnitude of the ΔpH generated could be varied from 0.80 to 0 pH units; no $\Delta\psi$ was observed. Carbon-14-labeled serotonin was added to these suspensions of ghosts after establishment of the ΔpH or $\Delta\psi$, and the distribution of the biogenic amine was measured after apparent steady-state levels had been reached. In the representative experiment illustrated in Fig. 12, curve A, the ^{14}C-serotonin accumulation under these conditions varied directly with the magnitude of the ΔpH, with a proportionality constant of 0.4. In each instance, the net ^{14}C-serotonin distribution after subsequent addition of 50 mM NH_4^+ was close to zero (date not shown). Data from 16 separate ghost preparations in which the ΔpH was the only driving force for biogenic amine accumulation, when plotted as amine distribution (mV) versus the ΔpH (mV), yielded a slope of 0.38 ± 0.09.

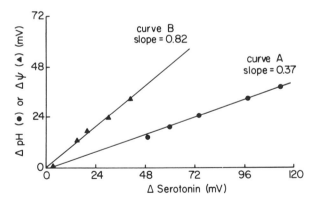

FIG. 12. Relationship between ΔpH or $\Delta\psi$ and steady-state accumulation of ^{14}C-serotonin. **A:** ΔpH versus ^{14}C-serotonin accumulation. Chromaffin vesicles, formed in 185 mM sodium isethionate, 20 mM ascorbate plus 4 mM Tris/maleate at pH 7.0, were suspended in the incubation medium of similar composition with one-half of the samples containing ^{14}C-methylamine (to measure ΔpH), and the other half ^{14}C-serotonin and various concentrations of NH_4^+ (0 to 30 mM). **B:** Experimental conditions were identical to **A**, except that ^{14}C-SCN$^+$ was substituted for methylamine and sodium thiocyanate (0 to 60 mM) for NH_4^+. [From Johnson et al. (15), with permission.]

Conversely, when chromaffin ghosts were incubated in the presence of ATP and 60 mM ammonia, a $\Delta\psi$ of up to 50 mV and no ΔpH was measured. Thus, in the second series of experiments, only a $\Delta\psi$ was present to drive amine uptake. The addition of varied external concentrations of thiocyanate to the incubation medium resulted in a range of values for the $\Delta\psi$ from 50 to 0 mV. In the representative experiment shown in Fig. 12, curve B, ^{14}C-serotonin distribution measured under identical steady-state conditions was proportional to the magnitude of the $\Delta\psi$ with a constant of proportionality of 0.8. The observation that amine distribution slightly exceeds the magnitude of the $\Delta\psi$ is probably secondary to limited internal binding or oxidation of the accumulated amine. Data from 10 separate ghost preparations in which the $\Delta\psi$ was the only driving force for amine accumulation yielded an average slope of $0.76 + 0.06$.

These results not only demonstrate that amine accumulation can be driven by either the ΔpH or $\Delta\psi$ but, in addition and of particular significance, indicate that over the same range of magnitudes of the driving force (0 to 50 mM), the accumulation of biogenic amines in the presence of a ΔpH alone is twice that observed in the presence of the $\Delta\psi$ of the same magnitude. These findings are consistent with an apparent driving force equal to $\Delta\psi - 2Z\Delta pH$.

Due to the fact that the biogenic amines tested possess two ionizable groups, an amine group (pK_a of 10) and a phenolic group (pK_a of 8.9), four amine species exist at physiological pH and are capable, at least theoretically, of being transported (Fig. 13): (a) the cationic species, the predominant species at neutral pH; (b) the neutral species, which is highly lipophilic, formed by deprotonation of the nitrogen; (c) the zwitterionic species, formed by deprotonation of one of the ring hydroxyl groups; and (d) the anionic species, formed by deprotonation of the nitrogen and a ring hydroxyl group.

Although at physiological pH the predominant percentage of amine species

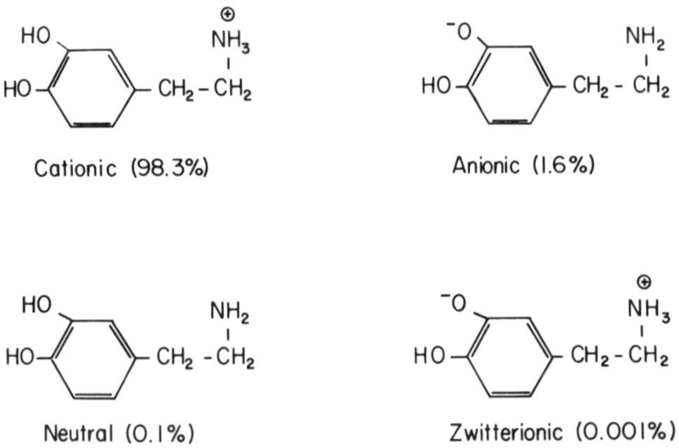

FIG. 13. Calculation of the molecular species of catecholamine existing in solution at physiological pHs.

is in the cationic form and positively charged (approximately 98%), the positive potential existing across the chromaffin granule membrane would render energetically unfavorable the translocation of this species.

Most of the available evidence to date suggests that the uncharged species is the species being transported. This evidence includes (a) a change in K_m of various amines transported, consistent with their pK_a and, therefore, their true concentration of undissociated species present in solution; (b) a change of K_m for the transport of the same catecholamine by changing the external pH, quantitatively consistent with a change in the percentage of the undissociated species available for the carrier.

The simplest model consistent with these and other experiments is illustrated in Fig. 14. The essence of the model is that a negatively charged carrier molecule is capable of binding an uncharged catecholamine. The complex moves vectorially inward due to the presence of an inwardly directed positive potential. At the inside membrane face, dissociation of catecholamine from the carrier occurs.

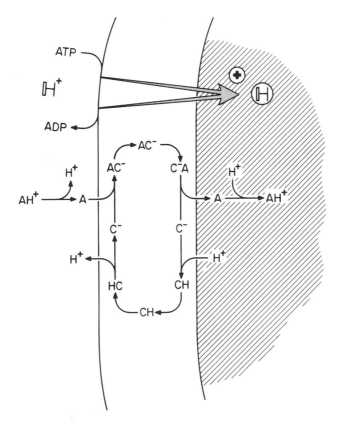

FIG. 14. Model for the coupling between electrochemical H$^+$ gradients and catecholamine transport in chromaffin granules.

Due to the high intragranular H^+ concentration, both the catecholamine and the carrier become protonated. Movement of the protonated carrier in the opposite vectorial plane then takes place, unaffected by the $\Delta\psi$ but at the expense of the H^+ gradient.

Although more remains to be elucidated on the molecular mechanism of this reaction, the evidence accumulating in the literature overwhelmingly indicates that catecholamine uptake into chromaffin granules occurs through a chemiosmotic mechanism coupled to an electrogenic H^+-ATPase.

This overall mechanism for amine uptake is not confined to the chromaffin granules, but is probably generally applicable to a variety of other storage vesicles. Evidence to date strongly indicates that in serotonin-dense granules of platelets (4), histamine granules of mast cells (14), and posterior pituitary granules (32), amine accumulation based on electrochemical proton gradients also occurs.

REFERENCES

1. Berneis, K. H., Pletscher, A., and DaPrada, M. (1970): Phase separation in solutions of noradrenaline and adenosine triphosphate: Influence of bivalent cations and drugs. *Br. J. Pharmacol.,* 39:382–389.
2. Carlsson, A., Hillarp, N-A., and Waldeck, B. (1963): Analysis of the Mg^{++}-ATP dependent storage mechanism in the amine granules of the adrenal medulla. *Acta Physiol. Scand.,* 59 (Suppl. 215):1–38.
3. Carty, S., Johnson, R. G., and Scarpa, A. (1980): The isolation of intact adrenal chromaffin granules using isotonic Percoll gradients. *Anal. Biochem.,* 106:438–445.
4. Carty, S., Johnson, R. G., and Scarpa, A. (1981): Serotonin transport in isolated platelet granules: Coupling to the electrochemical proton gradient. *J. Biol. Chem.,* 256:11244–11250.
5. Carty, S., Johnson, R. G., and Scarpa, A. (1982): Electrochemical proton gradient in dense granules isolated from anterior pituitary. *J. Biol. Chem.,* 257:7269–7273.
6. Casey, R. P., Njus, D., Radda, G. K., and Sehr, P. A. (1977): Active proton uptake by chromaffin granules: Observation by amine distribution and phosphorus-31 nuclear magnetic resonance techniques. *Biochemistry,* 16:972–976.
7. De Oliveira Filgueiras, O. M., van den Bosch, H., Johnson, R. G., Carty, S., and Scarpa, A. (1981): Phospholipid composition of some amine storage granules. *FEBS Lett.,* 129:309–313.
8. Dolais-Kitabgi, J., and Perlman, R. L. (1975): The stimulation of catecholamine release from chromaffin granules by valinomycin. *Mol. Pharmacol.,* 11:745–750.
9. Greville, G. D. (1969): A scrutiny of Mitchell's chemiosmotic hypothesis of respiratory chain and photosynthetic phosphorylation. *Curr. Top. Bioenerg.,* 3:1–79.
10. Hayflick, S., Johnson, R. G., Carty, S., and Scarpa, A. (1982): Kinetic and quantitative measurements of catecholamine transport in chromaffin ghosts using a glassy carbon electrode. *Anal. Biochem.,* 126:58–66.
11. Holz, R. W. (1978): Evidence that catecholamine transport into chromaffin vesicles is coupled to vesicle membrane potential. *Proc. Natl. Acad. Sci. USA,* 75:5190–5194.
12. Johnson, R. G., Beers, M., and Scarpa, A. (1982): H^+ ATPase of chromaffin granules: Kinetics, regulation, and stoichiometry. *J. Biol. Chem.,* 257:10701–10707.
13. Johnson, R. G., Carlson, N., and Scarpa, A. (1978): ΔpH and catecholamine distribution in isolated chromaffin granules. *J. Biol. Chem.,* 253:1512–1521.
14. Johnson, R. G., Carty, S., Fingerhood, B., and Scarpa, A. (1980): The internal pH of mast cell granules. *FEBS Lett.,* 120:75–79.
15. Johnson, R. G., Carty, S., and Scarpa, A. (1981): Proton:substrate stoichiometries during active transport of biogenic amines in chromaffin ghosts. *J. Biol. Chem.,* 256:5773–5780.
16. Johnson, R. G., Pfister, D., Carty, S., and Scarpa, A. (1979): Biological amine transport in chromaffin ghosts. *J. Biol. Chem.,* 254:10963–10972.
17. Johnson, R. G., and Scarpa, A. (1976): Ion permeability of isolated chromaffin granules. *J. Gen. Physiol.,* 68:601–631.

18. Johnson, R. G., and Scarpa, A. (1979): Protonmotice force and catecholamine transport in isolated chromaffin granules. *J. Biol. Chem.,* 254:3750–3760.
19. Kaback, H. R. (1972): Transport across isolated bacterial cytoplasmic membranes. *Biochim. Biophys. Acta,* 265:367–417.
20. Kanner, B. I., Sharon, I., Maron, R., and Schuldiner, S. (1980): Electrogenic transport of biogenic amines in chromaffin granule membrane vesicles. *FEBS Lett.,* 111:83–86.
21. Knoth, J., Handloser, K., and Njus, D. (1980): Electrogenic epinephrine transport in chromaffin granule ghosts. *Biochemistry,* 19:2938–2942.
22. Mitchell, P. (1963): Molecule, group, and electron translocation through natural membranes. *Biochem. Soc. Symp.,* 22:142–169.
23. Mitchell, P. (1966): *Chemiosmotic Coupling in Oxidative and Photosynthetic Phosphorylation.* Glynn Research, Bodmin, Cornwall.
24. Nicholls, D. (1982): *"Bioenergetics," an Introduction to the Chemiosmotic Theory.* Academic Press, London.
25. Njus, D., and Radda, G. K. (1978): Bioenergetic processes in chromaffin granules: A new perspective on some old problems. *Biochim. Biophys. Acta,* 463:219–244.
26. Njus, D., Sehr, P. A., Radda, G. K., Ritchie, G., and Seeley, P. J. (1978): Phosphorus-31 nuclear magnetic resonance studies of active proton translocation in chromaffin granules. *Biochemistry,* 17:4337–4343.
27. Phillips, J. H. (1978): 5-Hydroxytryptamine transport by the bovine chromaffin granule membrane. *Biochem. J.,* 170:673–679.
28. Pollard, H. B., Shindo, H., Creutz, C. F., Pazoles, C. J., and Cohen, J. S. (1979): Internal pH and state of ATP in adrenergic chromaffin granules determined by P-31 nuclear magnetic resonance spectroscopy. *J. Biol. Chem.,* 254:1170–1177.
29. Ramos, S., and Kaback, H. R. (1979): pH-Dependent changes in proton:substrate stoichiometries during active transport in *Escherichia coli* membrane vesicles. *Biochemistry,* 16:4271–4275.
30. Rottenberg, H. (1975): The measurement of transmembrane electrochemical gradients. *J. Bioenerg.,* 7:61–64.
31. Salama, G., Johnson, R. G., and Scarpa, A. (1979): Spectrophotometric measurements of transmembrane potential and pH gradients in chromaffin granules. *J. Gen. Physiol.,* 75:109–140.
32. Scherman, D., Nordmann, J., and Henry, J. P. (1982): Existence of an adenosine 5′-triphosphate dependent proton translocase in bovine neurosecretory granule membrane. *Biochemistry,* 21:687–694.
33. Taugner, G. (1971): The membrane of catecholamine storage vesicles of adrenal medulla: Catecholamine fluxes and ATPase activity. *Naunyn Schmiedebergs Arch. Pharmacol.,* 270:392–396.
34. Winkler, H. (1976): The composition of adrenal chromaffin granules: An assessment of controversial results. *Neuroscience,* 1:65–80.

Electrogenic Transport: Fundamental Principles and Physiological Implications, edited by Mordecai P. Blaustein and Melvyn Lieberman. Raven Press, New York © 1984.

Electric Aspects of Co- and Countertransport

E. Heinz and S. M. Grassl

Cornell University Medical College, New York, New York 10021

Rheogenicity of cotransport depends on the electric net charge transported, but not directly on the electric charge of the unloaded translocator. The kinetic manifestations of rheogenicity and their relationship to rate-limiting steps(s) are discussed using the model of a gated (conformational) channel. Most of the biological Na-linked cotransport systems appear to be positive rheogenic, i.e., associated with the translocation of excess cationic charge. This also applies to anionic substrates, even those bearing two or more negative charges, as exemplified by citrate transport in renal brush-border membrane vesicles. There is evidence that in addition to Na^+, protons may also participate in this cotransport directly, and not merely by neutralizing negative charges of the substrate. The involvement of K^+ in this transport is likely, as it stimulates equilibrium exchange of citrate, but a K-gradient in either direction does not add to the driving force. A possible reason for the predominance of positive rheogenicity in Na-linked cotransport systems may be a regulatory advantage, because such systems can be driven directly by an electric potential difference (PD), which can be raised (or dropped) by an electrogenic pump much faster than can an equivalent chemical PD of Na^+.

Secondary active transport of nutrients and other solutes is considered to involve true coupling between the flow of the solute concerned with that of a cation, such as Na^+ or H^+, through a biological barrier (7,11,25). True coupling, by definition, requires the transient formation of a new species, for example by binding of the species concerned to a translocator site. It differs, therefore, from mere electrostatic interaction between separate ion flows, such as to maintain electroneutrality. The coupling may be positive, as in cotransport (symport), or negative, as in countertransport (antiport). Since at least one ion species participates in the coupling, secondary active transport may be associated with electric phenomena, due to the net translocation of electric charge, depending on stoichiometry and electric valences of the ions concerned (9). There is not complete agreement as to whether the term rheogenic or electrogenic should be used in such cases. Until a consensus on the definition of these terms is reached, we shall reserve the term electrogenic for primary ion pumps, which

TABLE 1. *Test for rheogenicity*

The following partial differentials, which are derived in the Appendix, should be non-zero if the (coupled) co- or countertransport between A and B is rheogenic. For nonrheogenic systems they become zero, but only under condition that the chemical PDs of the species indicated in the subscripts are maintained constant, and that leakage of the substrate (A) is negligible or has been corrected for.

1. $\dfrac{1}{F}\left(\dfrac{\partial J_a}{\partial \Delta\psi}\right)_{\Delta\mu_A,\ \Delta\mu_B} = -(n_a + n_b)\, v_a L_r$

2. $F\left(\dfrac{\partial \Delta\psi}{\partial \Delta\mu_A}\right)_{\Delta\mu_B,\ \Delta\mu_i} = -\dfrac{(n_a + n_b)\, v_a L_r}{(n_a + n_b)^2\, L_r + \Sigma z_i^2 L_{ii}}$

J_a is the (coupled) flow of solute A.
$\Delta\mu_A$ and $\Delta\mu_B$ the chemical PD of solutes A and B, respectively.
n_a and n_b the number of charges moved into the cell (or, with opposite sign, out of the cell) by A and B, respectively, with each cycle of coupled transport.

---

actually generate electric energy, and the term rheogenic for passive and secondary active transport, which tends to dissipate (primary) electric energy.[1]

It may be convenient to characterize rheogenicity also with respect to the direction of the charge transfer. We propose to call rheogenic transport positive or negative, depending on whether a positive or negative charge, respectively, is moved into the cell, vesicle, or corresponding compartment, while the system functions in its physiologically normal direction. The physiologically normal direction may sometimes require an arbitrary definition. Accordingly, we would call the Na^+-linked transport of neutral nutrients as well as countertransport, such as between Na^+ and Ca^{2+} (17), positive rheogenic, under the assumption that the physiological function of the former system is the uptake of nutrients, and that of the latter, the extrusion of Ca^{2+}.

It is often experimentally difficult to verify rheogenicity. The most commonly used methods are to demonstrate the mutual effects between coupled flow, and electric PD or current, i.e., to show that an imposed electric PD affects transport rate and static head distribution of the substrate, and that, vice versa, an imposed change in substrate flow induces an electric PD or current, respectively (4,9). This is illustrated in Table 1 in which the partial differentials, which are derived from some basic flow equations, should be different from zero whenever $n_+ \neq n_-$, i.e., when the (coupled) translocation is rheogenic. Sources of error in these procedures are the leakage flows of these ions, which should be eliminated or corrected for, as is sometimes feasible by using inhibitors specific for the coupled transport to be tested (4).

Before turning to the details of electric effects more specifically, let us briefly

---

[1] In contrast to the definition adopted here, the term rheogenic is also used for pumps treated as constant-current sources in equivalent networks.

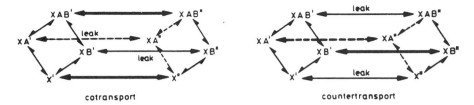

FIG. 1. Secondary active transport: X = unloaded translocator; A, B = solute species transported; AX, XB, XAB = complexes between translocator and solutes. Prime and double prime refer to left and right side orientation, respectively, of translocator and its complexes. *Thick arrows:* transport-effective pathways; *thin arrows:* (inner) leakage pathways.

look at some general features of the coupling mechanism in co- and countertransport. As illustrated in Fig. 1, the same model with small alterations of some crucial parameters may serve to represent both cotransport and countertransport (7). Each involves chemical and osmotic steps including binding and unbinding of the solutes concerned to the respective translocator sites, and the translocation of the resulting complexes through the barrier, no matter whether this occurs by a mobile carrier or by conformational changes of a gated channel. In cotransport (left side) there are two transport-effective translocation routes: the complete complex (AXB) and the unloaded site (X), as indicated by the thick arrows. The other possible routes, those of the translocation of the incomplete complexes (AX, BX), indicated by the thin arrows, are (internal) leakage pathways, which waste energy and therefore tend to reduce or abolish the effectiveness of the system. In countertransport (right side) these relations are inverted. The pathways of AX and BX are the transport-effective ones, and those of AXB, the inner leakages. In either model it is irrelevant how the preponderance of the transport-effective routes over the leakage routes is manifested, by an affinity effect, i.e., through cooperation between A and B, or by a velocity effect, i.e. through differential probability of translocation of the various "complexes" (11). Theoretically it should be possible to interconvert the two systems, simply by altering one or a few crucial parameters. Whether nature makes use of this possibility is not known yet.

In the following, we shall concentrate on cotransport, in line with the topic of our latest research; but from the foregoing it should be easy to extend what holds for the one system to the other by analogy. Assuming that cotransport systems are efficient enough, we simplify the model by omitting the pathways of internal leakage (Fig. 2). We see that on each side the translocatable "complex" (XAB) can be formed through two parallel binding sequences, i.e., through XA and XB. The law of detailed balance demands that the following relationships between all dissociation constants hold:

$$\frac{K_a}{K_{ba}} = \frac{K_b}{K_{ab}} = r \text{ (cooperativity)}$$

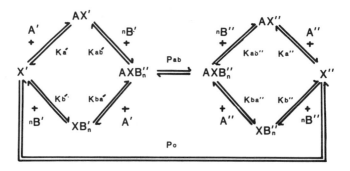

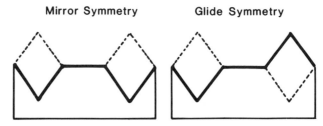

FIG. 2. Cotransport of A and B. **Top:** Random reaction sequence A, B, X, AX, XB, AXB as in Fig. 1. K_a, K_b, K_{ab}, K_{ba} = Dissociation constants; P_{ab}, P_o = Probability of translocation of AXB and X, respectively. **Bottom:** Ordered reaction sequence. Mirror symmetry: $K'_{ab} = 0$, $K''_{ab} = 0$, $K'_a = \infty$, $K''_a = \infty$. Glide symmetry: $K'_{ab} = 0$, $K''_{ba} = 0$, $K'_a = \infty$, $K''_b = \infty$.

and

$$\frac{K_a}{K_b} = \frac{K_{ba}}{K_{ab}} = s \text{ (preference)}$$

In the random system (Fig. 2, top) r and s have finite values. In the ordered system (Fig. 2, bottom) which is now much under consideration, r is infinite, whereas s is either zero (A before B) or infinite (B before A). An ordered system with "glide symmetry" (first in, first out), which seems to fit a channel model better than a mobile-carrier model, has been postulated for Na-linked sugar transport on the basis of kinetic evidence (12, 24). Other workers, however, have found strong evidence against an ordered system (G. Semenza, personal communication). Since it cannot yet be determined whether such discrepancies are due to real differences between different systems or to inadequacies of one or the other approach, we shall leave this question open for the present. It is nonetheless tempting to speculate on the possible significance of the two alternatives. Would one have a distinct advantage over the other? The answer clearly depends on which is the rate-limiting step of the overall process. If this is the translocation, as is usually assumed, the binding processes would be close to equilibrium in either system, so there should not be much difference between the random and the ordered model. If, however, the binding processes are rate-limiting, the random mechanism would, under otherwise equal conditions, likely be more expeditious than an ordered one, as two pathways instead of one would be available to form the translocatable complex. The question of rate limitation is often linked to that of whether a carrier or a channel is involved, based on the assumption that in the former the translocation steps, and in the latter the binding and unbinding processes, are rate-limiting (Fig. 3). This assumption may be plausible, but it is by no means established, especially not for a biological Patlak-Läuger-type gated channel (16,20), which appears to be of a protean nature, as it may sometimes resemble a mobile carrier and at other times a channel.

Let us now see where the electric potentials would become effective within our models, if, for instance, the system depicted were positive rheogenic (e.g., B being a cation). In other words, which steps are electrosensitive? The electrosensitive steps need not always be the same as the rate-limiting steps, although such may be so with mobile carrier systems. This more simple situation is shown in Fig. 3 (left). Here, the rate-limiting translocational pathways, i.e., the translocation of loaded and unloaded translocator, respectively, proceed essentially along the electric field. The kinetics of the various possibilities of such a model have been extensively treated (3,10). It is obviously important

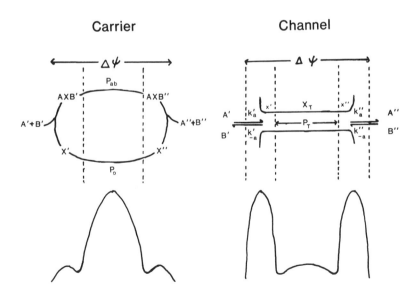

FIG. 3. Rate-limiting and electrosensitive steps in carrier and fixed-channel model, respectively. Rate-limiting barriers are defined by *dashed lines*. X, A, B, XB, AXB same as in Fig. 1.

whether the unloaded or loaded translocator site carries the charge and which of them is rate-limiting, because the electrical PD will affect the kinetic parameters, J_{max} and K_m, differently in the four possible cases. This was derived for a mobile carrier, but it also applies to a gated channel, e.g., that of the Patlak-Läuger type, provided its conformational changes are slow enough to limit the overall transport rate, and involve a charge translocation through the barrier.

Figure 3 (right) shows a model, the rate-limiting steps of which are located in the boundaries, as is usually assumed for channel mechanisms. As has been previously mentioned, substrate movement across these particular barriers, although rate-limiting, is not necessarily electrosensitive, unless there is a major drop in electric PD between the bulk solution and the binding site within the channel or gate. But even if this were not so, i.e., if only translocation of the charge within the channel, and not the (rate-limiting) binding steps were electrosensitive, as might be the case in any model, especially with channels, the translocation of the solutes across the barrier might still be affected by an electric PD, although not by the acceleration of a rate-limiting step but, rather, by the maintenance of a Boltzman-like distribution of the charged species. For example, in a Patlak-Läuger-type gated channel mechanism, an electric field will favor the distribution of the conformational states of the loaded or unloaded species, whichever carries the electric charge, in the one or the other direction, and thereby affect the transport, kinetically and energetically. If the movement of a charged translocator, be it a carrier or a gate, rate-limiting or not, happens to be associated with movement of charge through the barrier, an electric field may even affect the translocation of a neutral solute. It could of course not contribute to the driving force but only affect the kinetics by introducing an additional asymmetry to the system (Fig. 4).

A more general treatment of the effect of electric forces of rheogenic transport has more recently been presented (6), to which the reader may refer. It may suffice here to say that from the fact that transport is affected by an electric PD, little can be said about the underlying model, but more refined analyses of the kinetic details may sometimes help to arrive at the right model.

To return to the experimental observations made thus far, it is striking that most of the Na-dependent cotransport systems, especially those concerned with the reabsorption of nutrients, appear to be predominantly electropositive (13,18). Electroneutral systems seem to be less frequent, although there are differences between different systems, and electronegative ones have hardly ever been unequivocally demonstrated. This is more surprising in the case of anions, including even those with two or three negative charges, many of which appear to be transported in a cationic form with resulting positive rheogenicity (5,14,19, 23,28). For renal and intestinal brush-border vesicles, this is true for glutamate and for anions of the tricarboxylic acid cycle, such as succinate and citrate (14,21). Here, obviously, more cations must be cotransported than would be electrically equivalent to the transported anion, e.g., more than 1 Na^+ for glutamate and more than 2 Na^+ has been shown to be transported with each succinate molecule (15,27).

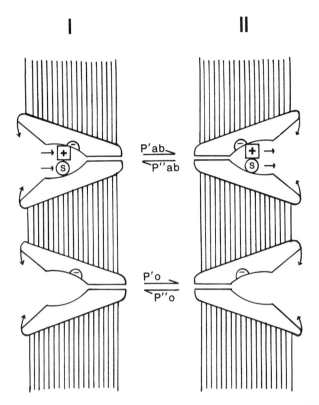

FIG. 4. Patlak-Läuger-type gated channel. Two conformational states, I and II, of transport proteins, in each of which the binding sites are accessible from the left or right side, respectively. Note that any charge located near the binding site of the translocator will be moved relative to the barrier during the conformational change. Transition between the two conformational states is assumed to be so fast that intermediate states can be neglected. This does not exclude that the transitions could be infrequent enough to limit the rate of the translocation.

The question arises whether, in addition to Na^+, other cations may participate in the cotransport. The observation that some such systems are stimulated by lowering the pH would suggest that protons are involved (1,22). For succinate, the Wright group reports that its transport is pH-independent and that an "excess of 2 Na^+," presumably 3, are moved per succinate (15,26,27) as would make the participation of H^+ in this transport system less likely. This is different with citrate transport into these vesicles, which is strongly stimulated by lowering pH (5,26). Since at physiological pH, citrate is predominantly present as trivalent anion, we would have to postulate that at least 4 cations per citrate are cotransported. Since citrate appears to be transported by the same system as succinate (8), we would expect that with either anion 3 Na^+ are required and in addition 1 H^+ per citrate. However, whenever H^+ is involved in catalysis with weak acids, in enzymatic and transport processes alike, the question arises of whether the H^+ attaches to the substrate, modifying it, or directly to the catalyst. Also, in the present case, the question is whether the H⁺ acts by modifying the substrate, i.e., by converting the citrate to a divalent anion, thereby making it acceptable to the succinate carrier, or by functioning as a co-solute in addition to the Na⁺ (Fig. 5). This question is difficult to answer but has been approached by us on the following basis. If the protonated, divalent citrate were the adequate species of this transport system, the pH changes should be ineffective to the extent that the concentration of this species remains constant. This has been tested by readjusting the total citrate concentration at each pH, according to the extended Henderson-Hasselbach equation, so as to keep the divalent species

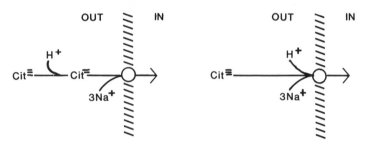

FIG. 5. Effect of H⁺ on cotransport, by substrate modification **(left)** or by cotransport **(right)**.

CITRATE

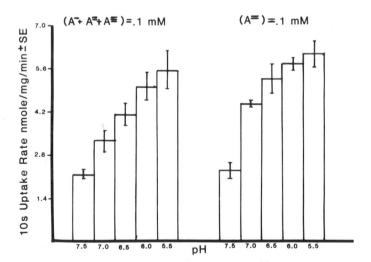

FIG. 6. pH-dependence of citrate transport. **Left:** Total citrate concentration is constant (0.1 mM) at each pH. **Right:** Concentration of divalent citrate is constant (0.1 mM) at each pH. Vesicles of calf renal brush border membrane; the approximate initial rate (10 sec) of citrate uptake at 25°C.

of citrate unchanged. It is seen (Fig. 6) that, nonetheless, the stimulation by H^+ remains undiminished. The assumption that both the divalent and the monovalent citrate may serve as substrate would not make much difference, since, by keeping the sum of these two species constant, the pH effect cannot be abolished. Hence, we tend to believe that the H^+ ions stimulate by cotransport rather than by substrate modification.

Another ion often related to transport is K^+, but since the early work of Crane (2) it is thought of more in terms of acting through ancillary countertransport. In other words, an outward directed K-gradient could add driving force to the inward transport of substrate, even in the absence of a diffusion PD. This has indeed been verified for glutamate (1,22). In terms of our model (Fig. 2), this could easily be explained by a preferential binding of K^+ to the empty translocator. To the extent that outward movement of the latter limits the overall transport rate, it should stimulate inward transport of the substrate. At the same time, however, such an effect should reduce, abolish, or even invert electropositive rheogenicity of the transport. In the present citrate system, any such effect of K^+ can be dismissed, since in the absence or presence of Na an outward K gradient, except through a valinomycin-induced diffusion PD, does not stimulate the citrate uptake (Fig. 7). Peculiarly, however, we also found that the mere presence of K^+ at equal concentrations inside and outside stimulates equilibrium exchange of citrate (7). This suggests a Na-dependent binding of K^+ to the citrate-loaded translocator, thereby raising its mobility. But in this case, an inward directed K^+ gradient should have a cotransport effect and stimulate uptake that has not been confirmed (Fig. 7). The only plausible, although unwarranted, explanation seems to be that K^+ binds to, and thereby accelerates, both loaded and unloaded carriers.

Finally we may ask the obvious question: why are so many cotransport systems for substrates positively rheogenic? At the moment we can only speculate. One reason may be that positive rheogenic transport can be driven by any PD of whatever origin, whereas an electroneutral Na-driven transport would depend specifically on a Na-concentration gradient. Since cellular membranes are usually negative on the inside, such a PD might favor any positive rheogenic process, as the uptake of nutrients or the extrusion of Ca^{2+}. More important could be another advantage in this context, which concerns the expeditiousness of the transport. On initiation of an electrogenic pump (Na^+ or H^+) the electric PD is likely to rise more than 1,000 times faster than the equivalent Na^+ chemical PD (8). This is because the generation of an appropriate Na-concentration gradient is a slow process: it requires the electroneutral transport of millions of Na^+ ions per cell and mV, whereas (considering the low electric capacity of the membrane), the generation of an equivalent electric PD by such a pump requires the electrogenic movement of much fewer ions, on the order of a few hundreds of Na^+ ions per cell and per millivolt. Provided that the pump is fast enough as compared to the passive ion movements, it may raise the electric PD at least 1,000 times faster than it would the equivalent chemical PD of

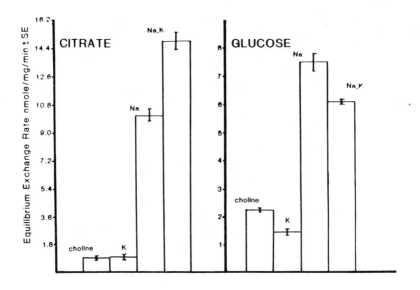
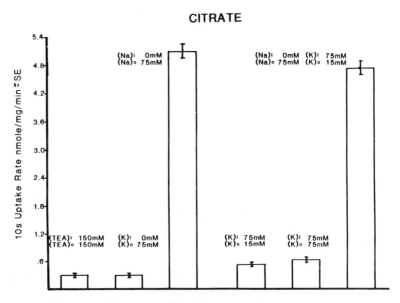

FIG. 7. K^+ effect on citrate transport. **Top:** Equilibrium exchange of citrate at equal concentration of all solutes inside and outside calf renal brush border membrane vesicles at 25°C and pH 7.5; $[Na^+] + [K^+] + [choline] = 100$ mM throughout. It is seen that the presence of 50 mM K^+ on both sides significantly stimulates unidirectional citrate influx, in contrast to glucose influx. It is also seen that without Na^+, neither K^+ (50 mM) nor choline significantly affects citrate exchange. Citrate concentration = 2 mM inside and outside. **Bottom:** It is seen that an outward K^+ gradient does not significantly alter the Na^+-linked net uptake of citrate into calf renal brush border membrane vesicles at pH 7.5 and 25°C. The figure shows also that in the absence of Na^+, the presence of K^+, with or without an inward or outward gradient, does not stimulate citrate influx beyond the control, in which tetraethylamonium (TEA) is the only cation present. Citrate concentration 0.1 mM outside throughout.

Na⁺. Hence, in case of need, a positive rheogenic cotransport could be turned on (or off) with a delay of only milliseconds simply by stepping up (or down) an electrogenic pump.

APPENDIX

Basic equations:
1. Coupled flow of A and B (J_r)

(a) $J_r = -L_r[\nu_a \Delta\mu_A + \nu_b \Delta\mu_B + (\nu_a z_a + \nu_b z_b) F \Delta\psi]$

where ν_a, ν_b = stoichiometric coefficients of A and B, respectively, taken as positive in the forward, and as negative, in the backward direction; z_a, z_b = electrical valencies of A and B, respectively; $\Delta\mu_A$, $\Delta\mu_B$ = chemical PD of A and B, respectively; $\Delta\psi$ = electric PD; F = Faraday constant; and L_r = rate coefficient at given electrochemical PDs of A and B.

(b) $J_a = \nu_a J_r + J_a^{\text{leak}}$ (c) $J_b = \nu_b J_r + J_b^{\text{leak}}$

From these equations it follows, provided that J_a^{leak} can be eliminated:

$$\frac{1}{F}\left(\frac{\partial J_a}{\partial \Delta\psi}\right)_{\Delta\mu_A, \Delta\mu_B} = -\nu_a(n_a + n_b) L_r$$

where n_a, n_b are $\nu_a z_a$ and $\nu_b z_b$, respectively.

2. Electroneutrality requires that

$$z_a J_a + z_b J_b + \Sigma z_i J_i = 0$$

where z_i, J_i are electric valencies and passive flows of the ions present other than A and B. Substituting from 1, we obtain:

$$F\left(\frac{\partial \Delta\psi}{\partial \Delta\mu_A}\right)_{\Delta\mu_B, \Delta\mu_i} = -\frac{(n_a + n_b)\nu_a L_r}{(n_a + n_b)^2 + \Sigma z_i^2 L_{ii}}$$

where L_{ii} is the rate coefficient of flow at a given electrochemical PD of ion i.

ACKNOWLEDGMENTS

Supported by grants of USPHS NIH GM No. 26554–04 and NIH GM No. 27859.

REFERENCES

1. Burckhardt, G., Kinne, R., Stange, G., and Murer, H. (1980): The effects of potassium and membrane potential on sodium-dependent glutamic acid uptake. *Biochim. Biophys. Acta,* 599:191–201.
2. Crane, R. F. (1965): Na⁺-dependent transport in the intestine and other animal tissues. *Fed. Proc.,* 24:1000–1006.

3. Geck, P., and Heinz, E. (1976): Coupling in secondary transport. Effect of electrical potentials on the kinetics of ion linked co-transport. *Biochim. Biophys. Acta*, 443:49–53.
4. Geck, P., Pietrzyk, C., Burckhardt, B. C., Pfeiffer, B., and Heinz, E. (1980): Electrically silent cotransport of Na^+, K^+ and Cl^- in Ehrlich cells. *Biochim. Biophys. Acta*, 600:432–447.
5. Grassl, S. M., Heinz, E., and Kinne, R. (1982): Potential-sensitive Na^+-dependent citrate uptake in renal brush border membrane vesicles (abstract). *Fed. Proc.*, 41:1115.
6. Hansen, U. P., Gradmann, D., Sanders, D., and Slayman, C. L. (1981): Interpretation of current-voltage relationships for "active" ion transport systems: 1. Steady-state reaction-kinetic analysis of class-I mechanisms. *J. Membr. Biol.*, 63:165–190.
7. Heinz, E. (1978): *Mechanics and Energetics of Biological Transport.* Springer, Heidelberg, New York, London.
8. Heinz, E. (1980): Electrogenic and electrically silent proton pumps. In: *Hydrogen Ion Transport in Epithelia*, edited by I. Schulz, pp. 41–46, Elsevier/North-Holland Biomedical Press, New York.
9. Heinz, E. (1981): *Electric Potentials in Biological Membrane Transport.* Springer, Berlin, Heidelberg, New York.
10. Heinz, E. and Geck, P. (1978): The electrical potential difference as a driving force in Na^+-linked cotransport of organic solutes. In: *Membrane Transport Processes, vol. 1*, edited by J. D. Hoffman, pp. 13–30. Raven Press, New York.
11. Heinz, E., Geck, P., and Wilbrandt, W. (1972): Coupling in secondary active transport. Activation of transport by co-transport and/or counter-transport with the fluxes of other solutes. *Biochim. Biophys. Acta*, 255:442–461.
12. Hopfer, U., and Groseclose, R. (1980): The mechanism of Na^+-dependent D-glucagon transport. *J. Biol. Chem.*, 255:4453–4462.
13. Kinne, R., Barac, M., and Murer, H. (1980): Sodium cotransport systems in the proximal tubule: Current developments. *Curr. Top. Membranes & Transport*, 13:303–313.
14. Kippen, I., Hirayama, B., Klinenberg, J. R., and Wright, E. M. (1979): Transport of tricarboxylic acid cycle intermediates by membrane vesicles from renal brush border. *Proc. Natl. Acad. Sci. USA*, 76:3397–3400.
15. Kippen, I., Wright, S., Hirayama, B., and Klinenberg, J. (1980): Direct measurement of Na: Citrate cotransport in rabbit renal brush border membrane vesicles (abstract). *Fed. Proc.*, 39:734.
16. Läuger, P. (1980): Kinetic properties of ion carriers and channels. *J. Membr. Biol.*, 57:163–178.
17. Mullins, L. J. (1977): A mechanism for Na/Ca transport. *J. Gen. Physiol.*, 70:681–695.
18. Murer, H., and Kinne, R. (1980): The use of isolated membrane vesicles to study epithelial transport processes. *J. Membr. Biol.*, 55:81–95.
19. Murer, H., Leopolder, A., Kinne, R., and Burckhardt, G. (1980): Recent observations on the proximal tubular transport of acidic and basic amino acids by rat renal proximal tubular brush border vesicles. *Int. J. Biochem.*, 12:223–228.
20. Patlak, C. S. (1957): Contributions to the theory of active transport. II. The gate-type non-carrier mechanism and generalizations concerning tracer flow efficiency, and measurement of energy expenditure. *Bull. Math. Biophys.*, 19:209–235.
21. Samarzija, I., and Fromter, E. (1982): Electrophysiological analysis of rat renal sugar and amino acid transport. V. Acidic amino acids. *Pfluegers Arch.*, 393:215–221.
22. Schneider, E. G., and Sacktor, B. (1980): Sodium gradient-dependent L-glutamate transport in renal brush border membrane vesicles. *J. Biol. Chem.*, 255:7645–7649.
23. Schneider, E. G., Hammerman, M. R., and Sacktor, B. (1980): Sodium gradient-dependent L-glutamate transport in renal brush border membrane vesicles. *J. Biol. Chem.*, 255:7650–7656.
24. Turner, R. J. (1981): Kinetic analysis of a family of cotransport models. *Biochim. Biophys. Acta*, 649:269–280.
25. West, I. C. (1980): Energy coupling in secondary active transport. *Biochim. Biophys. Acta*, 604:91–126.
26. Wright, S. H., Kippen, I., and Wright, E. M. (1982): Effect of pH on the transport of Krebs cycle intermediates in renal brush border membranes. *Biochim. Biophys. Acta*, 684:287–290.
27. Wright, S. H., Kippen, I., and Wright, E. M. (1982): Stoichiometry of Na^+-succinate cotransport in renal brush-border membranes. *J. Biol. Chem.*, 257:1773–1778.
28. Wright, S. H., Kippen, I., Klinenberg, R., and Wright, E. M. (1980): Specificity of the transport system for tricarboxylic acid cycle intermediates in renal brush borders. *J. Membr. Biol.*, 57:73–82.

Contributions of Electrogenic Pumps to Resting Membrane Potentials: The Theory of Electrogenic Potentials

R. A. Sjodin

Department of Biophysics, University of Maryland School of Medicine, Baltimore, Maryland 21201

Bioelectric potentials were given theoretical consideration long before sophisticated methods were developed for their experimental study. Bernstein (2,3) was the first to formulate a quantitative theory for the origin of bioelectric potentials expressed in terms of the theory for the electrical behavior of electrolytic solutions. According to this theory, membrane potentials of excitable cells were not due to mysterious life forces nor did they depend on complicated phenomena taking place within the membrane. In the resting state at least, they could, in large part, be understood by assuming that the cell membrane imposes a barrier on aqueous ionic diffusion but does not profoundly alter the electrodiffusional nature of the process itself. Thus, the resting potential depends primarily on the concentration gradient of K ions across the membrane, an insight that nearly a century of work has proven correct.

The theory of bioelectric potentials was put on a firmer and more realistic basis by Goldman (11) who solved the Nernst-Planck equation for a biological membrane by assuming a constant electrical field within the membrane. In this case, the current carried across the membrane by each kind of ion depends only on the permeability coefficient, the membrane potential, and the aqueous concentrations for that ion. Each individual ionic current is independent of the currents carried by the other ions present. The equation obtained for the flux of an ion across the membrane is (11,13)

$$j = P \frac{zE_m F}{RT} \left(\frac{C_i e^{zE_m F/RT} - C_o}{e^{zE_m F/RT} - 1} \right) \qquad (1)$$

where P is the permeability coefficient, E_m the membrane potential, Z the ionic valence, F the Faraday, C_i inside aqueous concentration, C_o outside aqueous concentration, and where R and T have their usual significance. Application of the principle of electrical neutrality to the ions Na, K, and Cl is made by setting $j_{Na} + j_K = j_{Cl}$ in which case the following equation is obtained for the membrane potential (11,13).

$$E_m = \frac{RT}{F} \ln \frac{P_K[K]_o + P_{Na}[Na]_o + P_{Cl}[Cl]_i}{P_K[K]_i + P_{Na}[Na]_i + P_{Cl}[Cl]_o} \tag{2}$$

This equation is usually called the Goldman-Hodgkin-Katz equation and, for easy reference, will be denoted here as the GHK equation. The equation is also termed the constant-field equation, but this terminology is avoided here due to some possible confusion. Equation (2) can be derived without the explicit assumption of a constant electrical field in the membrane (31,32). The general solution of the Nernst-Planck differential equation for a system of different ions and a membrane with fixed electrical charges reduces to Eq. (2) in the special case in which the total ionic concentrations on either side of the membrane are equal (31,32). As biological membranes operate in accordance with this condition because of the need for osmotic equilibrium, they can be expected to have a constant electrical field and to have a membrane potential governed by Eq. (2).

The GHK equation has been successfully applied to the membrane potential of nerve and muscle cells (12,13). It has since found application to a wide variety of cells by many investigators. Although it would be difficult to overestimate the influence of the GHK equation on the theory and experimental investigation of bioelectric potentials, the equation is inadequate if transport mechanisms and electrodiffusion both are present within the membrane and if these additional ion-transport mechanisms produce net ionic currents across the membrane. It is now known that at least two membrane ion-transport processes that produce net ionic current are present in excitable cells. One is the electrogenic Na pump (1,10,14,19) and the other is Na/Ca countertransport (4,18,24).

A notable failure of the GHK equation to predict correct results is in the case of skeletal muscle fibers recovering from the Na-enriched state in the presence of an adequate external K concentration (1,10,14,19). In such cases, the fiber membrane potential becomes more negative than the K ion equilibrium potential. Equation (2) can never predict a membrane potential that is more negative than the K equilibrium potential. It is clear that the previous approaches to the theory of bioelectric potentials do not take into account a sufficient number of important membrane processes. The GHK equation works well when other electrogenic transport mechanisms are either absent, as in the case of poisoned ionic pumps, or operate at a relatively low rate, as can occur when electrodiffusional ion movements account for the greater part of the ionic traffic across the membrane.

The purpose of this work is to add the missing parts to the theory of bioelectric potentials by incorporating electrogenic transport mechanisms into the GHK equation. Added stipulations are that, whatever changes and additions are introduced, the general form of the GHK equation be preserved and the additional terms involve readily accessible quantities such as ionic concentrations and quantities that can be obtained by well-defined experimental measurements using standard methods. Transcendental solutions to equations are accepted. Ruling them out would preclude much of the work.

GENERALIZED THEORETICAL METHODS

The general method applied is an extension of the approach of Mullins and Noda (23). In addition to electrodiffusional fluxes, j, that are assumed to obey Eq. (1), the pumped fluxes of Na and K are introduced as m_{Na} and m_K, respectively. The convention used is that the positive direction for ionic flux is outward. The coupling ratio is defined as

$$r = -\frac{m_{Na}}{m_K} \quad (3)$$

It is always a positive quantity, as the coupled Na and K pumped fluxes oppose each other. The coupling ratio can be a constant, in the case of a rigidly coupled pump or it can vary in accordance with Eq. (3).

In the steady state, both electrodiffusive leakages of Na and K must be exactly balanced by respective pumped fluxes of Na and K so that

$$j_{Na} = -m_{Na} \quad (4)$$

and

$$j_K = -m_K \quad (5)$$

Dividing Eq. (4) by Eq. (5) gives

$$j_{Na} = \frac{m_{Na}}{m_K} j_K = -r j_K \quad (6)$$

Equation (2), for Na and K terms only, is obtained by setting $j_{Na} = -j_K$ and substituting using Eq. (1). It is clear that Eq. (6) will give a similar equation with P_K multiplied by r:

$$E_m = \frac{RT}{F} \ln \frac{r P_K[K]_o + P_{Na}[Na]_o}{r P_K[K]_i + P_{Na}[Na]_i} \quad (7)$$

which is the steady-state equation of Mullins and Noda (23). This illustrates the general method. Any linear multiplier of any one of the permeability coefficients, say P_K, will simply yield a GHK-type equation that contains the linear flux or current multiplier in an appropriate position in the equation. It should be noted that the linear multiplier need not be a constant, as no integration of a differential equation is involved.

The method just illustrated can be generalized to non-steady-state conditions. Although the results have been previously reported (27,30), they will be briefly summarized. For the non-steady state, it is convenient to define pump/leak ratios as follows

$$f_{Na} = -\frac{m_{Na}}{j_{Na}} \tag{8}$$

and

$$f_K = -\frac{m_K}{j_K} \tag{9}$$

These quantities are clearly not constants, as electrodiffusive fluxes will vary in accordance with the ionic concentrations and membrane potential, and pumped fluxes will vary in accordance with separately determined pump activation parameters involving activating ion concentrations and pump rate constants.

It is simplest to obtain solutions for the membrane potential when Cl ions are assumed to be at equilibrium, as in muscle fibers in which Cl ions redistribute very rapidly. Although individual pumped fluxes no longer balance individual passive fluxes, electrical neutrality must hold with $j_{Cl} = 0$ so that

$$j_{Na} + j_K + m_{Na} + m_K = 0 \tag{10}$$

To obtain an equation for E_m, Eq. (9) is divided by Eq. (8) giving

$$j_{Na} = -r\frac{f_K}{f_{Na}} j_K \tag{11}$$

According to the principle of linear multiplication of flux, previous considerations permit the solution for E_m to be immediately written as

$$E_m = \frac{RT}{F} \ln \frac{r\left(\frac{f_K}{f_{Na}}\right) P_K[K]_o + P_{Na}[Na]_o}{r\left(\frac{f_K}{f_{Na}}\right) P_K[K]_i + P_{Na}[Na]_i} \tag{12}$$

This equation predicts potentials more negative than the K equilibrium potential whenever the electrodiffusive K flux, j_K, is in the inward direction, as f_K becomes negative in this case.

Electrogenic-pump equations having this form are useful whenever the pumped and passive ionic fluxes are known, as in cases where they have been measured by tracers or net ionic movements. Equation (12) has been applied to frog skeletal muscle in this way by Sjodin and Ortiz (30). It is not convenient, however, to use equations such as Eq. (12) when the value of E_m is known and one wants to infer something quantitative about individual pumped fluxes, or when pump-rate parameters and ion-activator concentrations are known and one wants to calculate the value of E_m.

To obtain relations in which the pump parameters appear in explicit form, Eq. (10) for electrical neutrality is solved to yield a parameter Y given by

$$E_m = \frac{RT}{F} \ln \frac{YP_K[K]_o + P_{Na}[Na]_o}{YP_K[K]_i + P_{Na}[Na]_i} \tag{13}$$

where

$$Y = \frac{f_K - 1}{f_{Na} - 1} = r \frac{f_K}{f_{Na}} \tag{14}$$

Solution of Eq. (14) using the definition of r gives

$$Y = \frac{1}{1 - nf_{Na}} \tag{15}$$

where n is a number defined by

$$n = \frac{r-1}{r} \tag{16}$$

Applying the defining equation for f_{Na} (Eq. 8), the more useful form of Y becomes

$$Y = \frac{1}{1 + \frac{nm_{Na}}{j_{Na}}} = \frac{j_{Na}}{j_{Na} + nm_{Na}} \tag{17}$$

From Eq. (1), the value of j_{Na} is

$$j_{Na} = P_{Na}\bar{f}([Na]_i e^{E_m F/RT} - [Na]_o) \tag{18}$$

where

$$\bar{f} = \frac{E_m F}{RT}\left(\frac{1}{e^{E_m F/RT} - 1}\right) \tag{19}$$

For most negative values of the membrane potential and for most values of $[Na]_i$, the passive Na efflux can be neglected, which results in

$$j_{Na} = -P_{Na}\bar{f}[Na]_o \tag{20}$$

The expression to be used for the pumped Na flux depends on the kind of pump-activation equation that fits pumped-flux data. In the case of skeletal muscle fibers operating in the linear region of $[Na]_i$ (22,26) and for squid giant axons (7,28), a simple and useful expression for m_{Na} is

$$m_{Na} = k_{Na}\left(\frac{[K]_o}{[K]_o + k_m}\right)[Na]_i \tag{21}$$

This equation is based on the three assumptions: that (a) the pump operates in the linear region for $[Na]_i$, (b) S-shaped K activation kinetics are ignored, and (c) the inhibitory effect of $[Na]_o$ is ignored. It is clear that these assumptions need not be made, but that more complicated equations for m_{Na} will result if they are not made.

The general procedure is next to insert the explicit values for j_{Na} and m_{Na} from Eqs. (20) and (21) into Eq. (17) to obtain a value of Y that can be inserted into Eq. (13), which can then be solved for E_m. The resulting equation will give E_m in terms of the usual permeability coefficients and ionic concentrations and the Na pump parameters as given in Eq. (21).

As the general equations hold for non-steady-state conditions, ionic concentrations will change with time and the membrane potentials will be time dependent. The equations developed are for a point in time at which ionic concentrations have particular values. The time dependence can always be worked out from the kinetics of the ion movements. This need not be done, however, to understand the basic applications of the equations.

SOME APPLICATIONS TO ELECTROGENIC SODIUM PUMPS

So that potential users of electrogenic potential equations will not become discouraged by some mathematical complexities to soon follow, a simplified but easily realizable special case will be considered in which particularly simple results are obtained. At very high Na pumping rates occurring at saturating concentrations of $[K]_o$, as observed in Na-enriched skeletal muscle fibers, for example, the pumped Na flux is many times the electrodiffusive Na flux and, approximately, $Y = -P_{Na}\bar{f}\,[Na]_o/nk_{Na}[Na]_i$. Substitution into Eq. (13) gives

$$E_m = \frac{RT}{F} \ln \frac{\bar{f}P_K[K]_o - nk_{Na}[Na]_i}{\bar{f}P_K[K]_i - nk_{Na}[Na]_i^2/[Na]_o} \quad (22)$$

The equation is transcendental, as $\bar{f}$ is a function of membrane potential by Eq. (19). It is, nevertheless, easy to use if one knows the pump parameters, n and k_{Na}. For skeletal muscle fibers from the frog, r is approximately 1.5 (30) and $n = 0.33$ from Eq. (16).

To evaluate k_{Na}, use is made of the fact that a steady state of the muscle Na is obtained at around $[Na]_o = 120$ mM, $[Na]_i = 10$ mM, $[K]_o = 5$ mM, and $E_m = -84$ mV (26). A steady state is also reached at around $[Na]_o = 110$ mM, $[Na]_i = 10$ mM, $[K]_o = 2.5$ mM, and $E_m = -92$ mV (23). In the steady state, $J_{Na} = -m_{Na}$ and

$$P_{Na}\bar{f}\,[Na]_o = k_{Na}\left(\frac{[K]_o}{[K]_o + k_m}\right)[Na]_i \quad (23)$$

The value of k_m is approximately 5 mM for K activation in frog sartorius muscle (26). Application of these values sets the value of k_{Na} at around 100 P_{Na}. As $P_{Na} = 0.01\,P_K$ in muscle (12), k_{Na} is approximately equal to P_K. It is evident that k_{Na} contains the surface-to-volume ratio of the fibers so that it is equivalent to an effective "permeability coefficient" for the pump since, when multiplied by a concentration, a flux is obtained.

Values of E_m calculated from Eq. (22) are plotted against the magnitude of $[Na]_i$ in Fig. 1 on the basis that the sum of $[Na]_i + [K]_i$ in muscle fibers is constant at about 150 mM (29). A constant value of $[K]_o = 20$ mM is used, and $[Na]_i$ values ≥ 50 mM are employed to comply with the simplifying assumption of a high pumping rate at saturating values of $[K]_o$. The value of E_K, the potassium equilibrium potential, is plotted as open circles. The values of E_m calculated from Eq. (22), plotted as solid dots, indicate the additional negativity created by the electrogenic Na pump. The electrogenic pump contribution to E_m is between -10 and -50 mV, increasing with increasing values of $[Na]_i$. Pump contributions to E_m in this range have been observed in skeletal muscle fibers (1,10,14,19). It can be noted from Fig. 1 that the electrogenic pump serves as an effective buffer for E_m, keeping its value relatively constant over a range where severe depolarization would result in its absence.

The neglect of passive Na efflux in deriving electrogenic potential equations is not essential, but considerable mathematical simplification results from the

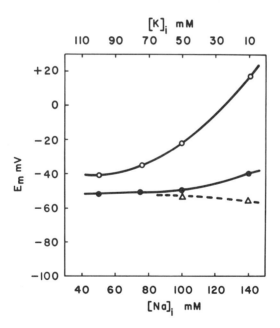

FIG. 1. Calculated values of the resting membrane potential, E_m, are plotted versus the Na and K ion concentrations within the skeletal muscle fiber. Values of E_m calculated from Eq. (22) appear as *solid circles*. All values are calculated with $[K]_o = 20$ mM and $[Na]_o = 120$ mM. Values of the potassium equilibrium potential, E_K, are plotted as *open circles*. *Open triangles* are from solutions of an equation not requiring the assumption that passive Na efflux is of negligible magnitude. The value of k_{Na} is equal to P_K as calculated in the text.

assumption that passive Na efflux is negligible. At high values of $[Na]_i$, as in the right-hand part of Fig. 1, the assumption becomes progressively less valid. For the two highest values of $[Na]_i$, a more complicated equation in which electrodiffusive Na efflux is not neglected was also used. The equation is not presented but points calculated from it appear in Fig. 1 as open triangles. Equation (22) is seen to be valid up to $[Na]_i = 100$ mM. At higher values of $[Na]_i$, the more accurate equation predicts values of E_m that are somewhat more hyperpolarized. This is expected, since the real passive inward Na leak against which the pump operates becomes less at very high values of $[Na]_i$. The approximation of neglecting passive unidirectional Na efflux is numerically valid up to $[Na]_i = 100$ mM. Qualitative conclusions are not altered above this value of $[Na]_i$, however. The buffering action of the pump on the resting potential is even more pronounced with passive Na efflux taken into account.

The only approximation made in deriving the generally valid equation for E_m is that the electrodiffusive unidirectional Na efflux is of negligible magnitude so that j_{Na} is given by Eq. (20). Substitution of Eqs. (20) and (21) into Eq. (17) and substitution of the resulting value of Y into Eq. (13) yields the general electrogenic equation for zero chloride current, i.e., Cl ions at equilibrium:

$$E_m = \frac{RT}{F} \ln \frac{\bar{f}(P_K[K]_o + P_{Na}[Na]_o) - nk_{Na}([K]_o/[K]_o + k_m)[Na]_i}{\bar{f}(P_K[K]_i + P_{Na}[Na]_i) - nk_{Na}([K]_o/[K]_o + k_m)[Na]_i^2/[Na]_o} \quad (24)$$

This equation provides the resting membrane potential with an active electrogenic Na:K pump when ionic concentrations, permeability coefficients, and the two pump parameters n and k_{Na} are known. As n depends only on r, Eq. (16), its value is known if r is known. The value of k_m is usually known.

Equation (24) reduces to the GHK equation when the pump current is zero. Some conditions that produce this are: (a) The pump is highly inhibited with ouabain or is unfueled of adenosine triphosphate (ATP), reducing k_{Na} to zero or a very low value. (b) The pump operates at zero or very low rate because either $[K]_o$ or $[Na]_i$ has been made zero or reduced to a very low value. (c) The pump operates in an electrically neutral mode with $r = 1$ making $n = 0$.

More complicated kinetic expressions for the Na pumping rate can easily be incorporated into Eq. (24). Some useful features to incorporate would be S-shaped kinetics for activation by $[K]_o$, the inhibitory effect of $[Na]_o$ on the pumping rate, and pump operation out of the linear region (22,26). For the purposes of this discussion, however, nothing is lost by neglecting these higher-order effects.

It is instructive to plot values of E_m calculated from Eq. (24) versus the logarithm of $[K]_o$ in the usual way. To amplify the effect due to the electrogenic Na pump, the equation is plotted for the case of Na-enriched frog muscle using permeability and pump parameters previously given. The following values are used in the plot: $[Na]_i = 50$ mM; $[K]_i = 100$ mM; $[Na]_o = 120$ mM; $P_{Na} =$

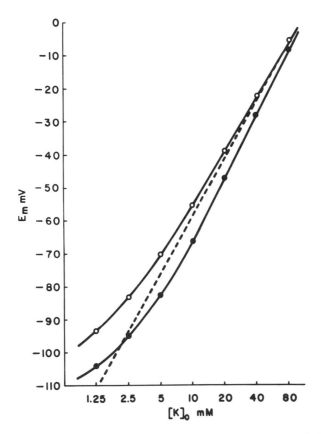

FIG. 2. Calculated values of E_m are plotted against $[K]_o$ semilogarithmically. *Solid circles* are calculated from Eq. (24) *open circles* are calculated from Eq. (2). The *dotted line* is a plot of the potassium equilibrium potential. Calculated values are for frog muscle with $[Na]_i = 50$ mM; $[K]_i = 100$ mM; and other parameters as given in the text.

0.01 P_K; $k_{Na} = P_K$; $n = 0.33$; $k_m = 5$ mM. The results are plotted in Fig. 2 as solid dots. The values calculated from the GHK equation (Eq. (2) with Cl ions at equilibrium) are plotted as open circles for comparison. The dotted line represents the K equilibrium potential and has a slope of 58 mV per 10-fold concentration change. In all cases, the points plotted are calculated values at the indicated $[K]_o$ values. The points are then fit by the lines drawn by best eye fit.

The electrogenic membrane hyperpolarization is evident at all $[K]_o$ values used and, for values of $[K]_o \leq 20$ mM, amounts to about 10 mV. Some other features can be noted. At large depolarizations, less electrogenic hyperpolarization occurs. This cannot be due to lessened pump activity since the saturation value is approached as $[K]_o$ is elevated and the pump fluxes will remain high and rather constant. The increment in extra potential negativity developed per unit of pump current diminishes at large depolarizations. This effect will be explored further in the next section of text.

Another interesting feature relates to the direction of the electrodiffusive K flux. At values of $[K]_o$ above about 2 mM, j_K is inward and K ions are both pumped inwardly and pulled inwardly by the potential. At values of $[K]_o < 2$ mM, j_K is outward and the K pump must oppose the electrodiffusive K loss. At around 2 mM $[K]_o$, K ions are at equilibrium, and the pump current now balances only the Na leak. Potassium net movement is still inward due to the pump.

It should be noted, however, that although absolute values of E_m differ significantly from both the GHK and Nernst equations, the slopes of the E_m versus log $[K]_o$ lines are rather similar. At high values of $[K]_o$, both the GHK equation and the electrogenic potential equation have slopes that approach the Nernst slope, the electrogenic slope being somewhat steeper than Nernst, and the GHK slope being somewhat less steep than the Nernst slope. At low values of $[K]_o$, the GHK equation slopes and the electrogenic potential equation slopes are similar, and it would be unlikely that they could be experimentally differentiated. At intermediate $[K]_o$ levels, the electrogenic equation E_m versus log $[K]_o$ slope is Nernstian over a wide range. It is clear that E_m versus log $[K]_o$ slopes measured experimentally do not provide either an accurate or a sensitive test for the presence of electrogenic pumps.

The experimental data available on the recovery of Na-enriched muscle fibers in the presence of external K ions provide a convenient test for the validity of Eq. (24). It was first found by Kernan (14) that frog muscle fibers made rich in Na by immersion in cold K-free Ringer's solution had membrane potentials more negative than the K equilibrium potential during recovery in the presence of external K. As the increased internal negativity persisted only during the period of Na extrusion brought on by the increased value of $[Na]_i$, the effect was attributed to the action of a Na pump that was not electrically neutral. The kinetics of this process were further studied by Frumento (10), who found a correlation between the magnitude of Na extrusion rate and the increased internal negativity. The availability of Na pumped fluxes for frog muscle fibers makes it possible to calculate the value of E_m expected during the process of Na extrusion using Eq. (24). The values used for equation parameters are those previously given except that $P_{Na} = 0.016\ P_K$. At the time $t = 0$ the muscle fibers are assumed to be Na loaded to the point where $[Na]_i = [K]_i = 75$ mM. The value of E_m calculated from Eq. (24) is plotted against time in Fig. 3. The values used for $[Na]_i$ and $[K]_i$ were obtained by applying calculated flux values for Na and K to a 75 μm diameter muscle fiber to calculate $[Na]_i$ and $[K]_i$ versus time. The plot in Fig. 3 is for a constant value of $[K]_o = 2.5$ mM. The value of E_m obtained from Eq. (24) is in reasonable agreement with experimental results (10). To obtain agreement of E_m with experimental measurements at the end of the period of Na extrusion, P_{Na} was chosen at the value

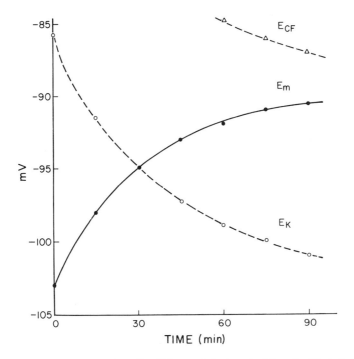

FIG. 3. Values of E_m calculated from Eq. (24) are plotted against time (*solid circles*) for frog sartorius muscle fibers extruding Na during recovery from Na enrichment. Values of parameters used and the method of calculation are presented in the text. Included in the graph are the equilibrium potential for K ions, E_K (*open circles*), and 3 values of E_m calculated from the constant-field or GHK Eq. (2) with Cl⁻ ions at equilibrium (*open triangles*).

of 0.016 P_K. This value is about 1.5 times higher than the value estimated by Hodgkin and Horowicz (12), because the electrogenic action of the Na pump was not taken into account at that time.

PUMP CURRENT-VOLTAGE RELATIONS

It is the pump current, the Faraday times the difference between the absolute magnitudes of pumped Na efflux and pumped K influx, that leads to electrogenic voltage changes across the membrane. It is of interest to examine the relations that exist between the pump current magnitude and the electrogenic change in membrane potential produced.

Denoting pump current by I_P, it is evident that $I_P = F(m_{Na} + m_K)$, or, defining net pumped flux as $\overline{m} = m_{Na} + m_K$, $I_P = F\overline{m}$. In terms of previously defined quantities, it follows that

$$I_P = nFm_{Na} = nFk_{Na}\left(\frac{[K]_o}{[K]_o + k_m}\right)[Na]_i \qquad (25)$$

To obtain the relation between I_P and E_m, the above value of I_P is substituted into Eq. (24). Defining the quantities P_i and P_o by the relations

$$P_i = P_K[K]_i + P_{Na}[Na]_i \tag{26}$$

and

$$P_o = P_K[K]_o + P_{Na}[Na]_o \tag{27}$$

and solving for I_P gives

$$I_P = \frac{\bar{f}F(P_o - P_i e^{E_m F/RT})}{1 - \dfrac{[Na]_i}{[Na]_o} e^{E_m F/RT}} \tag{28}$$

Equation (28) can be used to plot the current–voltage relation for the pump. A convenient point of reference is the reversal potential at which $I_P = 0$ and through which I_P changes sign. The reversal potential is clearly simply the value of E_m calculated from the GHK equation (Eq. (2) without the Cl terms). Equation (28) is plotted in Fig. 4 for the same conditions used for Fig. 2 but

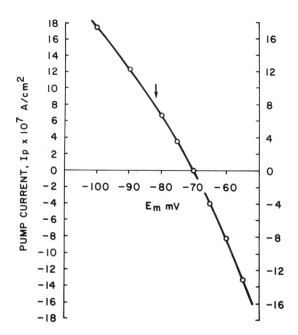

FIG. 4. The Na pump current, I_p, is plotted versus E_m using Eq. (28). The *arrow* points to a potential of −82 mV, which is the resting membrane potential calculated from Eq. (24) and which represents the resting potential set by the Na pumping rate for the following given fixed conditions: frog muscle with parameters given in the text; $[K]_i = 100$ mM; $[Na]_i = 50$ mM; $[K]_o = 5$ mM; and $[Na]_o = 120$ mM.

with $[K]_o$ fixed at 5 mM, i.e., $[K]_o = 5$ mM; $[K]_i = 100$ mM; $[Na]_o = 120$ mM; and $[Na]_i = 50$ mM. Again, the example is for frog muscle with all parameters as before. As expected from the form of Eq. (28), the current-voltage relation for the pump has an exponential bend characteristic of the constant-field flux equation kind of rectification (compare with Eq. (1)). Nevertheless, the I–V relation is linear for ± 5 mV from the reversal potential and fairly linear for ± 10 mV from the reversal point. The pump current is positive (outward) for hyperpolarizations since a pump electromotive force (EMF) is involved in producing the potential changes. The I_P versus E_m relation thus has a negative slope. It can be noted that the negative pump current part of the graph is only hypothetical, in the sense that negative pump currents with K pumping exceeding Na pumping are not normally observed. The positive I_P side of the curve is experimentally realizable by changing the pumping rate in various ways. For the data parameters given, the actual pump rate under these conditions moves the potential to -82 mV shown by the arrow. It is convenient to refer to this potential as the set point for the pump. The set point is clearly always given by Eq. (24), or its equivalent with altered pump kinetics. The set point would move to more hyperpolarized values if the pump were stimulated, say, with insulin (16), and would move to more depolarized values if the pump were partially inhibited.

A final point is that the slope of the I_P versus E_m curve at the reversal point is 70 μMho/cm², which is about equal to the K conductance for frog muscle fibers. Recall that k_{Na}, the pump-rate factor having the units of the permeability coefficient, is approximately equal to P_K for frog muscle.

Pump current-voltage relations are useful because they tell how much pumping is required to produce a given value of the membrane potential. Recall from Fig. 2, for example, that for increasing membrane depolarization, continuously less hyperpolarization is produced by about the same pumping rate. This is now seen to be due to the rectification present in the I_P–E_m relation evidenced in Fig. 4. For less negative values of the potential, the pump I–V relation is steeper and more pump activity is required to produce the same change in potential.

Since actual set points for the resting potential met in practice usually do not lie more than about 10 mV from the reversal or GHK potential, a knowledge of $\partial I_P/\partial E_m$ at the reversal potential is a good index of how much pump current will be required in view of the approximate linearity over this range (Fig. 4). For this reason, it is useful to have an equation for $\partial I_P/\partial E_m$. A simplified mathematical approach is to solve the basic equation for electrical neutrality using only the net pump flux $\overline{m}$ without splitting into m_{Na} and m_K terms. Solving the condition $j_{Na} + j_K + \overline{m} = 0$ using substitutions from Eq. (1) gives the following equation for E_m:

$$E_m = \frac{RT}{F} \ln \frac{P_K[K]_o + P_{Na}[Na]_o - \overline{m}/\bar{f}}{P_K[K]_i + P_{Na}[Na]_i} \qquad (29)$$

Neglecting small higher-order exponential terms, one obtains

$$E_m = \frac{RT}{F} \ln \frac{P_K[K]_o + P_{Na}[Na]_o + (RT/E_mF)\overline{m}}{P_K[K]_i + P_{Na}[Na]_i} \tag{30}$$

An equation similar to Eq. (30) has been obtained by Moreton (17). Writing Eq. (30) in terms of I_P, solving for I_P, and differentiating with respect to E_m gives

$$\frac{\partial I_P}{\partial E_m} = \frac{F^2}{RT} \left\{ P_i e^{E_m F/RT} \left(1 + \frac{E_m F}{RT}\right) - P_o \right\} \tag{31}$$

where P_i and P_o are as previously defined. At the reversal potential, this relation simplifies to

$$\frac{\partial I_P}{\partial E_m} = \frac{F^2}{RT} \left(\frac{E_m F}{RT} P_i e^{E_m F/RT}\right) \tag{32}$$

Equation (32) is simple, easy to use, and gives answers accurate to within 4%.

To obtain a relation that follows directly from the main approach applied from the outset, Eq. (28) can be differentiated with respect to E_m. The equation obtained at the reversal potential is

$$\frac{\partial I_P}{\partial E_m} = \frac{F^2}{RT} \left(\frac{E_m F}{RT}\right) \left(\frac{P_i}{(e^{-E_m F/RT} - 1) + ([Na]_i/[Na]_o)(1 - e^{E_m F/RT})}\right) \tag{33}$$

which reduces to Eq. (32) on making some approximations. Values of $\partial I_P/\partial E_m$ at the reversal potential are plotted against the reversal potential using Eq. (33) solved for different ionic concentrations (Fig. 5). Three curves are plotted at different values of $[K]_i$ and $[Na]_i$ with $[K]_i + [Na]_i = 150$ mM held constant. The reversal potential is calculated from the GHK equation at different values of $[K]_o$ and $[Na]_o$. The values of $[K]_i$ and $[Na]_i$ remain constant for any given curve. To avoid the inconvenience of using negative axes on the graph, $-\partial I_P/\partial E_m$ is plotted against minus the reversal potential.

The curves again all illustrate the fact that more pump current is required per unit of potential change the more depolarized the membrane. It is also evident from an examination of the curves that, at a given reversal potential value, less and less pump current is required to produce a given potential change the lower the K content and the higher the Na content of the fibers. Thus, as fibers gain more Na, not only do they produce more pump current due to increased activation of the pump, but it is inherently easier for the pump to hyperpolarize the membrane. This is clearly a valuable mechanism for the fibers that works to maintain a healthy value of the membrane potential under conditions favoring large depolarizations. This is obviously the mechanistic basis for the buffering effect of the pump current on the membrane potential so apparent in Fig. 1.

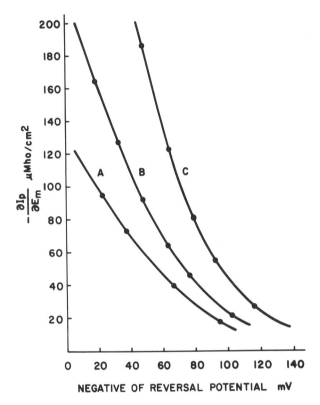

FIG. 5. The absolute value of the Na pump current required to hyperpolarize the membrane a given amount is plotted versus the absolute value of the reversal potential calculated from Eq. (2). The points plotted are calculated from Eq. (33) for frog muscle for three internal ionic states: curve **A:** $[K]_i = 50$ mM; $[Na]_i = 100$ mM; **B:** $[K]_i = 75$ mM; $[Na]_i = 75$ mM; **C:** $[K]_i = 140$ mM; $[Na]_i = 10$ mM.

ELECTROGENIC PUMP EQUATIONS WITH A Cl ION CURRENT PRESENT

In the theory previously developed, it was assumed that there was no Cl current so that $j_{Cl} = 0$. This is equivalent to assuming that Cl ions are not pumped and that P_{Cl} is very high so that Cl ions redistribute rapidly to reach an equilibrium. This situation holds to a good approximation for the skeletal muscle fiber examples given (30). In this case, the chloride term can be introduced into equations by adding the quantity $P_{Cl}[Cl]_i$ to the sum in the numerator and $P_{Cl}[Cl]_o$ to the sum in the denominator of the argument of the logarithm in the electrogenic pump equations. As the chloride concentrations give an equilibrium ratio, the equations predict the same value for the potential with or without the Cl terms. For this reason, Cl terms have been left out of the equations.

For nerve cells, and other cell types, elimination of Cl terms in this way is not a valid procedure, as Cl ions are often not in an equilibrium distribution and may even be pumped. If Cl ions are pumped in an electrogenic manner (i.e., the pump produces a net anion current), such current needs to be incorporated into equations in a way similar to the method used for the Na:K pump. This will not be done here, but j_{Cl} and, hence, P_{Cl} will be taken into account.

The basic condition for electrical neutrality now becomes

$$j_{Na} + j_K - j_{Cl} + m_{Na} + m_K = 0 \tag{34}$$

Substituting from Eqs. (1), (8), and (9) and applying the definitions of r and n previously given, the following equation is obtained for the membrane potential when, as before, the passive Na efflux is assumed to be negligible:

$$E_m = \frac{RT}{F} \ln \frac{AP_K[K]_o + BP_{Na}[Na]_o + CP_{Cl}[Cl]_i + nk_{Na}^2 a^2[Na]_i^2}{AP_K[K]_i + BP_{Na}[Na]_i + CP_{Cl}[Cl]_o + nk_{Na}^2 a^2[Na]_i^3/[Na]_o} \tag{35}$$

where

$$A = \bar{f}(P_{Na}\bar{f}[Na]_o - k_{Na}a[Na]_i) \tag{36}$$

$$B = \bar{f}(P_{Na}\bar{f}[Na]_o - (n+1)k_{Na}a[Na]_i) \tag{37}$$

$$C = \bar{f}(P_{Na}\bar{f}[Na]_o - nk_{Na}a[Na]_i) \tag{38}$$

and

$$a = \frac{[K]_o}{[K]_o + k_m}, \quad (k_m \text{ squid} \cong 1.5 \text{ m}M) \tag{39}$$

When $P_{Cl} = 0$ or Cl ions are in equilibrium, Eq. (35) reduces to Eq. (24). When the Na pump current is zero (electroneutral pump, $n = 0$; $[K]_o$ or $[Na]_i = 0$ so that pump is inactivated; or pump is unfueled or completely inhibited, $k_{Na} = 0$), then Eq. (35) reduces to Eq. (2), the GHK equation.

Equation (35), with its auxiliary Eqs. (36)–(39), is generally valid for an electrogenic Na pump and any distribution whatever of K, Na, and Cl ions. The only approximation used in its derivation is that the passive Na efflux is low enough to neglect. This restricts its usage to membrane potentials and values of $[Na]_i$ that permit this assumption to be made. The restriction is not burdensome, however, as by far the majority of cases examined in experimental practice meet this condition. For example, Fig. 1 can be consulted to see that the assumption is valid even when $[Na]_i$ is approximately equal to $[Na]_o$ in frog muscle, as long as E_m is approximately -50 mV. The same theory can be applied without approximation but considerably more terms appear in the equation.

APPLICATIONS TO THE SODIUM PUMP OF SQUID GIANT AXONS

Much work has been done on the Na:K pump in squid giant axons and much pertinent permeability information is available for this preparation. For this reason, it is logical to test the electrogenic potential equations developed on data from squid giant axons.

Permeability coefficients for ions can be calculated from Eq. (1) provided that net passive ionic fluxes and the membrane potential are known. Sodium influx has been measured in squid giant axons by several workers (6,7,25). The average value from these sources is 54 pmol/cm²sec. Neglecting a small passive backflux of Na of about 1 pmol/cm²sec, P_{Na} is calculated from Eq. (1) using $[Na]_i = 50$ mM, $[Na]_o = 425$ mM, and $E_m = -60$ mV. The value of P_{Na} so calculated is 0.048×10^{-6} cm/sec. The K fluxes in intact squid giant axons were measured by Caldwell and Keynes (8), who obtained for K efflux, 38 pmol/cm²sec, and for K influx, 16.3 pmols/cm²sec. The residual K influx after maximal pump poisoning with cyanide (CN) is about 5 pmol/cm²sec (28). Using K passive influx as 5 pmol/cm²sec, net passive efflux is 33 pmol/cm²sec. Using $[K]_i = 350$ mM and $[K]_o = 10$ mM (8), Eq. (1) gives $P_K = 0.57 \times 10^{-6}$ cm/sec with a membrane potential of -60 mV. These calculations give $P_{Na} = 0.08$ P_K. Keynes (15) has measured $P_{Cl} = 0.19$ P_K with $[Cl]_i = $ mM and $[Cl]_o = 560$ mM.

Pumped Na and K fluxes have also been measured in squid giant axons. Adjusted to a value of 50 mM for $[Na]_i$, $m_{Na} = 24$ pmol/cm²sec and $m_K = 12$ pmol/cm²sec, using average values of flux (20,28). The value of r for the squid giant axon Na pump is thus 2.0, and $n = 0.5$. The product of $k_{Na} \times a = 0.48 \times 10^{-6}$ cm/sec from this data. A summary of the values used in solving Eq. (35) for the squid giant membrane appears in Table 1.

The results of the calculations are that, for the squid giant axon, Eq. (35) with electrogenic Na pumping predicts a value of -60.8 mV for E_m, whereas Eq. (2), the GHK equation without electrogenic transport, predicts $E_m = -55.1$ mV.

The value of -60.8 mV is eminently reasonable and corresponds nicely with normal measured values of the resting potential for the squid giant axon membrane. The electrogenic hyperpolarization produced by the Na pump according to Eq. (35) and the above data amounts to about 5 mV. This is greater than the strophanthidin-induced depolarization observed in squid giant axons. De-Weer and Geduldig (9) report values from about 1 to 4 mV for this quantity under various conditions. The normally observed depolarization observed for

TABLE 1. *Data used for squid giant axon calculations*

Ion	P (cm/sec)	P_{REL}	$[C]_o$ (mM)	$[C]_i$ (mM)	m (pmol/cm²sec)
K	0.57×10^{-6}	1.0	10	350	−12
Na	0.048×10^{-6}	0.08	425	50	24
Cl	0.11×10^{-6}	0.19	560	41	—
r		n		$k_{Na}a$	
2.0		0.5		0.48×10^{-6} cm/sec	

fresh axons was somewhat over 1 mV. The strophanthidin-induced depolarization may not be the best index of the contribution of the Na pump to the membrane potential, however. It has been reported for internally dialyzed squid giant axons that strophanthidin reduces Na efflux to a residual value that is still some 10 times greater than the value observed for axons unfueled of ATP (7). The action of strophanthidin on Na pump electrogenic hyperpolarization may, therefore, be incomplete. It would clearly be of interest to know the value of the ATP-sensitive component of the membrane potential in internally dialyzed axons.

Physiologically, the axon must be in an overall steady state. For the measured value of r (approximately 2), a steady state would be predicted at $E_m = -65$ mV by Eq. (7) using the concentration and permeability data in Table 1. Since measured values of E_m are seldom this negative and are usually around -60 mV, the Na pump is evidently not activated enough to provide this steady state in a freshly dissected axon, in the face of the heightened Na leak that most likely occurs. Accordingly, $[Na]_i$ rises and $[K]_i$ falls with the time during which dissected axons are stored in seawater. A new steady state is predicted when $[Na]_i$ has risen high enough to activate the Na/K pump until Na and K leaks are now compensated. From Eq. (7), the new steady state is predicted at $E_m = -60.2$ mV, with $[Na]_i = 110$ mM; $[K]_i = 290$ mM; and $[Cl]_i = 50$ mM. These calculations show that a dissected squid giant axon stored in normal seawater will approach this new steady state at about constant membrane potential (-60 mV), as observed.

ELECTROGENIC Na/Ca TRANSPORT

It is possible to incorporate any electrogenic transport process in the membrane into the theory for the membrane potential in much the same way as applied to the Na pump. Only possible mathematical complexity sets practical limitations on the extent to which useful results can be obtained. The membrane process that uses energy stored in the electrochemical gradient for Na ions to extrude Ca ions from excitable cells is electrogenic, as more than two Na ions are transported inwardly per transport cycle for every Ca ion transported outwardly (4,5,18,21).

The present theoretical methods have been applied to Na/Ca transport to obtain the contribution of this process to the membrane potential of excitable cells (27). The results will be summarized. Following the previous methods for non-steady-state conditions, pump/leak ratios can be defined as follows:

$$f_{Ca} j_{Ca} = -t_{Ca} \qquad (40)$$

where t_{Ca} is the outward flux of Ca ions produced by Na/Ca transport. Similarly,

$$f'_{Na}(t_{Na} + j_{Na}) = -m_{Na} \qquad (41)$$

where t_{Na} is the inward Na flux driving Ca transport. As before, $f_K j_K = -m_K$ (see Eq. (9)). Defining the stoichiometric factor of Na/Ca transport as it is possible to show that

$$q = -\frac{t_{Na}}{t_{Ca}} \qquad (42)$$

$$q f_{Ca} j_{Ca} + \frac{f_K}{f'_{Na}} r j_K + j_{Na} = 0 \qquad (43)$$

and that E_m is given by

$$E_m = \frac{RT}{F} \ln \frac{2q f_{Ca} P'_{Ca}[Ca]_o + r(f_K/f'_{Na}) P_K[K]_o + P_{Na}[Na]_o}{2q f_{Ca} P''_{Ca}[Ca]_i + r(f_K/f'_{Na}) P_K[K]_i + P_{Na}[Na]_i} \qquad (44)$$

where

$$P'_{Ca} = P_{Ca} (e^{E_m F/RT} + 1)^{-1} \qquad (45)$$

and

$$P''_{Ca} = P'_{Ca} e^{E_m F/RT} \qquad (46)$$

Equation (44) reduces to a steady-state equation when $f_{Ca} = f'_{Na} = f_K = 1$.

Equation (44) can be used as it stands if the components of flux appearing in Eqs. (40)–(42) are known. For example, assume that all flux components are known initially in the steady state. Then assume that an increase in Na/Ca transport activity occurs so that t_{Ca} increases and f_{Ca} assumes a value greater than 1. As t_{Ca} increases, t_{Na} must also increase as indicated by Eq. (42). The net increase in positive-charge delivery to the fiber interior depolarizes the membrane so that, for constant Na and K pumping rates, an increase in passive charge movement out of the fiber ensues to balance the increased charge entry due to Na/Ca transport. Changes in the ratio f_K/f'_{Na} and in j_{Ca}, j_K, and j_{Na} are then selected so that Eqs. (43) and (44) are both obeyed. Although this must be done by trial and error, in the absence of a computer program, it is not difficult to arrive at the correct result in two or three tries. The reason for this is that P_K is generally higher than P_{Na} and P_{Ca}, and most of the compensation in charge movement is due to K ions. Thus, one can assume as a first approximation that j_{Ca} and j_{Na} remain constant and that the change in j_K can be reckoned from the change in Na/Ca charge movement. In this way, Eq. (43) can be used to calculate a first value for the ratio f_K/f'_{Na} which can be used via Eq. (44) to calculate a first value for E_m. Then Eq. (1) is tested to see if the estimated depolarization really does produce the desired change in j_K. The change will usually deviate a bit. Then the first value of the potential is used to calculate changes in j_{Ca} and j_{Na}. These changes adjust j_K to a new second value. Usually, one more change in the potential leads to values that satisfy all equations.

Were it not for mathematical complexity, one could proceed as in the derivation of Eqs. (24) and (35) by inserting explicit values of the ion-transport terms.

Due to the presence of the term f_{Ca}, however, it is not possible to insert values for all of the transport terms without increasing the transcendental nature of the final equation severalfold. As it is not practical to solve the resulting equation by trial and error, it would not be useful to present the final form of the equation with explicit values without providing a computer program. It is clear that the barrier to further application is of computational nature and is not due to theoretical limitations.

A particularly simple case of Na/Ca transport can be treated without mathematical difficulty. When $[Ca]_o$ is equal to zero, j_{Ca} can be assumed to be zero due to the minute value of electrodiffusional Ca efflux at most values of $[Ca]_i$. In this case, the condition for electrical neutrality is that $j_K + j_{Na} + m_{Na} + m_K + 2m_{oCa} = 0$. Writing the sum $m_{Na} + m_K$ as the pumped net cation flux given by $\bar{m}$, and recalling that m_{oCa} is already a net Ca efflux, the solution of the above electroneutrality condition gives for the potential:

$$E_m = \frac{RT}{F} \ln \frac{P_K[K]_o + P_{Na}[Na]_o + (1/\bar{f})(\bar{m} + 2m_{oCa})}{P_K[K]_i + P_{Na}[Na]_i} \qquad (47)$$

where

$$\bar{m} = nm_{Na} = nk_{Na}\left(\frac{[K]_o}{[K]_o + k_m}\right)[Na]_i \qquad (48)$$

and

$$m_{oCa} = k_7\beta \frac{k_{-8}[Na]_o^4[Ca]_i[X]_T}{k_{-8}[\alpha_o + \beta[Na]_o^4[Ca]_i] + \alpha_i[k_7\beta[Na]_o^4[Ca]_i + k_8]} \qquad (49)$$

Also, $t_{Ca} = m_{oCa} - m_{iCa} = m_{oCa}$ for zero influx of Ca. Equation (49) is the result of Mullins' work (18). The k's, α's, and β's refer to rate constants. Equation (47) can be directly applied whenever Ca influx can be neglected, and Na pump and Na/Ca exchange net currents are known.

SUMMARY AND CONCLUSIONS

Pumped and transported components of ionic flux have been added to passive electrodiffusive components. This permits the derivation of equations for the resting membrane potential that take account of electrogenic mechanisms in which the transport mechanism or pump itself produces a net ionic current. Such equations are general in that they apply to non-steady-state conditions in which intracellular ionic concentrations are changing.

The equations developed allow calculation of resting membrane potentials in terms of ionic concentrations, membrane permeability to ions, and kinetic relations for pumped ionic fluxes. When applied to skeletal muscle fibers, the equations predict a buffering effect of the Na/K pump on the membrane potential over a wide range in the values $[K]_i$ and $[Na]_i$ such that a fairly constant membrane potential occurs under conditions in which the passive ionic fluxes themselves would produce increasing degrees of depolarization.

A plot of the membrane potential versus log $[K]_o$ with an electrogenic Na pump present gives a curve with slopes both greater than and less than 58 mV per 10-fold concentration change. Over a middle range of $[K]_o$ values, the slope is 58 mV. The slope of E_m versus log $[K]_o$ curves is, therefore, not a very sensitive test for the presence of an electrogenic pump.

For the same internal ionic concentrations, less electrogenic increment in membrane potential is observed the higher the value of $[K]_o$, and the more depolarized the membrane. This is due to a rectification present in the pump current-voltage curve, which requires that more pump current be present to produce a given membrane hyperpolarization at depolarized values of the potential than at hyperpolarized values of the potential. A gain in Na and a loss of K by the fibers affects the rectification curve in such a way that less pump current is required to produce the same degree of hyperpolarization. This mechanism ensures that adequate internal negativity will be maintained at high values of $[Na]_i$ and $[K]_o$ where saturation of the pumping rates might be expected.

In the non-steady state of Na extrusion, the condition for which these equations were developed, it is clearly possible for the Na pump to generate potentials considerably higher than those generated under steady-state conditions. For steady-state conditions, both skeletal muscle and squid giant axon Na pumps generate additional internal negativity amounting to a few millivolts. During net Na extrusion, on the other hand, the skeletal muscle Na pump has been observed to produce additional internal negativity of up to -40 mV (10), and hyperpolarizations from about 20 to 80 mV are apparent in plots of the present equations (Fig. 1). It is of interest to inquire if any limits exist for pump-generated potentials in the two cases—steady and non-steady state. The question of a limit for steady-state conditions has already been adressed. It has been shown that the pump-generated contribution cannot exceed the value (RT/F) ln $(1/r)$ which, for $r = 1.5$, is about -10 mV (33). This value is clearly exceeded for non-steady-state conditions. The question of a limit for these conditions is best examined by considering the nature of Eq. (24). The argument of the logarithm may never become negative. Thus, the limiting value for the argument is zero. This limiting condition is easily obtained from Eq. (24) and yields

$$E_m \text{ (MAX)} = -\frac{RT}{F} \left\{ \frac{nk_{Na}([K]_o/[K]_o + k_m)[Na]_i}{P_K[K]_o + P_{Na}[Na]_o} \right\} \quad (50)$$

This equation has no theoretical maximum as long as $[Na]_i$ can rise indefinitely. There are practical limits to how high $[Na]_i$ can rise, however. In skeletal muscle fibers, $[Na]_i$ cannot rise beyond 140 mM, which is complete replacement of intracellular K by Na. For this limiting case, E_m (MAX) $= -210$ mV, using equation parameters previously given. For non-steady-state conditions, therefore, the maximum contribution of the pump to the potential depends on how much

Na loading can take place and also on the membrane permeability to K and to Na. Lowering these permeability values at maximum Na loading increases the limiting negativity.

The equations can be amplified to take into account P_{Cl} when Cl ions are not at equilibrium. Applying the theory to data from squid giant axons predicts a value of -60 mV for the normal resting potential and a value of -55 mV for the resting potential in the complete absence of Na pumping.

Equations can be extended to include other electrogenic transport mechanisms such as Na/Ca transport. In this case, however, explicit statement of the Na/Ca transport fluxes results in equations with increased transcendental nature so that computer-programmed solutions become necessary except for a particularly simple case occurring in the nominal absence of external Ca ions.

REFERENCES

1. Adrian, R. H., and Slayman, C. L. (1966): Membrane potential and conductance during transport of sodium, potassium and rubidium in frog muscle. *J. Physiol. (Lond.)*, 184:970–1014.
2. Bernstein, J. (1902): Untersuchungen zur thermodynamik der Bioelektrischen ströme. *Pfluegers Arch.*, 92:521–562.
3. Bernstein, J. (1912): *Elektrobiologie*, F. Vieweg, Braunschweig.
4. Blaustein, M. P. (1974): The interrelationship between sodium and calcium fluxes across cell membranes. *Rev. Physiol. Biochem. Pharmacol.*, 70:33–82.
5. Blaustein, M. P., Russell, J. M., and DeWeer, P. (1974): Calcium efflux from internally dialyzed squid axons: The influence of external and internal cations. *J. Supramol. Struct.*, 2:558–581.
6. Brinley, F. J., Jr., and Mullins, L. J. (1967): Sodium extrusion by internally dialyzed squid axons. *J. Gen. Physiol.*, 50:2303–2331.
7. Brinley, F. J., Jr., and Mullins, L. J. (1968): Sodium fluxes in internally dialyzed squid axons. *J. Gen. Physiol.*, 52:181–211.
8. Caldwell, P. C., and Keynes, R. D. (1960): The permeability of the squid giant axon to radioactive potassium and chloride ions. *J. Physiol. (Lond.)*, 154:177–189.
9. DeWeer, P., and Geduldig, D. (1978): Contribution of sodium pump to resting potential of squid giant axon. *Am. J. Physiol.*, 235:C55–C62.
10. Frumento, A. S. (1965): Sodium pump—its electrical effect in skeletal muscle. *Science*, 147:1442–1443.
11. Goldman, D. E. (1943): Potential, impedance, and rectification in membranes. *J. Gen. Physiol.*, 27:37–60.
12. Hodgkin, A. L., and Horowicz, P. (1959): The influence of potassium and chloride ions on the membrane potential of single muscle fibres. *J. Physiol. (Lond.)*, 148:127–160.
13. Hodgkin, A. L., and Katz, B. (1949): The effect of sodium ions on the electrical activity of the giant axon of the squid. *J. Physiol. (Lond.)*, 108:37–77.
14. Kernan, R. P. (1962): Membrane potential changes during sodiumn transport in frog sartorius muscle. *Nature*, 193:986–987.
15. Keynes, R. D. (1963): Chloride in the squid giant axon. *J. Physiol. (Lond.)*, 169:690–705.
16. Moore, R. D. (1973): Effect of insulin upon the sodium pump in frog skeletal muscle. *J. Physiol. (Lond.)*, 232:23–45.
17. Moreton, R. B. (1969): An investigation of the electrogenic sodium pump in snail neurones, using the constant-field theory. *J. Exp. Biol.*, 51:181–201.
18. Mullins, L. J. (1977): A mechanism for Na/Ca transport. *J. Gen. Physiol.*, 70:681–695.
19. Mullins, L. J., and Awad, M. Z. (1965): The control of the membrane potential of muscle fibers by the sodium pump. *J. Gen. Physiol.*, 48:761–775.
20. Mullins, L. J., and Brinley, F. J., Jr. (1967): Some factors influencing sodium extrusion by internally dialyzed squid axons. *J. Gen. Physiol.*, 50:2333–2355.
21. Mullins, L. J., and Brinley, F. J., Jr. (1975): Sensitivity of calcium efflux from squid axons to changes in membrane potential. *J. Gen. Physiol.*, 65:135–152.

22. Mullins, L. J., and Frumento, A. S. (1963): The concentration dependence of sodium efflux from muscle. *J. Gen. Physiol.*, 46:629–654.
23. Mullins, L. J., and Noda, K. (1963): The influence of sodium-free solutions on the membrane potential of frog muscle fibers. *J. Gen. Physiol.*, 47:117–132.
24. Nelson, M. T., and Blaustein, M. P. (1981): Effect of Na_o-dependent calcium efflux on the membrane potential of internally perfused barnacle muscle fibers. *Biophys. J.*, 33:61a.
25. Shanes, A. M., and Berman, M. D. (1955): Kinetics of ion movement in the squid giant axon. *J. Gen. Physiol.*, 39:279–300.
26. Sjodin, R. A. (1971): The kinetics of sodium extrusion in striated muscle as functions of the external sodium and potassium ion concentrations. *J. Gen. Physiol.*, 57:164–187.
27. Sjodin, R. A. (1980): Contribution of Na/Ca transport to the resting membrane potential. *J. Gen. Physiol.*, 76:99–108.
28. Sjodin, R. A., and Beaugé, L. A. (1968): Coupling and selectivity of sodium and potassium transport in squid giant axons. *J. Gen. Physiol.*, 51:152s–161s.
29. Sjodin, R. A., and Beaugé, L. A. (1973): An analysis of the leakages of sodium ions into and potassium ions out of striated muscle cells. *J. Gen. Physiol.*, 61:222–250.
30. Sjodin, R. A., and Ortiz, O. (1975): Resolution of the potassium ion pump in muscle fibers using barium ions. *J. Gen. Physiol.*, 66:269–286.
31. Teorell, T. (1951): Zur quantitativen behandlung der membranpermeabilität. *Z. Electrochem.*, 55:460–469.
32. Teorell, T. (1953): Transport processes and electrical phenomena in ionic membranes. In: *Progress in Biophysics and Biophysical Chemistry*, Vol. 3, edited by J. A. V. Butler and J. T. Randall, pp. 305–369. Pergamon Press, London.
33. Thomas, R. C. (1972): Electrogenic sodium pump in nerve and muscle cells. *Physiol. Rev.*, 52:563–594.

Electrogenic Transport: Fundamental Principles and Physiological Implications, edited by Mordecai P. Blaustein and Melvyn Lieberman. Raven Press, New York © 1984.

The Energetics and Kinetics of Sodium–Calcium Exchange in Barnacle Muscles, Squid Axons, and Mammalian Heart: The Role of ATP

Mordecai P. Blaustein

Department of Physiology, University of Maryland School of Medicine, Baltimore, Maryland 21201

Virtually all animal cells studied to date are characterized by resting intracellular free calcium concentrations ($[Ca^{2+}]_{in}$) on the order of 10^{-7} M—about 4 to 5 orders of magnitude lower than the free Ca concentrations in the extracellular milieu ($[Ca^{2+}]_{out} \simeq 1-5 \times 10^{-3}$ M). Furthermore, because of the inwardly directed voltage gradient across the plasmalemma of these cells (i.e., cytoplasm negative at rest), the total electrochemical gradients that favor the inward movements of Ca are especially steep.

Intracellular Ca ions play a critical role in numerous physiological processes. The low resting $[Ca^{2+}]_{in}$ levels in cells are, therefore, particularly important because the entry of relatively small amounts of Ca into the cytosol can increase $[Ca^{2+}]_{in}$ markedly (by an amount, $\Delta[Ca^{2+}]_{in}$). This modulation of $[Ca^{2+}]_{in}$ can be represented as a "signal-to-background" ratio, R_{Ca}:

$$R_{Ca} = \frac{\Delta[Ca^{2+}]_{in}}{[Ca^{2+}]_{in(rest)}} \quad (1)$$

R_{Ca} may be as large as 10:1, or even 50:1 or more, during cell activity. This accounts for the ability of Ca to serve as a "second messenger" in the regulation of various physiological processes, including muscle contraction and neurotransmitter release in the cells that we shall discuss in this chapter.

Clearly, information about the mechanisms by which cells control and modulate their $[Ca^{2+}]_{in}$ is required in order to understand how the Ca-activated processes are regulated. In general, four classes of mechanisms are involved:

(a) **Cytoplasmic Ca buffers** are usually Ca-binding proteins with very high affinity for Ca; they may include the Ca "sensors" such as calmodulin and troponin-C, which actually participate in the regulation of Ca-activated processes. The binding of Ca to these substances is usually very rapid, with time constants in the millisecond range.

(b) **Ca sequestering organelles** include endoplasmic reticulum (sarcoplasmic reticulum in muscle), mitochondria, and perhaps other vesicular structures such as secretory granules. These systems help to remove Ca^{2+} from the cytoplasm following a transient rise in $[Ca^{2+}]_{in}$. They can either release Ca slowly into the cytoplasm, so that it can be extruded across the plasma membrane or, as exemplified by the sarcoplasmic reticulum in vertebrate skeletal muscle, they can be triggered to release a pulse of Ca into the cytoplasm in order to initiate Ca-activated processes.

(c) **Gated Ca entry mechanisms** are either voltage-gated or chemosensitive divalent cation-selective channels that enable a pulse of Ca to enter the cytoplasm from the external medium.

(d) **Ca extrusion mechanisms** help to maintain cells in steady Ca balance. Two types have been recognized: those that utilize ATP as an immediate energy source (and involve Ca-dependent ATPases), and those that utilize energy stored in the Na electrochemical gradient (and involve an exchange of Na for Ca).

This chapter is focused on the Ca extrusion mechanisms, and particularly on the properties of the Na-Ca exchange system and its physiological significance.

PARADIGMS FOR Ca EXTRUSION

Figure 1 illustrates the two main Ca extrusion systems that have been observed in animal cells: the adenosine triphosphate (ATP)-driven Ca pump, and the Na-Ca exchange carrier system. The ATP-driven Ca pump is mediated by a plasma membrane Ca-ATPase, analogous to the ubiquitous $Na + K - ATPase$ (see the Na pump at the top of Fig. 1). It was first clearly identified in human red blood cells, in which it appears to be the only plasma membrane Ca transport system present (51; and see 50). Subsequent work has established that similar Ca-dependent ATPases are present in a number of other types of cells (cf. 50) although, as discussed below, their physiological role remains uncertain. An interesting and important property of the red cell Ca-ATPase and Ca transport is its modulation by calmodulin (11,32,55).

The Na-Ca exchange carrier system was first recognized in mammalian cardiac muscle (48) and squid giant axons (1,7). A similar transport system has now been identified in numerous other types of cells (cf. 5,8). Many of these cell types, including cardiac muscle and squid axons, have both Ca-ATPase and Na-Ca exchange mechanisms; this brings up the crucial question of what role each of the transport systems plays in the control of $[Ca^{2+}]_{in}$—a question that we shall address below.

As illustrated in Fig. 1, the Na-Ca exchange system can apparently operate in any one of three modes: (a) a forward mode, in which Na influx is coupled to Ca efflux; (b) a reverse mode, in which Ca influx is coupled to Na efflux; and (c) a Ca-Ca exchange mode, in which there is no net movement of Ca. In many types of cells, available evidence indicates that the Na-Ca exchange can drive Ca either out of, or into the cells, depending on the prevailing Na electrochemical gradient (see below).

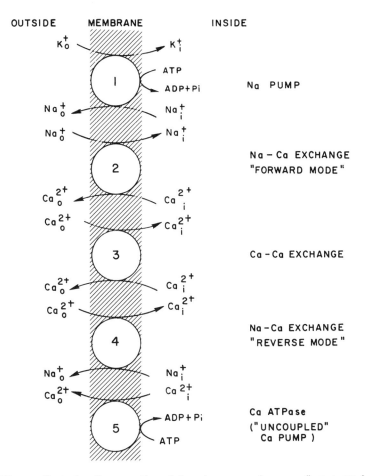

FIG. 1. Diagram illustrating the operation of the plasma membrane sodium pump (**1**), the "uncoupled," ATP-driven Ca pump (**5**), and the three modes of operation of the Na-Ca (and Ca-Ca) exchanger (**2–4**). See text for details.

Canine erythrocytes present an interesting situation. These cells have both Na-Ca exchange and ATP-driven Ca transport systems, but no sodium pump (to extrude Na; cf. Fig. 1, top). It has, therefore, been suggested that the Ca-ATPase maintains a large electrochemical gradient for Ca, and that the Na-Ca exchange is used to extrude Na in exchange for entering Ca, thereby maintaining the (small) Na electrochemical gradient and controlling cell volume in these cells (40,41; and see 50).

THE ENERGETICS OF Na-Ca EXCHANGE

The ability of the Na-Ca exchange system to mediate the net movement of Ca (or Na), mentioned in the preceding section, implies that the electrochemical gradients for Ca and Na, $\Delta\bar{\mu}_{Ca}$ and $\Delta\bar{\mu}_{Na}$, respectively, are coupled through the exchange carrier system. The relationship between the two gradients is then given by (cf. 5–7):

$$\Delta\bar{\mu}_{Ca} = n\Delta\bar{\mu}_{Na}, \tag{2}$$

where n is the stoichiometry of the exchange (i.e., the number of Na^+ ions exchanged for 1 Ca^{2+}). This expression can be expanded in terms of the membrane potential, V_M, and the equilibrium potentials for Ca and Na, E_{Ca} and E_{Na}, respectively:

$$2F(V_M - E_{Ca}) = nF(V_M - E_{Na}), \tag{3}$$

where

$$E_{Ca} = \frac{RT}{2F} \ln \frac{[Ca^{2+}]_{out}}{[Ca^{2+}]_{in}}, \tag{4a}$$

and

$$E_{Na} = \frac{RT}{F} \ln \frac{[Na^+]_{out}}{[Na^+]_{in}}, \tag{4b}$$

where R, T, and F are the gas constant, absolute temperature, and Faraday number, respectively. Then, from Eqs. (3), (4a), and (4b), we obtain the relationship between the Na and Ca ion concentrations:

$$\frac{[Ca^{2+}]_{out}}{[Ca^{2+}]_{in}} = \left(\frac{[Na^+]_{out}}{[Na^+]_{in}}\right)^n \cdot \exp\left[(2-n)(V_M F/RT)\right] \tag{5}$$

As the aforementioned relationships indicate, with appropriately tight coupling between Na and Ca fluxes through the Na-Ca exchanger, it should be possible to use the energy available from one of the ion gradients (for example, Na) to move the other ion (in this example, Ca), against a large electrochemical gradient. Moreover, in this case (for example), there could be considerable leakage influx of Na, so long as the efflux of Ca is tightly coupled to Na influx (i.e., there is little leakage efflux of Ca; cf. 30).

With the stage thus set, it is now possible to focus on three of the critical questions involving Na-Ca exchange: (a) What direct role, if any, does ATP play in the exchange? (b) What is the stoichiometry of the exchange? (c) What are the relative roles of the ATP-driven Ca transport system and Na-Ca exchange in Ca extrusion and the control of $[Ca^{2+}]_{in}$?

Na-Ca EXCHANGE IN SQUID AXONS AND BARNACLE MUSCLE FIBERS: THE EFFECT OF ATP

Giant nerve and muscle fibers have proven very useful for the detailed analysis of some of the kinetic properties of the Na-Ca exchange transport system. In particular, these large cells can be conveniently used to study unidirectional tracer fluxes. Moreover, with internal dialysis or perfusion methods, the solute composition on both sides of the plasma membrane can usually be well controlled.

The observation that Ca efflux from ^{45}Ca-injected squid axons is, in part, dependent on external Na (7), provided early evidence for an Na-Ca exchange process. The nerve fibers also exhibited an external Ca-dependent Na efflux, and showed a net gain of Ca when the Na electrochemical gradient across the plasma membrane was reduced (1). These findings indicated that the Na-Ca exchange could move Ca in either direction across the membrane; they raised the possibility that the maintenance of the Ca gradient was dependent, at least in part, on energy stored in the Na electrochemical gradient.

The fact that the external Na-dependent Ca efflux took place even in cyanide-poisoned axons that were, presumably, depleted of ATP (7), implied that Na-Ca exchange was not directly fueled by ATP. However, when cyanide was applied to the axons, a small, transient decline in Ca efflux was usually noted (Fig. 2) before the Ca efflux rose; the rise in Ca efflux was presumed to be a consequence of the release of Ca from intracellular buffer sites.

Baker and Glitsch (2) called attention to the cyanide-mediated transient fall in Ca efflux. They raised the possibility that, under normal resting conditions, the Na-Ca exchange system was under metabolic control. Since ATP hydrolysis was not required for this exchange, they suggested that phosphorylation of the carrier might increase the affinity for external Na and, perhaps, internal

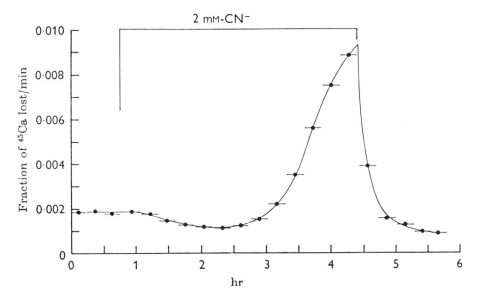

FIG. 2. Effects of 2 mM cyanide (CN$^-$) on the rate constant for Ca efflux (ordinate) from a ^{45}Ca-injected squid axon. Note the transient fall in Ca efflux that begins shortly after the application of cyanide. [From Blaustein and Hodgkin (7), with permission.]

Ca, thereby facilitating the exchange but not altering the steady-state distribution of Ca. Subsequently, DiPolo (20,21), Blaustein (6), and Requena (44), tested the influence of ATP on Ca efflux in internally dialyzed squid axons, in which the concentrations of small cytoplasmic solutes (<1,000 to 2,000 MW) could be controlled. They all found that, at low $[Ca^{2+}]_{in}$, ATP promotes the external Na-dependent efflux of Ca. As illustrated in Fig. 3A, ATP increases the apparent affinity of the carriers for internal Ca; that is, it reduces the apparent half-saturation, K_{Ca_i}. A similar effect was seen in internally perfused giant barnacle

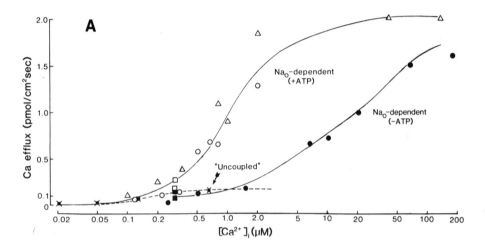

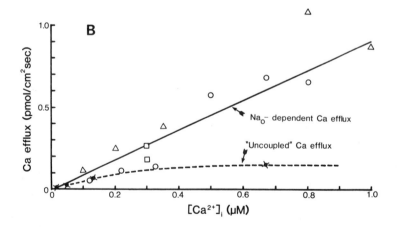

FIG. 3. A: The relationship between $[Ca^{2+}]_{in}$ and Ca efflux in squid giant axons. External Na-dependent Ca efflux in the presence (*open symbols*) or absence (*solid symbols*) of ATP, and the ATP-driven uncoupled Ca efflux ✶. △, ○, ●: Data from the Squid, *Loligo pealii* (6). □, ■: Data from the Squid, *Dorytheutis plei* (20). ✶: Data from *Loligo pealii* (24). **B:** A linear expansion of the external Na-dependent Ca efflux (in ATP-fueled fibers) and the uncoupled Ca efflux over the expected dynamic range of physiological free Ca^{2+} concentrations.

muscle fibers (39; and see 26). This shift in affinity should effectively increase the rate of exchange when $[Ca^{2+}]_{in}$ is in the low (physiological) range.

Although ATP also reduced the apparent half-saturation for external Na ($\bar{K}_{Na_o}$) in squid axons (Fig. 4; and see Refs. 2 and 6), this effect was not seen in barnacle muscle (39). However, a shift in $\bar{K}_{Na_o}$ may not be very important for the exchange, because the carrier should always be saturated with external Na under normal physiological conditions.

The fact that only ATP or its hydrolyzable analogs are able to promote the external Na-dependent Ca efflux in squid axons (21) is consistent with the idea that the carrier must be phosphorylated in order to alter the kinetic parameters. Additional evidence that a phosphorylation step is involved comes from the observation that vanadate, an inhibitor of many phosphorylation reactions, blocks the action of ATP on the Na-Ca exchange system in barnacle muscle (Fig. 5; and see 39). In the next chapter of this volume, Caroni et al. provide direct evidence that the Na-Ca exchanger in heart muscle plasma membrane—which is also influenced by ATP (43)—is modulated by a calmodulin- and cyclic AMP-dependent phosphorylation reaction. Whether or not the phosphorylation is similarly modulated in the giant nerve and muscle fibers has not yet been ascertained.

Ca-Ca EXCHANGE

The Na-Ca exchange carrier is also involved in another mode of exchange: Ca-Ca exchange (Fig. 1; and see 6,9,39). Surprisingly, as illustrated in Fig. 6, ATP markedly *inhibits* this mode of exchange (which does not result in any net Ca transport, because the stoichiometry is 1 Ca^{2+}:1 Ca^{2+}). Thus, in effect, phosphorylation by ATP also appears to shift the carriers from a (nonproductive) Ca-Ca exchange mode to the Na-Ca exchange mode. However, the physiological significance of this shift is not readily apparent because, at normal (saturating) concentrations of external Na, the carriers mediate Na-Ca exchange primarily, even when ATP is depleted (Fig. 7; and see Refs. 6,39).

Although the Ca-Ca exchange mode appears to involve an electroneutral 1:1 exchange of Ca (9,34), knowledge of its properties may be particularly informative if the same carriers can mediate both Na-Ca exchange and Ca-Ca exchange. Indeed, while not conclusive, there is strong circumstantial evidence that the same carrier system is involved in both modes of exchange (6):

(a) The affinity for internal Ca is similar for both modes; in ATP-depleted squid axons, K_{Ca_i} is about 3 to 8 μM.

(b) In both modes of exchange, the affinity of the internal binding sites for Ca is increased by ATP.

(c) Both Na_o-dependent and Ca_o-dependent Ca effluxes are inhibited by internal Na.

(d) Both modes of Ca efflux are reduced by external Sr and Mn.

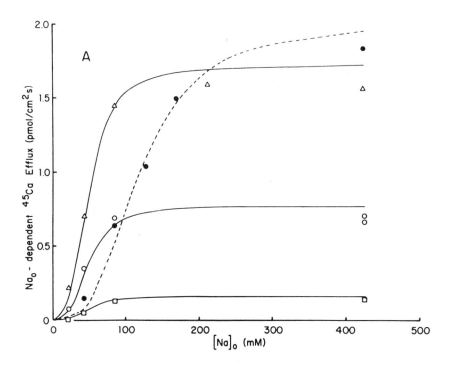
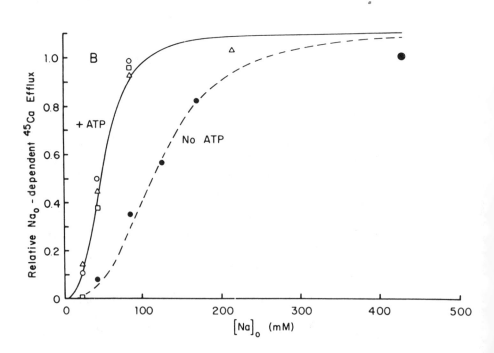

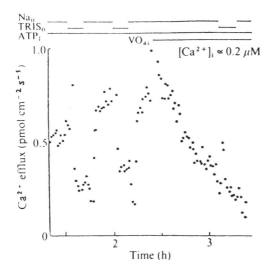

FIG. 5. Effects of external Na and Tris, and internal ATP and vanadate, on Ca efflux from an internally perfused giant barnacle muscle fiber. The periods during which external Na was removed, or 1 mM vandate was added to the internal solution, are indicated at the top. Vanadate reduced the Ca efflux to about the level observed before the introduction of ATP (data not shown). The $[Ca^{2+}]_{in}$ was $\simeq 0.2$ μM. [From Nelson and Blaustein (39), with permission.]

(e) Under conditions that should be optimal for both modes of exchange to operate simultaneously, Ca-Ca exchange is "occluded," and the Na-Ca exchange mode appears to predominate (Fig. 7), perhaps because external Na inhibits Ca-Ca exchange while activating Na-Ca exchange (external Na-dependent Ca

---

FIG. 4. A: The external Na-dependent Ca efflux from ATP-fueled (*open symbols*) and ATP-depleted (*solid circles*) squid axons, graphed as a function of the external Na concentration. The Ca efflux into Ca-free, Na-free 10 K (choline) or Ca-free, Na-free 12 K (choline) + CN was used as a reference (○ on the ordinate). When [Na]$_{out}$ was reduced below 425 mM, it was replaced mole for mole by choline. Each symbol refers to a different axon; those represented by *open symbols* were dialyzed with ATP-containing fluids; axon represented by the *solid circles* was ATP-depleted. The calculated free Ca levels were: $[Ca^{2+}]_{in} \simeq 100$μ M, (●); $[Ca^{2+}]_i \simeq 2.5$ μM (△); $[Ca^{2+}]_{in} \simeq 0.50$ μM, (○); $[Ca^{2+}]_{in} \simeq 0.31$ μM, (□). The curves were drawn to fit the equation:

$$J_{Na\text{-}Ca} = J^*_{Na\text{-}Ca}/[1 + (\bar{K}_{Na_o}/[Na]_o)^n],$$

where $J_{Na\text{-}Ca}$ is the efflux Na-dependent Ca efflux at external Na concentration, [Na]$_o$. The maximal efflux, $J^*_{Na\text{-}Ca}$ had values of 1.95 (●), 1.72 (△), 0.77 (○), and 0.16 (□) pmol/cm²/sec. The apparent mean half-saturation constant for external Na, $\bar{K}_{Na_o}$, had value of 50 mM for the continuous curves (representing the axons dialyzed with ATP-containing fluids) and 120 mM for the *broken curve* (representing the ATP-depleted axon). **B:** Relative Na$_o$-dependent Ca efflux graphed as a function of [Na]$_{out}$. The data from Fig. 4A have been normalized to a value of 1.0 for the external Na-dependent Ca efflux into Ca-free media with 425 mM Na. The curves were drawn to fit the equation given above with $J^*_{Na\text{-}Ca} = 1.1$. The calculated (least squares) slope of the Hill plot (not shown) of the external Na-dependent Ca efflux for the ATP-fueled axons was 3.07 ± 0.23 (SE); the slope of the Hill plot for the ATP-depleted axon was 2.74 ± 0.58 (SE). [From Blaustein (6), with permission.]

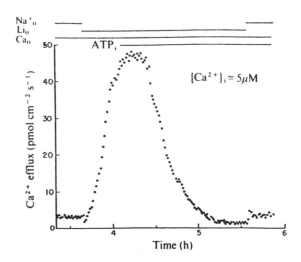

FIG. 6. Effects of external Na and Li, and internal ATP, on ^{45}Ca efflux from a barnacle muscle fiber perfused with high $[Ca^{2+}]_{in}$. Internal perfusion with fluid containing $[Ca^{2+}]_{in} \simeq 5$ μM labeled with ^{45}Ca was begun 200 min before data shown here were obtained. After 250 min of perfusion, 2 mM ATP and an ATP-regenerating system were added to the perfusion fluid. The muscle fiber was exposed to 456 mM Na seawater, except during the period indicated, when the seawater contained 456 mM Li. [From Nelson and Blaustein (39), with permission.]

efflux). Clearly, a reasonable explanation for the "occlusion" is that a single set of carriers mediates both modes of exchange.

Another interesting feature of the Ca-Ca exchange mode is that it is activated by external (9) *and* internal (6) alkali metal cations (Li, K, Rb and, perhaps, low concentrations of Na). *Both* external and internal cation binding sites need to be appropriately loaded *simultaneously,* in order for exchange to take place (6). Furthermore, during the Na-Ca exchange mode of operation, the kinetics of activation by external Na are not affected by the fractional saturation of the transport system by internal Ca (3,6). Both sets of observations suggest that the carrier may be a long molecule that traverses the plasma membrane and mediates the simultaneous translocation of ions in both directions across the membrane.

ATP-DRIVEN, UNCOUPLED Ca EFFLUX FROM SQUID AXONS

Adenosine triphosphate has yet another effect on Ca efflux from squid axons. Following up on an earlier observation by Baker and McNaughton (3), DiPolo and Beaugé (22–24) showed that, at very low $[Ca^{2+}]_{in}$ (<0.1 μM), ATP activates a small component of Ca efflux even in the absence of external Na and Ca. This so-called uncoupled Ca efflux appears to be analogous to the ATP-driven Ca efflux in human red blood cells (e.g., 50), mentioned above. Vanadate, which inhibits several transport ATPases, was also observed to inhibit this uncoupled Ca efflux in squid axons (25,28). In barnacle muscle, however, Nelson and

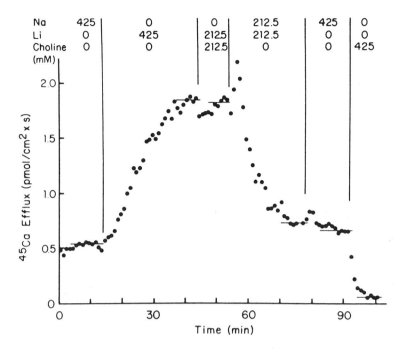

FIG. 7. "Occlusion," by external Na, of Ca efflux into Li-containing fluid, in a squid axon. The monovalent cation content of the superfusion fluid bathing the axon during each segment of the experiment is indicated at the top of the respective time segments. All external solutions contained 10 mM Ca, 50 mM Mg, 12 mM K, and 2 mM CN. The ^{45}Ca-containing dialysis fluid flow, with free $[Ca^{2+}]_{in} \simeq 1$ μM, was begun 30 min before 0 time on the graph. [From Blaustein and Hodgkin (7), with permission.]

Blaustein (39) could find no evidence for a vanadate-sensitive, ATP-driven ("uncoupled") Ca extrusion system at physiological $[Ca^{2+}]_{in}$ values, although vanadate inhibits the external Na-dependent Ca efflux [Fig. 5; but contrast DiPolo et al. (25,28)]. But in mammalian cardiac muscle sarcolemmal preparations, there is a Ca-dependent ATPase and ATP-driven Ca transport system (13,14,16), although its activity is only about one-tenth that of the Na-Ca exchange system in the same preparation.

These observations, concerning parallel Ca-dependent ATPases and Na-Ca exchange transport systems in the same plasma membranes, lead directly to the question: What are the relative physiological roles of the two Ca transport systems in the control of $[Ca^{2+}]_{in}$?

THE RELATIVE ROLES OF Na-Ca EXCHANGE AND ATP-DRIVEN Ca EXTRUSION

DiPolo and Beaugé (23) have concluded that, since the " 'uncoupled' Ca efflux . . . operates in the physiological range of $[Ca^{2+}]_{in}$, it plays a fundamental part in the regulation of intracellular ionised calcium" in squid axons. Also, Barry and Smith (4) have suggested that the Na-Ca exchange system may operate primarily as a Ca entry mechanism in mammalian cardiac muscle—with other systems (perhaps the ATP-driven Ca transport system) required to mediate Ca efflux. However, as we shall see, much of the available evidence seems consistent with the view that Na-Ca exchange plays a very important role in the regulation of resting $[Ca^{2+}]_{in}$.

Original estimates of $[Ca^{2+}]_{in}$ in squid axons, with aequorin and arsenazo III, were on the order of 20 to 50 nM (27). However, these values were probably too low by a factor of about 3 because of calibration errors (cf. ref. 36). Indeed, recent studies with Ca-selective microelectrodes have yielded a value of about $0.11 \mu M$ for resting $[Ca^{2+}]_{in}$ (29). This suggests that the dynamic physiological range for $[Ca^{2+}]_{in}$ may be about 0.1 to 1.0 μM or higher. As illustrated in Fig. 3, the uncoupled Ca efflux, with a maximal Ca transport rate (J_{Max}^{Ca}) of about 0.15 pmol/cm²/sec, is already nearly saturated at the lower end of the dynamic physiological $[Ca^{2+}]_{in}$ range (see Fig. 3B). Even at a $[Ca^{2+}]_{in}$ of 0.1 μM, the absolute value of the external Na-dependent Ca efflux, with a J_{Max}^{Ca} of about 2 pmol/cm²/sec, is greater (in the presence of ATP) than the uncoupled Ca efflux (Fig. 3B).

The recent, elegant experiments with Ca- and Na-specific ion-selective microelectrodes by Sheu and Fozzard, (53) have shown directly that, in mammalian cardiac muscle fibers, the Ca and Na electrochemical gradients appear to be coupled as expected from Eq. 2. When $[Na^+]_{in}$ is altered, $[Ca^{2+}]_{in}$ changes in parallel, so that the relationship, $\Delta \bar{\mu}_{Ca} = n \Delta \bar{\mu}_{Na}$, remains unchanged, with n relatively constant (between 2.5 and 3.0). The implication is that, in this tissue as well, $[Ca^{2+}]_{in}$ is regulated primarily by Na-Ca exchange. The exchange mechanism is known to operate in reverse mode (see Fig. 1), under some circumstances; that is, with Ca entering in exchange for exiting Na (e.g., 19). Nevertheless, it is difficult to imagine how the aforementioned observations on the coupling of the Na and Ca ion gradients could be explained by a system in which the Na-Ca exchange operates primarily to bring Ca into the cells (4), while the Ca-ATPase serves as the main Ca extrusion mechanism and regulator of $[Ca^{2+}]_{in}$.

Additional compelling information comes from net Ca-transport studies. One type of test that can be applied is to load the cells with Ca, and then to determine the conditions needed to extrude the Ca load. Indeed, Requena et al. (47) observed that squid axons failed to extrude a sodium load when the sodium electrochemical gradient across the axolemma was reduced (e.g., by replacing most of the external Na by choline); however, reduction of ATP levels by metabolic poisons did not inhibit the net extrusion of Ca in the presence of a normal Na gradient (Fig. 8). Clearly, these data show no evidence that an ATP-driven Ca pump plays a major role in net Ca efflux from squid axons; instead, they support the view that net Ca extrusion is governed primarily by the Na electrochemical gradient across the axolemma.

Various intracellular Ca-dependent physiological processes have often been

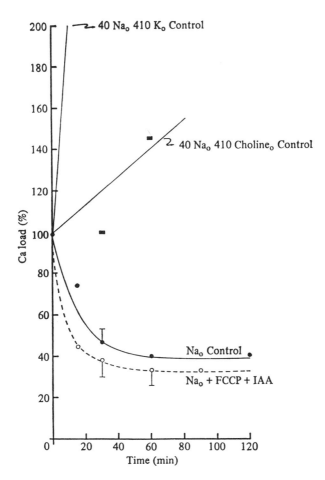

FIG. 8. The time course for unloading of Ca under different experimental conditions in *Dorytheutis plei* axons loaded with Ca by various means. The total analytical Ca content at the start of the unloading period was normalized to 100% Ca load for sets of control axons loaded either by stimulation in 100 mM Ca sea water, or by soaking for 1 to 2 hr, in 1 or 3 mM Ca, Na-free choline sea water. The Ca content of axons loaded by one of these methods and allowed to unload for the indicated periods were normalized accordingly. The extent of the Ca load (i.e., 100%) corresponds roughly to 1.5 mM for stimulated axon, and 500 or 200 μM for axons soaked in choline sea water. Axons were allowed to unload into different solutions all containing 3 mM Ca. Axons kept in 440 mM Na sea water (●) were able to lose the Ca load with a time constant of 20 min. Axons loaded in carbonylcyanide *p*-trifluoromethoxyphenylhydrazone (FCCP) + iodoacetic acid (IAA) containing solution and allowed to unload in Na sea water in the presence of FCCP + IAA (0) did so with a time constant of 10 min. Axons unloaded in 40 mM Na-410 mM choline (■) sea water, in which the chemical gradient for Na is negligible, could not lose the imposed Ca load. Axons unloaded in 40 mM Na-410 mM K sea water, in which the electrochemical gradient for Na is absent, continue to gain further Ca; the experimental points (not shown) correspond to a 400% Ca gain in 30 min. From the observed values for the total internal Ca, a maximal rate of efflux of 0.8 to 8 pmol/cm²/sec (depending on the extent of the Ca load) was calculated for the Ca efflux responsible for reducing the Ca load with the aforementioned time constants. [From Requena and Mullins (46), with permission.]

used to estimate relative $[Ca^{2+}]_{in}$ in cells. For example, the facilitation of evoked neurotransmitter release is a function of the residue of $[Ca^{2+}]$ in the cytoplasm remaining from prior nerve stimulation (cf. 31). At the squid synapse, there is strong evidence that facilitation is also a function of the intracellular Na concentration: The higher the $[Na^+]_{in}$, the greater the facilitation (17,18). The implication is that under normal physiological conditions, a small increase in $[Na^+]_{in}$, by reducing the Na electrochemical gradient, may inhibit net Ca extrusion because of a reduction in Ca efflux and/or increase in Ca influx mediated by the Na-Ca exchange. This conclusion is also consistent with the experiments of Requena et al. (45), in which light emitted by the Ca-sensitive photoprotein, aequorin (confined to a dialysis capillary in the center of a squid axon), was used to assess changes in $[Ca^{2+}]_{in}$. These axons could extrude a Ca load, and maintain a normal resting $[Ca^{2+}]_{in}$, so long as a large, inwardly directed Na electrochemical gradient across the axolemma was maintained.

Thus, the data from both cardiac muscle and squid axons clearly indicate that $[Ca^{2+}]_{in}$ and net Ca transport is controlled *primarily* by Na-Ca exchange in these cells under physiological conditions. The physiological role of the ATP-driven uncoupled Ca efflux remains an enigma, although its presence in the plasmalemma is not disputed. One suggestion (e.g., 24) is that, after a period of cell activity with Ca entry, the Na-Ca exchange, with its large J^{Ca}_{Max} (see Fig. 3) brings $[Ca^{2+}]_{in}$ back close to the resting value by extruding most of the excess Ca. Then, the ATP-driven uncoupled Ca extrusion system, with its high affinity (but low J^{Ca}_{Max}; see Fig. 3) takes over and drives out the last bit of Ca. The problem with this hypothesis is that there is no evidence from squid axons, or mammalian cardiac muscle, that $[Ca^{2+}]_{in}$ can be regulated near its normal resting level when the Na electrochemical gradient is significantly reduced. Quite the contrary! The $[Ca^{2+}]_{in}$ appears to be governed by the Na electrochemical gradient in these cells. Furthermore, as indicated by the data in Fig. 3B, the Na-independent (uncoupled) Ca efflux in squid axons accounts for only a small fraction of the (total) Ca efflux with $[Ca^{2+}]_{in}$ in the dynamic physiological range.

THE STOICHIOMETRY OF Na-Ca EXCHANGE

In the preceding section, we concluded that physiological evidence indicates that Na-Ca exchange plays an important role in the regulation of $[Ca^{2+}]_{in}$ in squid axons as well as mammalian cardiac muscle. This implies that there must be sufficient energy available from the Na electrochemical gradient in these cells to drive Ca out and maintain the large electrochemical gradient for Ca. Then, as noted above, the n in the relationship, $\Delta\bar{\mu}_{Ca} = n\Delta\bar{\mu}_{Na}$ (Eq. 2), gives a lower limit to the stoichiometry of Na-Ca exchange. In other words, with a known $\Delta\bar{\mu}_{Na}$, the n indicates how many Na^+ ions (at a minimum) must be extruded to maintain $[Ca^{2+}]_{in}$ and $\Delta\bar{\mu}_{Ca}$. In many types of cells, including cardiac muscle (35), barnacle muscle (49), and squid axons (6), a stoichiometry of 3 $Na^+:1\ Ca^{2+}$ seems sufficient to maintain $[Ca^{2+}]_{in}$ at its low, physiological resting level if Ca efflux is tightly coupled to a fraction of the Na influx

The observation that $n = 2.5 - 3.0$ in cardiac muscle directly implies that *more than* 2 Na^+ are exchanged for 1 Ca^{2+}. Calculations of $\Delta\bar{\mu}_{Ca}$ and $\Delta\bar{\mu}_{Na}$, respectively, using new, direct measurements of $[Ca^{2+}]_{in}$ (19) and $[Na^+]_{in}$ (54) from squid axons, and comparable data from barnacle muscle (cf. 49), are also consistent with an n of about 3. Earlier suggestions than an n of 4 might be required in squid axons (37) were based on an underestimate of $[Ca^{2+}]_{in}$ (see above) and an overestimate of $[Na^+]_{in}$.

Pitts (42) has directly determined a stoichiometric ratio of 3 Na^+ : 1 Ca^{2+} from measurements of tracer fluxes in cardiac sarcolemmal vesicles, and Bridge and Bassingthwaighte (12) obtained a similar value from net flux measurements in cardiac muscle. Although less direct, studies on the activation of Ca efflux by external Na in barnacle muscle (49) and squid axons (6,10) yield a Hill coefficient of 2.6 to 3.0—which also implies that the stoichiometry is *greater than* 2 Na^+ : 1 Ca^{2+}.

Additional evidence that Na-Ca exchange is electrogenic has come from the demonstration that external Na-dependent Ca efflux from squid axons is voltage-sensitive (10,38). And, in cardiac muscle sarcolemmal vesicles, Na-Ca exchange appears to be associated with net charge movement (15).

Although all of this evidence is consistent with a 3:1 stoichiometry, the available information does not provide conclusive proof that this is, indeed, the stoichiometry of the exchange. Surely, variable stoichiometries, or stoichiometries of >3:1, may be consistent with many of the observations mentioned above.

If the exchange is, indeed, electrogenic (i.e., associated with a net transfer of charge), as may occur with stoichiometries $\geq 3:1$, it should be possible to measure the current flow associated with the exchange. Elsewhere in this volume, Nelson and Lederer (see also ref. 33) describe experiments in which they attempted to determine the stoichiometry of Na-Ca exchange from simultaneous voltage clamp current and ^{45}Ca efflux measurements when the exchange was activated in barnacle muscle. Although the stoichiometry values they obtained were variable, and considerably greater than 3:1, they note that their methods have significant limitations. One particularly troublesome problem is that the conditions that activate Na-Ca exchange—high $[Ca^{2+}]_{in}$ and $[Na^+]_{out}$, and low $[Ca^{2+}]_{out}$—may also activate other, parallel, current sources (cf. ref. 52). Therefore, such voltage clamp measurements could provide an accurate indication of the Na-Ca exchange stoichiometry only if *all* the parallel current pathways were blocked or adequately accounted for, or if a selective blocker of the exchange were available to identify the specific currents and fluxes of interest.

Another potential source of error in the determination of the stoichiometry is the possibility that some of the carriers in a membrane may be mediating Na-Ca exchange while another fraction are mediating Ca-Ca exchange. In this case, measurement of Ca efflux and internal Ca-dependent Na influx, for example, might lead to an underestimate of the stoichiometry—if some of the Ca efflux is due to Ca-Ca exchange. This could give the appearance of a variable stoichiometry exchange, depending on the relative fractions of the carriers involved in each of the modes of exchange. Since these carriers mediate little, if any, Na-

Na exchange (cf. ref. 1), overestimates of the stoichiometry due to this mode of exchange would be unlikely.

If carriers must be correctly loaded at *both* sides of the membrane before exchange can occur, and if the ion movements in both directions are tightly coupled and occur simultaneously (see above), it is difficult to imagine how the stoichiometry could be truly variable (except for the mixture of Na-Ca and Ca-Ca exchange). The available evidence does not appear compatible with a carrier that has numerous ($\gg$ 3) sites available for Na^+ ions, so that exchange could occur when a variable number of these sites are loaded. The data concerning simultaneous exchange seem to mitigate against this possibility.

Another interesting feature of the Na-Ca exchange carrier may also have a bearing on this issue. In its Na-Ca exchange modes (forward and reverse), the carrier can mediate either the (net) exit or entry of Ca in exchange for Na, depending on the prevailing Na and Ca electrochemical gradients (e.g., refs. 1,47). A small change in the Na electrochemical gradient appears sufficient to shift the net flow of Ca from an outward to an inward direction (e.g., 35, 53; and see Mullins, this volume). If Na-Ca exchange is the dominant mechanism controlling $[Ca^{2+}]_{in}$, this implies that the exchange must operate close to the equilibrium predicted by Eq. (2), suggesting that the stoichiometry may, in fact, be close to 3:1. Then, as long as $[Ca^{2+}]_{in}$ remains close to the value expected from this relationship (i.e., in resting cells), the transport system could operate in an extremely efficient manner—with very little flow through the system, and very little dissipation of energy (much like the operation of a "trickle charger" in charging up a battery that is nearly fully charged). Whether or not this is normally the case, remains to be determined.

SUMMARY

Two classes of Ca extrusion mechanisms have been described in animal cells: ATP-driven Ca efflux mediated by a Ca-dependent ATPase, and Na-Ca exchange. The kinetics of the Na-Ca exchange system are modified by ATP. The increase in the carrier's affinity for intracellular Ca appears to be the most important action of ATP. This change in affinity probably involves a phosphorylation step.

The ATP-driven Ca efflux and the Na-Ca exchange systems are both present in plasma membranes of squid axons and mammalian cardiac muscle fibers. In these cells, the maximum transport rate through the Na-Ca exchanger is about one order of magnitude greater than the maximum rate mediated by the ATP-driven Ca pump. In barnacle muscle, too, at physiological $[Ca^{2+}]_{in}$, at most only a small fraction of the Ca efflux is mediated by an ATP-driven Ca pump. Thus, $[Ca^{2+}]_{in}$ appears to be controlled primarily by the Na-Ca exchange system in these cells. Available evidence indicates that the stoichiometry of the exchange system in these cells may be close to 3 Na^+:1 Ca^{2+}, so that the electrochemical gradients for the two ions are close to equilibrium. In cardiac muscle, changes in the Na gradient are followed by appropriate changes in the Ca gradient, so that this equilibrium is maintained, as would be predicted for tight chemiosmotic coupling.

ACKNOWLEDGMENTS

I thank S.-S. Sheu for helpful comments on the manuscript, M. Tate and A. Wilder for preparation of the typescript, and E. M. Santiago for assistance with the illustrations. Supported by NIH grant, NS-16106, NSF grant PCM-11704, and a grant from the Muscular Dystrophy Association.

REFERENCES

1. Baker, P. F., Blaustein, M. P., Hodgkin, A. L., and Steinhardt, R. A. (1969): The influence of calcium on sodium efflux in squid axons. *J. Physiol. (Lond.)*, 200:431–458.
2. Baker, P. F., and Glitsch, H. G. (1973): Does metabolic energy participate directly in the Na^+-dependent extrusion of Ca^{2+} ions from squid axons? *J. Physiol. (Lond.)*, 233:44–46P.
3. Baker, P. F., and McNaughton, P. A. (1976): Kinetics and energetics of calcium efflux from intact squid axons. *J. Physiol. (Lond.)*, 259:103–144.
4. Barry, W. H., and Smith, T. W. (1982): Mechanisms of transmembrane calcium movement in cultured chick embryo ventricular cells. *J. Physiol. (Lond.)*, 325:243–260.
5. Blaustein, M. P. (1974): The interrelationship between sodium and calcium fluxes across cell membranes. *Rev. Physiol. Biochem. Pharmacol.*, 70:33–82.
6. Blaustein, M. P. (1977): Effects of internal and external cations and of ATP on sodium-calcium and calcium-calcium exchange in squid axons. *Biophys. J.*, 20:79–111.
7. Blaustein, M. P., and Hodgkin, A. L. (1969): The effect of cyanide on the efflux of calcium from squid axons. *J. Physiol. (Lond.)*, 200:497–527.
8. Blaustein, M. P., and Nelson, M. T. (1982): Sodium-calcium exchange: Its role in the regulation of cell calcium. In: *Membrane Transport of Calcium*, edited by E. Carafoli, pp. 217–236. Academic Press, London.
9. Blaustein, M. P., and Russell, J. M. (1975): Sodium-calcium exchange in internally dialyzed squid giant axons. *J. Membr. Biol.*, 22:285–312.
10. Blaustein, M. P., Russell, J. M., and DeWeer, P. (1974): Calcium efflux from internally dialyzed squid axons: The influence of external and internal cations. *J. Supramol. Struct.*, 2:558–581.
11. Bond, G. H., and Clough, D. L. (1973): A soluble protein activator of $(Mg^{2+} + Ca^{2+})$-dependent ATPase in human red cell membranes. *Biochim. Biophys. Acta*, 323:592–599.
12. Bridge, J. H. B., and Bassingthwaighte, J. B. (1983): Uphill sodium transport driven by an inward calcium gradient in heart muscle. *Science*, 219:178–180.
13. Caroni, P., and Carafoli, E. (1981): The Ca^{2+}-pumping ATPase of heart sarcolemma. *J. Biol. Chem.*, 256:3263–3270.
14. Caroni, P., and Carafoli, E. (1981): Regulation of Ca^{2+}-pumping ATPase of heart sarcolemma by a phosphorylation-dephosphorylation process. *J. Biol. Chem.*, 256:9371–9373.
15. Caroni, P., Reinlib, L., and Carafoli, E. (1980): Charge movements during Na^+-Ca^{2+} exchange in heart sarcolemmal vesicles. *Proc. Natl. Acad. Sci. USA*, 77:6354–6358.
16. Caroni, P., Zurini, A., and Clark, A. (1983): The Ca^{2+}-pumping ATPase of heart sarcolemma. *Ann. N.Y. Acad. Sci.*, 402:402–420.
17. Charlton, M. P., and Atwood, H. L. (1977): Modulation of transmitter release by intracellular sodium in squid giant synapse. *Brain Res.*, 134:367–371.
18. Charlton, M. P., Thompson, C. S., Atwood, H. L., and Farnell, B. (1980): Synaptic transmission and intracellular sodium: Ionophore induced sodium loading of nerve terminals. *Neurosci. Lett.*, 16:193–196.
19. Deitmer, J. W., and Ellis, D. (1978): Changes in the intracellular sodium activity of sheep heart Purkinje fibers produced by calcium and other divalent cations. *J. Physiol. (Lond.)*, 277:437–453.

20. DiPolo, R. (1974): Effect of ATP on the calcium efflux in dialyzed squid giant axons. *J. Gen. Physiol.,* 64:503–517.
21. DiPolo R. (1977): Characterization of the ATP-dependent calcium efflux in dialyzed squid giant axons. *J. Gen. Physiol.,* 69:795–813.
22. DiPolo, R. (1978): Ca pump driven by ATP in squid axons. *Nature,* 274: 390–392.
23. DiPolo, R., and Beaugé, L. (1979): Physiological role of ATP-driven calcium pump in squid axons. *Nature,* 278:271–273.
24. DiPolo, R., and Beaugé, L. (1980): Mechanisms of calcium transport in the giant axon of the squid and their physiological role. *Cell Calcium,* 1:147–169.
25. DiPolo R., and Beaugé, L. (1981): The effects of vanadate on calcium transport in dialyzed squid axons. Sidedness of vanadate-cation interactions. *Biochim. Biophys. Acta,* 645:229–236.
26. DiPolo, R., and Caputo, C. (1977): The effect of ATP on calcium efflux in dialyzed barnacle muscle fibres. *Biochim. Biophys. Acta,* 470:389–394.
27. DiPolo, R., Requena, J., Brinley, F. J., Jr., Mullins, L. J., Scarpa, A., and Tiffert, T. (1976): Ionized calcium concentrations in squid axons. *J. Gen. Physiol.,* 67:433–467.
28. DiPolo, R., Rojas, H. R., and Beaugé, L. (1979): Vanadate inhibitors uncoupled Ca efflux but not Na-Ca exchange in squid axons. *Nature,* 281:228–229.
29. DiPolo, R., Rojas, H., Vergara, J., Lopez, R., and Caputo, C. (1983): Measurements of intracellular ionized calcium in squid giant axons using calcium-selective electrodes. *Biochim. Biophys. Acta,* 728:311–318.
30. Heinz, E., and Geck, P. (1974): The efficiency of energetic coupling between Na^+ flow and amino acid transport in Ehrlich cells—A revised assessment. *Biochim. Biophys. Acta,* 339:426–431.
31. Kretz, R., Shapiro, E., and Kandel, E. R. (1982): Post-tetanic potentiation at an identified synapse in *Aplysia* is correlated with a Ca^{2+}-activated K^+ current in the presynaptic neuron: Evidence for Ca^{2+} accumulation. *Proc. Natl. Acad. Sci. USA,* 79:5430–5434.
32. Larsen, F. L., and Vincenzi, F. F. (1979): Calcium transport across the plasma membrane: Stimulation by calmodulin. *Science,* 204:306–309.
33. Lederer, W. J., and Nelson, M. T. (1983): Effects of extracellular sodium on calcium efflux and membrane current in single muscle cells from barnacle. *J. Physiol. (Lond.),* 341:325–339.
34. Lederer, W. J., Nelson, M. T., and Rasgado-Flores, H. (1982): Electroneutral Ca_o-dependent Ca efflux from barnacle muscle single cells. *J. Physiol. (Lond.),* 325:34–35P.
35. Lee, C. O., and Dagostino, M. (1982): Effect of strophanthidin on intracellular Na ion activity and twitch tension of constantly driven canine cardiac Purkinje fibers. *Biophys. J.* 40:185–198.
36. Marban, E., Rink, T. J., Tsien, R. W., and Tsien, R. Y. (1980): Free calcium in heart muscle at rest and during contraction measured with a Ca^{2+}-sensitive microelectrode. *Nature,* 286:845–850.
37. Mullins, L. J. (1977): A mechanism for Na/Ca transport. *J. Gen. Physiol.,* 70:681–695.
38. Mullins, L. J., and Brinley, F. J., Jr. (1975): Sensitivity of calcium efflux from squid axons to changes in membrane potential. *J. Gen. Physiol.,* 50:2333–2355.
39. Nelson, M. T., and Blaustein, M. P. (1981): Effects of ATP and vanadate on calcium efflux from barnacle muscle fibers. *Nature,* 289:314–316.
40. Parker, J. R. (1978): Sodium and calcium movements in dog red blood cells. *J. Gen. Physiol.,* 71:1–17.
41. Parker, J. R. (1979): Active and passive Ca movement in dog red blood cells and resealed ghosts. *Am. J. Physiol.,* 237:C10–C16.
42. Pitts, B. J. R. (1979): Stoichiometry of sodium-calcium exchange in cardiac sarcolemmal vesicles. *J. Biol. Chem.,* 254:6232–6235.
43. Reinlib, L., Caroni, P., and Carafoli, E. (1981): Studies on heart sarcolemma: Vesicles of opposite orientation and the effect of ATP on the Na^+/Ca^{2+} exchanger. *FEBS Lett.,* 126:74–76.
44. Requena, J. (1978): Calcium efflux from squid axons under constant sodium electrochemical gradient. *J. Gen. Physiol.,* 72:443–470.
45. Requena, J., DiPolo, R., Brinley, F. J., Jr., and Mullins, L. J. (1977): The control of ionized calcium in squid axons. *J. Gen. Physiol.,* 70:329–353.
46. Requena, J., and Mullins, L. J. (1979): Calcium movement in nerve fibres. *Q. Rev. Biophys.,* 12:371–460.
47. Requena, J., Mullins, L. J., and Brinley, F. J., Jr. (1979): Calcium content and net fluxes in squid giant axons. *J. Gen. Physiol.,* 73:327–342.

48. Reuter, H., and Seitz, N. (1968): The dependence of calcium efflux from cardiac muscles on temperature and external ion composition. *J. Physiol. (Lond.)*, 195:451–470.
49. Russell, J. M., and Blaustein, M. P. (1974): Calcium efflux from barnacle muscle fibers. Dependence on external cations. *J. Gen. Physiol.*, 63:144–167.
50. Schatzman, H. J. (1982): The plasma membrane calcium pump of erythrocytes and other animal cells. In: *Membrane Transport of Calcium*, edited by E. Carafoli, pp. 41–108. Academic Press, London.
51. Schatzman, H. J., and Vincenzi, F. F. (1969): Calcium movements across the membranes of human red cells. *J. Physiol. (Lond.)*, 201:369–395.
52. Sheu, S.-S., and Blaustein, M. P. (1983): Effects of calcium on membrane potential and sodium influx in barnacle muscle fibers. *Am. J. Physiol.*, 244:C297–C302.
53. Sheu, S-S., and Fozzard, H. (1982). Transmembrane Na^+ and Ca^{++} electrochemical gradients in cardiac muscle and their relationship to force development. *J. Gen. Physiol.*, 80:325–351.
54. Vassort, G., Tiffert, T., Whittembury, J., and Mullins, L. J. (1982): The effects of internal Na^+ and H^+, external Ca^{++} and of membrane potential on Ca entry in squid axons. *Biophys. J.*, 37: 68a.
55. Vincenzi, F. F., and Hinds, T. R. (1980): Calmodulin and plasma membrane calcium transport. In: *Calcium and Cell Function, vol. I, Calmodulin*, edited by W. Y. Cheung, pp. 127–165. Academic Press, New York.

Electrogenic Transport: Fundamental Principles and Physiological Implications, edited by Mordecai P. Blaustein and Melvyn Lieberman. Raven Press, New York © 1984.

The Na^+/Ca^{2+} Exchanger of Heart Sarcolemma is Regulated by a Phosphorylation-Dephosphorylation Process

*Pico Caroni, Luciano Soldati, and Ernesto Carafoli

Laboratory of Biochemistry, Swiss Federal Institute of Technology (ETH), 8092 Zürich, Switzerland

Previous work in our laboratory has shown that the Ca^{2+}-pumping adenosine triphosphate (ATP)ase of heart sarcolemma is regulated by a phosphorylation-dephosphorylation cycle (6). The activating phosphorylation is catalyzed by a cyclic adenosine monophosphate (AMP) (and Ca^{2+}) dependent protein kinase, but the properties of the deactivating phosphatase have not yet been established. Work on the purified reconstituted ATPase (9,10) has shown that the ATPase proper is not the target of the phosphorylation-dephosphorylation process, and has suggested the existence of an accessory (protein) factor that mediates the regulatory function of the kinase-phosphatase couple.

Studies on axonal membranes (1,3,4,12,13) have demonstrated long ago that the other Ca^{2+}-ejecting system of (excitable) plasma membranes is activated by ATP. The suggestion has been advanced (12) that ATP may activate allosterically, by binding to a low-affinity site on the exchanger, but the observation that only hydrolyzable analogs of ATP are active indicates the involvement of a regulatory phosphorylation process.

We have studied the mechanism of the activation of the Na^+/Ca^{2+} exchanger by ATP in heart sarcolemmal vesicles (7). The experiments show that the activation is mediated by a membrane-bound, Ca^{2+} and calmodulin-dependent protein kinase. The deactivation of the activated exchanger is catalyzed by a protein phosphatase that is also dependent on Ca^{2+} and calmodulin.

It recently has been shown that one of the intracellular calmodulin-binding proteins, calcineurin, is a Ca^{2+} plus calmodulin-dependent protein phosphatase (23). We tested calcineurin on the ATP-activated Na^+/Ca^{2+} exchanger and found it to be active (i.e., deactivating). Recent work (20) has shown that the Na^+/Ca^{2+} exchanger of heart sarcolemma can be activated by a controlled proteolytic treatment. We have confirmed the activating proteolytic treatment, and shown it to be additive with that induced by ATP.

---

* Present address: Department of Biochemistry and Biophysics, University of California, San Francisco, Ca. USA.

MATERIALS AND METHODS

Materials

Digitoxigenin, ouabain, and arsenazo III were from Fluka AG, Buchs, Switzerland; valinomycin and protein kinase inhibitor (cyclicAMP-dependent, type II) from Sigma, St. Louis, Mo; γ-S-ATP from Boehringer Mannheim, FRG; R24571 from Janssen Pharmaceuticals, Beerse, Belgium. Calf hearts were obtained from the local slaughterhouse. Phosphorylase phosphatase was isolated from rabbit skeletal muscle (5,16), calmodulin from bovine brain (24).

Preparation of Calf Heart Sarcolemma

The method of Jones et al. (17) modified as indicated in (8), was used. Its specific application to calf heart has been described (7). The method involves a step in which the suspension is exposed to ATP, Ca^{2+}, and oxalate (17) to remove contaminating sarcoplasmic reticulum vesicles. At the end of the isolation procedure, the vesicles were suspended at a concentration of 10 mg protein/ml in either 160 mM NaCl, 20 mM Hepes, pH 7.4, or 160 mM KCl, 20 mM Hepes, pH 7.4, and stored frozen at $-80°C$.

Ca^{2+}-Transport by the Sarcolemmal Vesicles

The method of Reeves and Sutko (21) was modified as follows. Approximately 20 μg of Na^+-loaded vesicle protein (20 min of incubation in 160 mM NaCl at 37°C) were diluted 200 to 400-fold in a medium containing 160 mM KCl, 20 mM Hepes, pH 7.4, 1 μM valinomycin, and variable concentrations of free Ca^{2+}. The latter were obtained with Ca/EGTA or Ca/arsenazo III buffers, and checked with specific Ca^{2+} electrodes. The uptake of Ca^{2+} was monitored spectrophotometrically (660 to 680 nm) with arsenazo III, or isotopically. In both cases, the temperature of the uptake medium was 37°C. When the uptake was monitored with $^{45}Ca^{2+}$ (reaction volume 200 μl) the reaction was arrested with 100 μl of 160 mM KCl, 10 mM $LaCl_3$, 20 mM Hepes, pH 7.4. Aliquots of 75 μl were filtered and the filters (0.22 μM pore size) washed with 1 ml cold KCl, $LaCl_3$-Hepes. Initial rates of uptake were monitored with a metronome. The first point was at 1 sec. The zero time was obtained by diluting the vesicles directly in the stop solution. When the arsenazo III method was used, the zero time was obtained by diluting the Na-loaded vesicles in 160 mM NaCl, 20 mM Hepes, pH 7.4. The isotopic methods produced initial rates that were consistently lower (about 50%) than those obtained with the arsenazo III technique, probably due to incomplete recovery of the vesicles on the filters.

Phosphorylation and Dephosphorylation of the Sarcolemmal Vesicles

The compounds indicated below, dissolved in 100 μl of 160 mM NaCl, 20 mM Hepes, pH 7.4, were added at time zero to 10 μl of Na-loaded vesicles

(100 μg of protein) in the tube of a Beckman Airfuge. The suspension was incubated at 37°C for various times (see description of the experiments) treated with 10 mM Na$_2$-EDTA, pH 7.4 to arrest the reaction and then centrifuged (first preincubation). A second preincubation (see legends for tables and figures) was performed under the same conditions, but with different additions to the Na-medium. The vesicles were washed in NaCl-Hepes and then resuspended at a concentration of 5 mg protein/ml in the NaCl-Hepes medium. The Na$^+$/Ca^{2+} exchange of the Na$^+$-loaded vesicles was then determined as indicated above.

When bovine brain calcineurin was tested on the exchange activity, the conditions were the same as in the dephosphorylation experiments described above. Only one preincubation was carried out, as indicated in the legend for Table 4.

Activation of the Na$^+$/Ca^{2+} Exchanger by Trypsin

Sarcolemmal vesicles (approximately 5 mg protein per ml, 5 μl) were incubated with 1.5 μl of trypsin (1 mg/ml) for 20 min, at 0°C. Proteolysis was arrested with 3 μl of soybean trypsin inhibitor (5 mg/ml). The treatment with trypsin was applied during the second preincubation.

Protein Determination

The protein was precipitated with deoxycholate and trichloroacetic acid (2) and was determined on the precipitate (18).

RESULTS

Inhibition of the Na$^+$/Ca^{2+} Exchange by Phosphorylase Phosphatase, and Reversal of the Inhibition by an ATP-Dependent Process

Sarcolemmal vesicles preincubated with phosphorylase phosphatase and Mg^{2+} have a reduced Na$^+$/Ca^{2+} exchange activity (Table 1). The degree of inactivation, in vesicles preincubated with the phosphatase for 5 min at 37°C, is variable, depending on the sarcolemmal preparation used, and probably reflects the variable degree of endogenous phosphorylation induced in the vesicles by the ATP-Ca^{2+}-oxalate step applied during the isolation scheme. Treatment of the dephosphorylated vesicles with ATP, Ca^{2+}, and Mg^{2+}, reverses the inhibition (Table 1) and, in most of the cases, even raises the activity of the exchanger to levels that are higher than those of the control, nonphosphorylated vesicles. The observation suggests the involvement of a phosphorylation step, apparently catalyzed by a membrane-bound protein kinase. The results shown in Table 1 define some of the properties of the putative phosphorylation step. It requires Mg^{2+} and Ca^{2+}. The endogenous calmodulin present in the sarcolemmal vesicles is probably

TABLE 1. *Inhibition of Na^+/Ca^{2+} exchange activity by preincubation of the vesicles with phosphorylase phosphatase and its activation by an endogenous mechanism requiring Ca^{2+}, Mg^{2+}, endogenous calmodulin, and ATP*

Additions to the first preincubation medium	Additions to the second preincubation medium	Na^+/Ca^{2+} exchange activity (nmol Ca^{2+}/mg protein/sec)
PPase, Mg^{2+}	None	9.7
PPase, Mg^{2+}	Ca^{2+}, Mg^{2+}, ATP	28.1
PPase, Mg^{2+}	Ca^{2+}, Mg^{2+}	9.5
PPase, Mg^{2+}	Ca^{2+}, ATP	10.7
PPase, Mg^{2+}	EGTA, Mg^{2+}, ATP	9.9
PPase, Mg^{2+}	Ca^{2+}, Mg^{2+}, ATP, R24571	10.6
PPase, Mg^{2+}	Ca^{2+}, Mg^{2+}, ATP, protein kinase inhibitor	28.4

The first and the second preincubation of the vesicles were carried out at 37°C for 5 min in 160 mM NaCl, and 20 mM Hepes, pH 7.4 (for details see Materials and Methods). After washing, the vesicles were suspended at about 10 mg protein/ml in 160 mM NaCl, 20 mM Hepes, pH 7.4, and immediately preincubated a second time. Concentrations in the preincubation media were: phosphorylase phosphatase, 20 µg/ml; protein kinase (cyclic-AMP-dependent) inhibitor, 20 µg/ml; R24571 0.5 µM; $MgCl_2$, 1 mM; $CaCl_2$, 20 µM; EGTA, 0.2 mM. When ATP (1 mM) was present in the preincubation medium, 10 µM vanadate and 10 µM digitoxigenin were included to prevent the consumption of ATP and the loading of the (inside-out) vesicles with Na^+ by the Na, K-ATPase. The Na^+/Ca^{2+} exchange activity was measured with the arsenazo III technique, in the presence of 10 µM Ca^{2+}. PPase, phosphorylase phosphatase.

also involved, since the process is inhibited by the potent anticalmodulin compound R24271. Cyclic-AMP-dependent kinases are not involved, as shown by the lack of effect of an inhibitor of cyclic-AMP-dependent protein kinases. In the experiment shown in Table 1, phosphorylase phosphatase was used as the inactivating enzyme. Since it was considered unlikely that phosphorylase phosphatase is the enzyme involved in the inactivation of the exchanger *in vivo*, conditions were investigated where a membrane-bound, endogenous phosphatase could become activated, and mimic the effect of added phosphorylase phospha-

TABLE 2. *Inactivation of the Na^+/Ca^{2+} exchange process by an endogenous Ca^{2+} and calmodulin-dependent mechanism*

Additions to the first preincubation medium	Additions to the second preincubation medium	Na^+/Ca^{2+} exchange activity nmol Ca^{2+}/mg protein/sec
None	None	19.8
Ca^{2+}, Mg^{2+}, calmodulin	None	9.6
Ca^{2+}, Mg^{2+}, R24571, calmodulin	None	19.0
EGTA, Mg^{2+}, calmodulin	None	19.7

The first and the second preincubation of the sarcolemmal vesicles carried out as described in the legend for Table 1. When added, R24571 was incubated for 1 min at 37°C with the calmodulin-containing preincubation medium to allow full interaction of the drug with calmodulin. Concentrations in the preincubation media were: phosphorylase phosphatase, 20 µg ml; calmodulin 5 µg/ml; R24571, 0.5 µM; $MgCl_2$, 1 mM; EGTA, 0.2 mM; $CaCl_2$, 20 µM. The Na^+/Ca^{2+} exchange activity was measured with the arsenazo III technique in the presence of 10 µM $CaCl_2$.

tase. Table 2 shows that this is achieved by preincubating the sarcolemmal vesicles (5 min at 37°C) in the presence of Ca^{2+}, Mg^{2+}, and calmodulin. Also in this case, the degree of inactivation of the exchanger is variable with the preparation (see above), and in some cases it may be very small. A possible explanation here is the inactivation of the putative endogenous phosphatase during the preparation of the vesicles. Table 2 shows that the endogenous inactivating process requires both Ca^{2+} and added calmodulin. It requires Mg^{2+} as well. However, the response to the elimination of Mg^{2+} from the medium is highly variable, and is presently being investigated in detail.

The experiments shown in Tables 1 and 2 thus strongly support the suggestion that the Na^+/Ca^{2+} exchanger is activated by a membrane-bound Ca^{2+} and calmodulin-dependent protein kinase, and inactivated by a membrane-bound, Ca^{2+} and calmodulin-dependent protein phosphatase. The possibility that the inactivation by the treatment with Ca^{2+}, Mg^{2+}, and calmodulin in the absence of ATP merely increases the nonspecific permeability of the vesicle membrane, thus simulating a decreased Na^+/Ca^{2+} exchange activity, has been ruled out by a series of controls (see ref. 7 for a detailed description). They have shown that the ATP-dependent Ca^{2+} uptake is not decreased in vesicles treated with Ca^{2+}, Mg^{2+}, and calmodulin. They have also shown that the response of the exchanger in treated vesicles to valinomycin is not different from the controls, as would be the case if the electrogenic exchanger were operating in K^+-leaky vesicles.

The experiment outlined in Table 3 provides additional, and compelling evidence for the existence of the membrane-bound Ca^{2+} and calmodulin-dependent phosphorylation/dephosphorylation regulatory system. The ATP-analog, γ-S-ATP is not a substrate for ATPases; it can still function as a substrate for protein kinases, but subsequent phosphorylation is inhibited (11,14,15). Thus, if the ATP-dependent activating process involves a kinase-catalyzed phosphorylation it should still operate with γ-S-ATP. One the other hand, if the deactivating process is due to a phosphatase-catalyzed dephosphorylation step it should not act on the vesicles activated by γ-S-ATP. This is precisely what is observed

TABLE 3. *The activation and the inactivation of the Na^+/Ca^{2+} exchanger involve a phosphorylation/dephosphorylation process*

Nucleotide added to the activation medium	Additions to the final preincubation medium	Na^+/Ca^{2+} exchange activity nmol Ca^{2+}/mg protein/sec
ATP	None	28.5
	Ca^{2+}, Mg^{2+}, calmodulin	10.4
γ-S-ATP	None	29.1
	Ca^{2+}, Mg^{2+}, calmodulin	26.5

The exchanger was inactivated by preincubation in the presence of phosphorylase phosphatase (see legend to Table 1). In a second incubation, the exchange system was activated (see Table 1) with either 1 mM ATP or its analog γ-S-ATP (5 min, 37°C). A third preincubation (5 min at 37°C) was carried out either in 160 mM NaCl, 20 mM Hepes, pH 7.4, or in the presence of Ca^{2+}, Mg^{2+}, and calmodulin as described in the legend to Table 2. After each incubation, the vesicles were washed, loaded with NaCl, and the Na^+/Ca^{2+} exchange activity was measured with the arsenazo III method. (free Ca^{2+}, 10 μM).

in the experiment of Table 3. Thus, the suggestion that the Na$^+$/Ca^{2+} exchange system is regulated by a kinase/phosphatase couple seems convincingly corroborated.

Properties of the Kinase and the Phosphatase That Regulate the Na$^+$/Ca^{2+} Exchange Process

The finding that the activating kinase and the deactivating phosphatase are both dependent on Ca^{2+} and calmodulin seems, at first, paradoxical. A way out of this difficulty would be if the two opposing activities had different sensitivities to Ca^{2+} and calmodulin. It has already been shown (see Tables 1 and 2) that phosphatase requires added calmodulin, whereas kinase does not. The sensitivity of the two activities to Ca^{2+} was then explored. To this aim, sarcolemmal vesicles were either fully phosphorylated with Ca^{2+}, Mg^{2+}, and ATP (first preincubation) and then, in second preincubation, exposed to calmodulin and different Ca^{2+} concentrations. Or, they were fully dephosphorylated in the first preincubation with phosphorylase phosphatase and then, in the second preincubation, they were exposed to ATP and various concentrations of Ca^{2+}. In both cases, they were then loaded with NaCl as usual.

Figure 1 illustrates that the rates of deactivation and activation at saturating Ca^{2+} concentrations proceed almost linearly for about 20 sec. It is thus possible to conduct a study of the initial velocity of the two processes as a function of the added Ca^{2+}. Figure 2 shows that the activity of the kinase is half maximally stimulated at 1.2 μM Ca^{2+}. It also shows that added calmodulin, although not required for the maximal activity of the kinase, increases its affinity for Ca^{2+} to a K_m of about 0.8 μM. Figure 3 also shows that the affinity of the deactivating phosphatase for Ca^{2+} is lower than that of the kinase, and is positively influenced by Mg^{2+}.

Inactivation of the Na$^+$/Ca^{2+} Exchanger by Calcineurin

Recent research has shown that calcineurin is a Ca^{2+}-calmodulin-dependent phosphatase (23), which dephosphorylates phosphorylase kinase. Calcineurin is a soluble protein, whereas the endogenous phosphatase that inactivates the Na$^+$/Ca^{2+} exchanger is membrane bound. However, both activities require Ca^{2+} and calmodulin. It thus became of interest to study whether calcineurin could deactivate the Na$^+$/Ca^{2+} exchanger in the presence of Ca^{2+} and calmodulin. One problem here was the fact that the conditions for testing calcineurin (addition of Ca^{2+} and calmodulin in the absence of Mg^{2+}) also activate the endogenous phosphatase, although generally not to optimal levels. Thus, it was necessary to evaluate the effect of calcineurin from the increase in the deactivation of the Na$^+$/Ca^{2+} exchanger with respect to the background deactivation produced by the endogenous phosphatase. Despite the absence of Mg^{2+}, the latter may sometimes be well pronounced. The results shown in Table 4, which are the

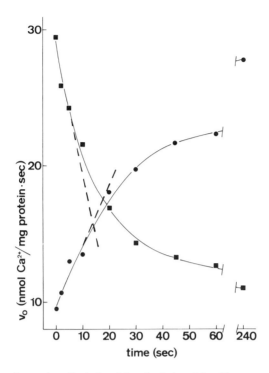

FIG. 1. Na$^+$/Ca^{2+} exchange in activated and deactivated vesicles. Time course of the activation and deactivation reactions. Vesicles were phosphorylated in the first preincubation and dephosphorylated in the second (■) or vice versa (●). The reactions were carried out at 15°C, and arrested as indicated by the addition of 1.5 mM EDTA (final concentration). The activation reaction was carried out in a medium containing 1 mM ATP, 1 mM MgCl$_2$, 20 μM Ca^{2+} and the deactivation reaction in a medium containing 1 mM MgCl$_2$, 20 μM Ca^{2+}, and 5 μg calmodulin per ml. After washing, the vesicles were loaded with NaCl, and the exchange activity was determined (arsenazo III method) in the presence of 10 μM Ca^{2+}.

average of two different experiments, show that calcineurin apparently deactivates the Na$^+$/Ca^{2+} exchanger beyond the level produced by the endogenous phosphatase.

Activation of the Na$^+$/Ca^{2+} Exchanger by Controlled Proteolysis: Its Additivity with the Effect of ATP Plus Ca^{2+}

As mentioned, it has been shown (20) that a controlled treatment with trypsin activates the Na$^+$/Ca^{2+} exchanger. It thus became of interest to study whether the modulating system, mediated by kinase, phosphatase, and the proteolytic treatment, share mode of action and target, or whether they are additive. Table 5 shows that the second alternative is correct. The Na$^+$/Ca^{2+} exchanger in fully phosphorylated vesicles can be activated to higher levels by exposure to trypsin. Trypsin also activates vesicles dephosphorylated with phosphorylase

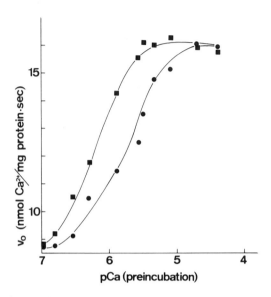

FIG. 2. Ca^{2+}-dependence of the activating reaction. The phosphorylation time was 20 sec, under the experimental conditions indicated in the legend to Fig. 1, in the presence of the amounts of Ca^{2+} indicated. In (●) no calmodulin was added, in (■) 5 μg per ml were present. The Na^+/Ca^{2+} exchange activity was followed with the arsenazo III method, in the presence of 10 μM Ca^{2+}.

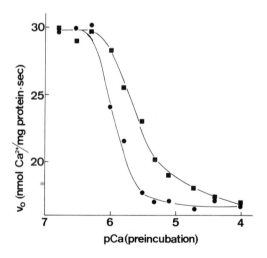

FIG. 3. Ca^{2+}-dependence of the deactivating reaction. The dephosphorylation time was 20 sec, in the presence of the amounts of Ca^{2+} indicated. (■) 1 mM $MgCl_2$, 5 μg calmodulin per ml; (●) 1 mM $MgCl_2$, 50 μg calmodulin per ml. The measurement of the Na^+/Ca^{2+} exchange was as in the legends to Figs. 1 and 2.

TABLE 4. *Inactivation of the Na^+/Ca^{2+} exchanger by calcineurin*

Additions to the preincubation	Initial rate of Na/Ca exchange (nmol Ca^{2+}/mg protein/sec)
None	15.9
Ca^{2+}, Mg^{2+}, ATP	19.6
Ca^{2+}, Mg^{2+}, calmodulin	12.8
Ca^{2+}, calcineurin, calmodulin	11.2

The preincubation of the vesicles was carried out at 37°C for 10 min in 160 mM NaCl and 20 mM Hepes pH 7.4. Concentrations in the preincubation medium were: Calmodulin, 5 μg/ml; calcineurin, 10 μg/ml; $MgCl_2$, 1 mM; $CaCl_2$, 20 μM. When ATP (1 mM) was present in the preincubation medium, 10 μM vanadate and 10 μM digitoxigenin were included to prevent ATP consumption by the sarcolemmal ATPases. The Na^+/Ca^{2+} exchange activity was measured with arsenazo III (10 μM Ca).

phosphatase. As expected, the final level of activity reached in this case is lower than that of fully phosphorylated vesicles exposed to trypsin.

DISCUSSION

The population of vesicles used in this study consists of approximately 40% inside-out, 45% right side-out vesicles. Since it is conceivable that the ATP-interacting site resides on the inner side of the sarcolemmal membrane, the kinase-mediated step probably activates only the exchanger in inside-out vesicles. In all likelihood, the same is also true of the phosphatase. As a result, then, a basal level of exchange activity due to the right side-out vesicles and unaffected

TABLE 5. *Additivity of the effects of phosphorylation and of proteolysis on the Na^+/Ca^{2+} exchanger*

Addition to the first preincubation	Addition to the second preincubation	Initial rate of Na/Ca exchange (nmol Ca^{2+}/mg protein/sec)
None	None	6.1
Ca^{2+}, Mg^{2+}, ATP	None	6.8
Phosphorylase phosphatase	None	5.0
None	Trypsin	10.3
Ca^{2+}, Mg^{2+}, ATP	Trypsin	10.8
Phosphorylase phosphatase	Trypsin	8.7

The first preincubation of the sacrolemmal vesicles was carried out as described in the legend to Table 1. The second preincubation was carried out as follows: After washing, the vesicles were resuspended in 160 mM NaCl and 20 mM Hepes, pH 7.4 at a concentration of about 5 mg protein/ml; 5 μl of the suspension were incubated with 1.5 μl trypsin (1 mg/ml) (about 1 μg trypsin per 20 μg sarcolemmal protein) for 20 min at 0°C. Proteolysis was arrested by adding 3 μl soybean trypsin inhibitor (5 mg/ml). The Na/Ca exchange activity was measured isotopically (10 μM Ca).

by the phosphorylation/dephosphorylation process, is present in the experiments presented here. This conclusion is based on the assumption that the Ca^{2+} affinity of the outer side of the Na^+/Ca^{2+} exchanger is adequate to sustain exchange activity under the experimental conditions (19).

The results presented have shown that the Na^+/Ca^{2+} exchange process is regulated by a phosphorylation/dephosphorylation system. One open problem, at this point, is whether the target of the phosphorylation/dephosphorylation system is the exchanger proper or some other protein, which would become phosphorylated and then regulate the exchanger. A comment is also in order regarding the finding that the phosphorylation leg of the regulation process, at variance with the other Ca^{2+}-ejecting system of heart sarcolemma, the specific ATPase (6), is not cyclic-AMP-dependent. Since both legs of the regulation process are Ca^{2+} and calmodulin-dependent, it is necessary to postulate that their different sensitivity to calmodulin and Ca^{2+} is a key element in determining their respective levels of activity. At saturating levels of Ca^{2+} and calmodulin, however, the concentration of ATP in the vicinity of the enzyme may also play an important role.

Both the endogenous kinase and the phosphatase are membrane-bound. The specificity of the phosphatase, however, is limited, since its effects can be mimicked by phosphatase. The inactivating (dephosphorylating) effect of calcineurin is potentially interesting, in view of the fact that the latter phosphatase is also Ca^{2+} and calmodulin dependent. Studies on the possible inactivation of the endogenous, membrane-bound phosphatase by anticalcineurin antibodies (kindly donated by Dr. C. Klee, Bethesda) are currently under way. Concerning the relationship of the kinase-mediated activating system to the activation induced by controlled proteolysis, it is clear from the experiments presented that the two phenomena are additive, and thus probably operate on independent molecular (or submolecular) targets.

Concerning the possible physiological meaning of the regulation system described, it may be postulated that the phosphorylation system responds to the increased Ca^{2+} influx into the cell consequent on hormonal stimulation (22). As a result, the rate of relaxation would be accelerated, provided that the exchanger operates only in the direction of Ca^{2+} efflux from the heart. If the exchanger also operates in the opposite direction, as it very likely does during part of the contraction/relaxation cycle, then the opposite effect would be achieved, i.e., a potentiation of the influx of Ca^{2+} into the cell on hormonal stimulation. These considerations, however, are based on the assumption that the Na^+/Ca^{2+} exchange system, in the resting state, is at least partially dephosphorylated, and thus susceptible to the activating phosphorylation. The low Ca^{2+} affinity of the deactivating phosphatase is difficult to reconcile with a dephosphorylating role *in vivo,* but it is possible that Ca^{2+} is compartmentalized in such a way as to produce the required high concentration in the vicinity of the phosphatase.

ACKNOWLEDGMENTS

The work was aided by the financial contribution of the Geigy Jubiläums Stiftung (Basel, Switzerland) and the Swiss Nationalfonds (Grant No. 3.634.0-80). Thanks are due to Dr. C. Klee, (Bethesda) for the kind gift of calcineurin and of anticalcineurin antibody, and to Dr. K. D. Philipson (Los Angeles) for making available to us a manuscript, prior to its publication, on the effect of trypsin on the exchanger.

REFERENCES

1. Baker, P. F., and Glitsch, H. G. (1973): Does metabolic energy participate directly in the sodium-dependent extrusion of calcium from squid giant axons? *J. Physiol. (Lond.)*, 233:44–46P.
2. Bensadoun, A., and Weinstein, D. (1976): Assay of proteins in the presence of interfering materials. *Anal. Biochem.*, 70:241–250.
3. Blaustein, M. P. (1977): Effect of internal and external cations and of ATP on sodium-calcium and calcium-calcium exchange in squid axons. *Biophys. J.*, 20:79–111.
4. Blaustein, M. P., and Hodgkin, A. L. (1969): The effect of cyanide on the efflux of calcium from squid axons. *J. Physiol. (Lond.)*, 200:497–527.
5. Brautigan, D. L., Picton, C., and Fischer, E. H. (1980): Phosphorylase phosphatase complex from skeletal muscle. Activation of one of two catalytic subunits by manganese ions. *Biochemistry*, 19:5787–5794.
6. Caroni, P., and Carafoli, E. (1981): Regulation of Ca^{2+}-pumping ATPase of heart sarcolemma by a phosphorylation-dephosphorylation process. *J. Biol. Chem.*, 256:9371–9373.
7. Caroni, P., and Carafoli, E. (1983): The regulation of the Na^+-Ca^{2+} exchanger of heart sarcolemma. *Eur. J. Biochem.*, 132:451–460.
8. Caroni, P., Reinlib, L., and Carafoli, E. (1980): Charge movements during the Na^+-Ca^{2+} exchange in heart sarcolemmal vesicles. *Proc. Natl. Acad. Sci. USA*, 77:6354–6358.
9. Caroni, P., Zurini, M., and Clark, A. (1982): The calcium pumping ATPase of heart sarcolemma. *Ann. NY Acad. Sci.*, 402:402–421.
10. Caroni, P., Zurini, M., Clark, A., and Carafoli, E. (1983): Further characterization and reconstitution of the purified Ca^{2+}-pumping ATPase of heart sarcolemma. *J. Biol. Chem.*, 258:7305–7310.
11. Cassidy, P., Hoar, P. E., and Kerrick, W. G. L. (1979): Irreversible thiophosphorylation and activation of tension in functionally skinned rabbit ileum strips by [^{35}S]ATPγS. *J. Biol. Chem.*, 254:11148–11152.
12. Di Polo, R. (1974): Effect of ATP on the calcium efflux in dialyzed squid giant axons. *J. Gen. Physiol.*, 64:503–517.
13. Di Polo, R. (1977): Characterization of the ATP-dependent calcium efflux in dialyzed squid giant axons. *J. Gen. Physiol.*, 69:795–813.
14. Eckstein, F. (1979): Phosphorothioate analogues of nucleotides. *Accounts Chem. Res.*, 12:204–210.
15. Goody, R. S., Eckstein, F., and Schirmer, R. H. (1972): The enzymatic synthesis of thiophosphate analogs of nucleotides. *Biochim. Biophys. Acta*, 256:155–163.
16. Hurd, S. S., Navoa, N. P., Hickenbottom, J. P., and Fischer, E. H. (1966): Phosphorylase phosphatase from rabbit muscle. *Methods Enzymol.*, 8:546–550.
17. Jones, L. R., Besch, H. R., Jr., Fleming, J. W., McConnaughey, M. M., and Watanabe, A. M. (1979): Separation of vesicles of cardiac sarcolemma from vesicles of cardiac sarcoplasmic reticulum. *J. Biol. Chem.*, 254:530–539.
18. Lowry, O. H., Rosenbrough, N. J., Farr, A. L., and Randall, R. J. (1951): Protein measurement with the folin phenol reagent. *J. Biol. Chem.*, 193:265–275.
19. Philipson, K. D., and Nishimoto, A. Y. (1982): Na^+-Ca^{2+} exchange in inside-out cardiac sarcolemmal vesicles. *J. Biol. Chem.*, 257:5111–5117.

20. Philipson, K. D., and Nishimoto, A. Y. (1982): Stimulation of Na^+-Ca^{2+}-exchange in cardiac sarcolemmal vesicles by proteinase pretreatment. *Am. J. Physiol.,* 243:C191–C195.
21. Reeves, J. P., and Sutko, J. L. (1980): Sodium-calcium exchange activity generates a current in cardiac membrane vesicles. *Science,* 208:1461–1464.
22. Reuter, H. (1974): Localisation of beta adrenergic receptors and effects of adrenaline and cyclic nucleotides on action potentials, ionic currents, and tension in mammalian cardiac muscle. *J. Physiol. (Lond.),* 242:429–451.
23. Stewart, A. A., Ingebretsen, T. S., Manalan, A., Klee, C. B., and Cohen, P. (1982): Discovery of a Ca^{2+} and calmodulin-dependent protein phosphatase. Probable identity with calcineurin (CaM BP80). *FEBS Lett.,* 137:80–84.
24. Watterson, D. M., Van Erdick, L. J., Smith, R. E., and Vanaman, T. C. (1976): Calcium dependent regulatory protein of cyclic nucleotide metabolism in normal and transformed chicken embryo fibroblast. *Proc. Natl. Acad. Sci. USA,* 73:2711–2715.

ns, edited by Mordecai P. Blaustein and Melvyn
Lieberman. Raven Press, New York © 1984.

An Electrogenic Saga: Consequences of Sodium-Calcium Exchange in Cardiac Muscle

L. J. Mullins

Department of Biophysics, University of Maryland School of Medicine, Baltimore, Maryland 21201

If the pumping of ions across the cell membrane is a necessary physiological activity, then the simplest pumping device would be a mechanism whereby metabolism drives machinery that either introduces or removes the particular ion from cell cytoplasm. This is clearly an electrogenic mechanism because it is moving a charge across the cell membrane. If there were a coupled mechanism whereby some ions are moved in one direction and others in the opposite direction, then it would only be coincidence that the coupling were such that equal quantities of electric charge were moved per cycle of the pump—thus, electrogenic ion pumping is not something of a rarity but ought to be the rule except in very special cases.

Thus far, this volume has reviewed the wide variety of electrogenic mechanisms that are known to occur in biological systems. If there exists one unifying principle in transport mechanisms, it would appear to be the fact that one can generate a substrate by metabolism building up an electrochemical gradient for hydrogen ions as shown in Fig. 1, and one can use this gradient to synthesize a high-energy compound, adenosine triphosphate (ATP) from adenosine diphosphate (ADP) and P_i. Furthermore, ATP itself can be used to drive another mechanism that will produce an ion gradient in the illustration shown as sodium ion gradient and, as previously noted in this volume, many subsidiary transport reactions can be run on the sodium gradient. The point to be made is that the energy in ion gradients can be converted into high energy substrates, and substrates can be used to produce ion gradients, and an extension of this idea is that the electrochemical gradient of one ion can be used to produce an electrochemical gradient in another ion so that electrochemical potentials of ions and substrates can be readily interconverted.

ELECTROCHEMICAL GRADIENTS FOR SODIUM AND CALCIUM

Mechanisms will be considered that use the electrochemical gradient for sodium in cells to carry out certain types of electrophysiological functions. The

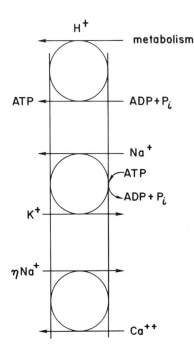

FIG. 1. Top: The electron transport chain produces H+ and hence, establishes a proton gradient across a membrane; this gradient is capable of synthesizing ATP from ADP and P_i. A separate transport system translocates this substrate across the membrane (not shown). The middle diagram uses ATP to energize a Na/K transport system whereby more Na ions are extruded than K taken up (electrogenic action). The ATP generated above is utilized in this form of transport. **Bottom:** the Na gradient formed by the Na/K pump is dissipated by a coupled transport of Na inward and Ca outward.

electrochemical gradient for sodium can be expressed as two terms, one, the membrane potential and the other the chemical potential for sodium, which is most conveniently expressed as E_{Na}. Let us consider E_{Na} first.

The question is whether or not E_{Na} changes during physiological activity. External sodium concentration is a highly regulated parameter in the animal body so that the question really being asked is whether or not internal sodium concentration changes during physiological activity. We may consider the case of nerve shown in Fig. 2. This is a comparison between the squid giant axon whose diameter is 500 μm and a nerve terminal where diameters may get down to 0.1 μm. The same amount of sodium must enter per unit area during the passage of a nerve impulse in order to overcome the membrane capacitance, so in the one case it is clear that the change in internal sodium concentration for a single impulse in a squid axon is utterly trivial, whereas in the example of the terminal the amount is clearly nontrivial. The effect of five or six nerve impulses into terminals of such a size as given above is that the inside sodium concentration would rise to high levels and would promote very substantially larger entries of calcium than would normally enter by a sodium-calcium exchange mechanism.

This exchange of internal sodium for external calcium is going to be the major theme of this discussion, and this example in the nerve terminal is by way of an introduction to the subject. Nerve terminals release transmitters from vesicles that are contained in the terminals just under the membrane. To do this, they need calcium, since it is calcium entry that is somehow involved

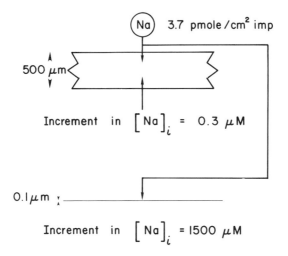

FIG. 2. In a large axon (the squid giant axon, **top**) the entry of Na in a single nerve impulse leads to a negligible increase in $[Na]_i$. In a nerve terminal where fiber diameter is 0.1 μm, a significant increase in $[Na]_i$ occurs with each action potential.

in the breakdown of the vesicles and transmitter release. If stimulation increases $[Na]_i$ and this in turn leads to an increased Ca entry on depolarization, then an increased supply of calcium to the terminals is dependent on past bioelectric activity. This is indeed a form of memory whereby the nerve terminal, by virtue of its accumulation of sodium as a result of past action potentials, is able to call for more calcium entry for subsequent depolarizations (or action potentials). Thus, $[Na]_i$ can change during a single bioelectric event and this change can have physiological consequences.

A second way by which the electrochemical gradient for sodium can be altered is by a change not in internal sodium (or E_{Na}) but by a change in membrane potential itself, since the total electrochemical potential for sodium is ($E - E_{Na}$). Action potentials last for a millisecond or less, but there is an action potential that lasts for hundreds of milliseconds, and this is the cardiac action potential. We suggest therefore that this long duration action potential might have been invented in order to reduce the sodium electrochemical gradient for an appreciable period of time. Why would one want to reduce the sodium electrochemical gradient? The answer to this question might be that if we supposed that the sodium electrochemical gradient is used to pump out calcium from an excitable cell, and if we reduce the gradient, we might slow or even stop this pumping. Extending the idea further, we might not only stop the pumping out of calcium by reducing the sodium electrochemical gradient, but we might actually cause the entry of calcium by the same mechanism that pumps out calcium running backwards.

To determine if this notion is feasible we must explore the thermodynamics of sodium-calcium exchange. Figure 3 shows that the sodium and calcium elec-

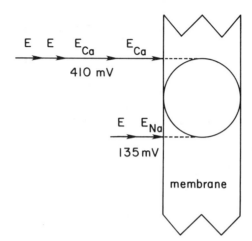

FIG. 3. Comparison of inward electrochemical gradients of both Na and Ca. These gradients are defined as $(E - E_{Na})zF$ for Na and $(E - E_{Ca})z'F$ for Ca. With $E_{Na} = +75$ mV, $E_m = -60$ mV, and $E_{Ca} = +145$ mV, the Ca electrochemical gradient is about 3 or more times that of the Na gradient.

trochemical gradients are disparate, and if sodium is to be used as an energy source for the extrusion of calcium, there must be many sodium ions acting cooperatively in order to bring this about. Notice that the electrochemical gradients of sodium and calcium are both directed inward, and the idea of countertransport is that the larger ion gradient will determine in which direction the transport moves. As I have it drawn here, the movement would clearly be in the form of calcium moving inward and the coupled outward movement of sodium. Since the primary purpose of sodium-gradient transport is the extrusion of calcium at the resting potential, it is not difficult to see that more than three sodiums would be required to effect the outward movement of calcium. Accordingly, in Fig. 4 I have shown for purposes of illustration only four sodium ions being coupled to the movement of one calcium ion. This gives the net difference between the sodium and calcium gradients of about 90 mV in favor of the extrusion of calcium; it is an empirical fact that large differences between sodium and calcium electrochemical gradients appear to favor large fluxes, although in this case changes in E_{Na} and changes in membrane potential are quite different in terms of magnitude of fluxes that each is capable of generating.

The plateau of the action potential in a cardiac cell is near zero, so that it is useful to look at what happens to the electrochemical gradients for sodium and calcium if the membrane potential is taken away, as it is during the plateau of the action potential. The lower part of the figure shows this and indicates that now the energy of the calcium gradient directed inward is much larger than that for sodium, so that the gradient transport mechanism will now run in a direction we might call backward and introduce substantial calcium into

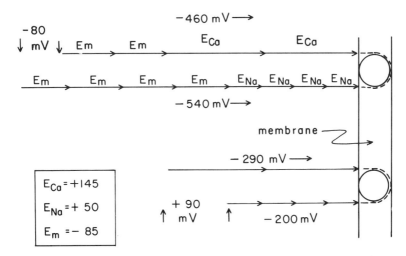

FIG. 4. A balance between Na and Ca electrochemical gradients at the resting potential is achieved if 4 Na are coupled to the movement of 1 Ca. In this condition there is an excess of gradient difference that ensures an extrusion of Ca. In the lower part of the figure, depolarizing the cell to zero membrane potential leads to an entry of Ca with a substantial gradient difference.

the cardiac cell. Here, then, is a calcium source resulting from depolarization that has been largely overlooked by workers in the field of cardiac electrophysiology.

If Ca extrusion proceeds at the resting potential and Ca entry takes place at the plateau where E_m is close to zero, then a potential must exist where there is no net flux of Na or Ca via the carrier; in short, a carrier reversal potential E_R. The value of this potential is obtained by setting the difference in electrochemical gradients for r (Na) and Ca equal to zero and solving for E_m, which for this condition will be E_R. The value r will be the number of Na$^+$ ions moved per Ca.

$$r(E_m - E_{Na}) - 2(E_m - E_{Ca}) = 0$$

hence,

$$r = 4 \quad E_R = 2E_{Na} - E_{Ca}$$
$$r = 3 \quad E_R = 3E_{Na} - 2E_{Ca}$$

Now the numbers given for E_{Na} and E_{Ca} in this diagram are somewhat arbitrary, as indeed is the coupling ratio. So, to be fair, I should include other values for both E_{Na} and for E_{Ca} as well as other values for the coupling ratios between the number of sodium ions per calcium. I do not believe that E_{Na} can be very much lower than 50 mV nor much higher than 70 mV, so Table 1 uses these two values for variations in this parameter. Similarly, for calcium I do not believe that, for muscle that is relaxed, E_{Ca} can be much lower than 120 mV (although of course, it can be much lower in the contractile state)

TABLE 1. *The reversal potential for the Na/Ca carrier and the electrochemical potential difference between Na and Ca if r = 3 or 4 and E_{Ca} and E_{Na} are as indicated*

| \multicolumn{4}{c}{r=4} | \multicolumn{4}{c}{r=3} |

E_{Ca}	E_{Na}	E_R	$4(E_m - E_{Na}) -$ $2(E_m - E_{Ca}) =$ gradient difference at resting E_m	E_{Ca}	E_{Na}	E_R	$3(E_m - E_{Na}) -$ $2(E_m - E_{Ca}) =$ gradient difference at resting E_m
145	50	−45	−80	145	50	−140	+55
145	70	−5	−160	145	70	−80	−5
120	50	−20	−130	120	50	−90	+5
120	70	+20	−210	120	70	−30	−55

nor do I believe that E_{Ca} can be much higher than +145 mV, so I have used this range of values. For the coupling ratio the numbers 3 or 4 are the two values suggested for an electrogenic sodium-calcium exchange, and thus are given in the table. It is clear that with $r = 4$, for all values of E_{Na} or E_{Ca}, at the resting potential there is a large gradient difference between the sodium and calcium gradients such that calcium will be pumped out of the fiber and the reversal potentials will vary from −45 to +20 mV. When the coupling ratio is 3, then there is only one value for E_{Ca} (a low one) and one value for sodium +70 mV (a high one) that gives any sort of gradient difference leading to calcium extrusion at the resting potential. Clearly, this is a major difficulty of 3:1 coupling; it is very close to lacking sufficient energy to extrude calcium from the fiber, or at least to extrude it rapidly.

One must also consider the possibility that there might be two or more populations of sodium-calcium exchangers each with different equilibrium-binding constants for both sodium and calcium and/or with that different coupling ratios. An even more complex consideration is that coupling may vary depending on experimental circumstances. It may, therefore, be safest to say that although the electrogenic nature of sodium-calcium exchange is beyond question, precise data about the absolute value of the coupling ratio are lacking. I shall proceed using the number 4 simply because I am used to this, but there is in principal no real difference between coupling ratios of 3 or 4 other than that the sensitivity of the system to voltage change is lower and, as indicated above, it lives on the edge of an energy lack. What does an electrogenic sodium-calcium exchange mean in physiological terms. I indicated above that it could in principle introduce calcium into a cell when either the sodium chemical or electrical gradients are reduced by physiological activity. Figure 5 shows that when the membrane is normally polarized and the concentrations of the sodium across it are 150 mM on the outside and 10 on the inside, there is a sodium flow inward that is coupled to the outward extrusion of calcium. With depolarization where the membrane potential reverses sign, there is an inward movement of calcium that is dependent on a coupling to an outward movement of sodium shown in

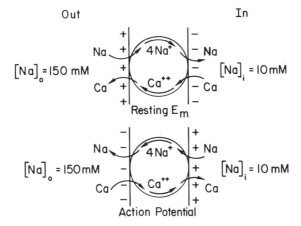

FIG. 5. Sodium-calcium exchange is a process that moves Ca inward during depolarization and outward during the level of membrane polarization occurring during diastole.

the lower half of this illustration. Let us examine the existing experimental support for this rather heavy dose of theory that I have given thus far.

Ca MOVEMENT IN RESPONSE TO MEMBRANE POTENTIAL

In 1975 Brinley and I (11) showed in squid axon that the calcium efflux measured from dialyzed axons is sensitive to membrane potential. Here we measured calcium efflux under conditions of physiological values of ionized calcium in the fiber and we varied the membrane potential. When we depolarized the cell a decrease in calcium efflux occurred. This recovered when polarization returned to normal. Even more interestingly, when we hyperpolarized the cell an increase in calcium efflux occurred, which again returned to control levels when hyperpolarization was turned off. This was, I believe, the first demonstration that hyperpolarization led to an enhancement of ion flux. These measurements suggested that the carrier that moved calcium outward in exchange for sodium moving inward was sensitive to membrane potential; the most likely reason for this sensitivity was that it transported more ionic charges in one direction than in another.

If the calcium efflux were sensitive to membrane potential, then symmetry considerations would make it likely that calcium influx was also sensitive to changes in membrane potential. Thus, DiPolo et al. (6) introduced aequorin contained within a dialysis tube inside a squid axon. The diameter of the dialysis tube was about one-fifth that of the axon, so the aequorin came to rest at a distance far from the membrane. When we depolarized such a squid axon with 100 mM potassium in seawater, we found that for about 1½ hr, the aequorin light continued to rise and, on repolarization, the aequorin light had a similar slow fall. This observation suggested that depolarization indeed leads to an increased calcium entry but, unfortunately, calcium is known to enter with depolarization by many different mechanisms. In 1957 Hodgkin and Keynes (8) showed that part of the calcium entry with stimulation could be accounted for by assuming that it moved through the sodium channels. Subsequently, Baker et al. (1) showed that calcium entry in squid axons as measured with aequorin, could be divided roughly into partial entry through sodium channels and partial entry through another mechanism, which they supposed to be calcium channels. Investigating this matter in detail Baker and co-workers (2) showed in 1973 that a steady depolarization with KCl, where all the sodium in seawater was replaced with KCl, gave a bright flash of aequorin light followed by a plateau. If all the sodium in seawater were replaced by choline, only an enhanced plateau results. These findings were offered as evidence that when all the sodium is removed, sodium-calcium exchange produces an enhanced calcium influx that has been known for some time, whereas if the membrane were additionally depolarized with potassium as well as having all the sodium removed, there was a transient entry of calcium that was assumed to be a calcium channel that opened with depolarization and then inactivated.

Requena and I (14) investigated this phenomenon in a somewhat different way and our results are as shown in Fig. 6. Here the left trace is one that is identical in form with those obtained by Baker et al. (2) and was produced

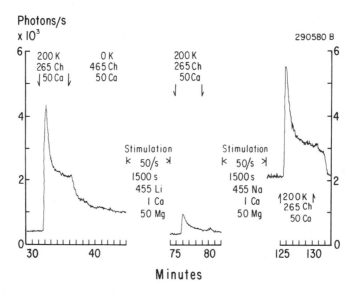

FIG. 6. A depolarization of an aequorin-injected squid giant axon with 200 mM K, choline solution leads to an increase in light emission as measured (photons/sec). Repolarization leads to a decline in the aequorin glow. The axon was then stimulated in Li-containing seawater for 1,500 sec (a treatment known to reduce $[Na]_i$ to about one-half) and the depolarization repeated **(middle trace)**. The axon was then stimulated in Na-seawater (to return $[Na]_i$ to initial values) and the **right** trace was recorded.

by changing from choline seawater to choline seawater with 200 mM potassium. After obtaining this record we changed the seawater surrounding the axon to one with all Na replaced by lithium and stimulated the axon for about half an hour. Analytical measurement on such axons showed that they lose about half of the internal sodium as a result of this stimulation, which in effect allows lithium to enter and sodium to exit with each nerve impulse.

When we tested the axon for its response to depolarization we were surprised to find that the response had virtually disappeared, as shown in the middle trace, and if after obtaining this response the axon were stimulated once again, but this time in sodium seawater where it could regain its lost sodium, it gave a trace very much like the initial one shown as the right hand trace of this illustration. We may suppose from this finding that the calcium channel requires internal sodium and that, in the absence of internal sodium, it fails to open. A more attractive possibility, however, is that there is no calcium channel and that the calcium entry with depolarization is almost exclusively caused by an entry of calcium in exchange for internal sodium. Calcium enters through sodium channels and, indeed, it seems through potassium channels as well. To illustrate the relative magnitudes of these sorts of entries, Fig. 7 shows that fully depolariz-

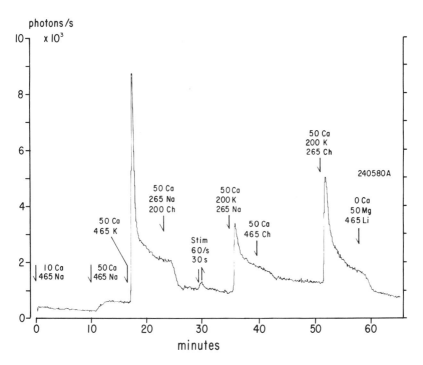

FIG. 7. Comparison between the light emission from an aequorin-injected axon on depolarization and a similar increase in light emission on stimulation at 60/sec for 30 sec. Note that this stimulated Ca entry is substantially smaller than that resulting from steady depolarization.

ing squid axons gives the first large trace, whereas stimulating the axon at 60 times for 30 sec gives a very small trace to the right of the large one. Thus, from a quantitative standpoint calcium entry from repetitive stimulation is very small compared with the entry produced by a full fiber depolarization. This does not mean that calcium entry with stimulation is not affected by or have a component of sodium-calcium exchange. Measurements show that most of the calcium entry with stimulation in the fresh axon is, in fact, entry through channels, and this entry is not influenced by reducing internal sodium or putting it back to control levels again. On the other hand, if $[Na]_i$ is doubled by stimulation in seawater, then there is a substantial increment of aequorin light with a test stimulation, suggesting that with each nerve impulse which is, after all, a phase of depolarization, there is an extra increment in Ca entry over that going through the channels. This result is important, because the increase in [Na] in nerve terminals that was mentioned earlier is undoubtedly related to this observed increment in Ca entry.

A strange aspect about the calcium entry with depolarization, as measured with aequorin, is its phasic nature. Light rises to a peak and then reaches a plateau. Because this process is not well understood, we decided to investigate it further by comparing the response to depolarization from an aequorin-injected and an arsenazo-injected axon. Results obtained in collaboration with Drs. Vassort, Whittembury, and Tiffert (13) show that an arsenazo-injected axon subjected to depolarization gives an absorbance versus time record that is a square wave. We must conclude, therefore, that the spike of the aequorin response occurs because the calcium concentration initially goes outside the linear range for aequorin-calcium interaction and that the bright flash is a reflection of a square or higher power law function. As the Ca concentration levels off, light emission returns to values where a linear relationship between calcium concentration and light is obtained. One may ask why the calcium concentration levels off in some kind of plateau when isotope measurements show clearly that an enhanced calcium entry is occurring continuously. It must be assumed that as calcium concentration rises, more and more buffers of calcium are activated so that eventually a concentration is reached in every element of axoplasmic volume where the rate of calcium entry into this element of volume and the rate of calcium uptake by the buffers are just equal, and there is therefore no further rise in concentration.

CONTROL OF Ca MOVEMENT BY $[Na]_i$

To make meaningful experimental measurements connecting Ca entry with depolarization, we have found it necessary to measure continuously internal sodium concentration and internal pH with electrodes and to measure membrane potential while also having arsenazo or aequorin in the axon as indicators of the ionized calcium in the axoplasm. When the extent of depolarization is constant (+5 mV), and internal sodium is varied over a range of values, we have

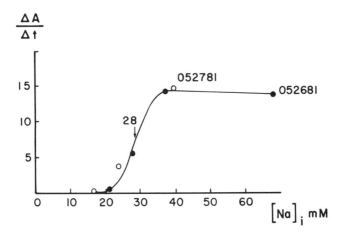

FIG. 8. The rate of increase of arsenazo III absorbance with time (proportional to Ca influx) is plotted against $[Na]_i$ as measured in the axon with a Na-sensitive electrode. Note the steep response of Ca entry to $[Na]_i$ (From Mullins, Tiffert, Vassort, and Whittembury, with permission.)

results shown in Fig. 8, from which it can be seen that calcium entry is half maximal when internal sodium concentration is 28 mM and it is saturated at a sodium concentration of about 38 mM. The function is very steep—it has a Hill coefficient of 7—and such a finding argues for a high degree of kinetic complexity of the carrier in binding as many particles as it must.

Na/Ca EXCHANGE IN THE HEART

Now, how does this apply to heart? Why select this particular tissue? The action potential of cardiac cells lasts for a long time; 300 msec is a representative value, but it can be longer than this. What does this mean? If a sodium-calcium exchange process enhances calcium entry when the membrane is depolarized, then there is a long time for available calcium to enter cardiac fibers because of the lengthy depolarization. It is important to recognize that heart muscle is very different from skeletal muscle in the sense that it is generally believed that every time an action potential passes over the cell it contracts, but this contraction may vary from being almost unmeasurable to maximal contraction, depending on the intervention of various regulatory phenomena that are in general poorly understood. The purpose of talking about sodium-calcium exchange in the context of heart is because until very recently it has been somewhat overlooked. Let us see what sense one can make out of cardiac electrophysiology by thinking in terms of sodium-calcium exchange. Figure 9 illustrates the events that follow depolarization of cardiac cells. There is calcium entry and a release of calcium from internal stores in many cardiac cell types. The calcium entry is generally thought to be by calcium channels. However, note that if they are calcium channels, then there must be some mechanism for pumping out

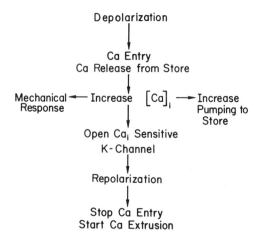

FIG. 9. Sequence of steps in which depolarization of cardiac cells leads to contraction. Note that the increase in [Ca]$_i$ leads to 3 events. The Ca$_i$-sensitive K channel is not the only process leading to repolarization, but it is included here because it has been described. Relaxation can occur because increased pumping to store brings down [Ca]$_i$ *or* because Na/Ca is activated by repolarization to bring down [Ca]$_i$. In any case, it is Na/Ca that brings about an unloading of the store during diastole.

the calcium, and sodium-calcium exchange is an obvious candidate. I have attempted to show that if sodium-calcium exchange exists, it works both ways—not only does it pump calcium out, but it pumps calcium in. Thus, we have calcium entry and calcium release from stores. Both of these increase the internal ionized calcium, which has two effects; it leads to the mechanical response of the heart cell and it also increases the rate of pumping to stores, since the sarcoplasmic reticulum (SR) pumps faster, the higher the level of ionized calcium. The SR in fact competes with the contractile machinery for calcium. Internal calcium can also act in heart muscle to open up a calcium-sensitive potassium channel. This is an easy and straightforward way of bringing about repolarization, although there are other kinds of potassium currents that can also do this.

Na$_o$-FREE EFFECTS

Repolarization leads to the halting of calcium entry and the starting of calcium extrusion. There is little disagreement that repolarization closes the calcium channel and in the generally favorable climate now for electrogenic sodium-calcium exchange, it would be difficult to argue against the idea that calcium extrusion starts when the membrane is repolarized. What evidence exists that sodium-calcium exchange has anything to do with cardiac contractility? Perhaps the clearest example of sodium-calcium exchange being involved is the well-known fact that, unlike skeletal muscle, cardiac cells contract when the sodium concentration surrounding them is reduced. It has been known for a long time and it has been studied a great deal but until very recently thought of as an electroneutral phenomenon that made some of the interpretations of the data difficult. Studies by Chapman (4) on frog cardiac cells are summarized in Fig. 10. In (a), external sodium was reduced, and the mechanical response is a contracture that gradually relaxes within minutes. This change in external sodium produces no change in membrane potential, so it is difficult to see how the calcium channels know that they should open. Indeed, they do not, nor is there any signal to the SR to release calcium; therefore, reversal of the sodium gradient produced by reducing external sodium must itself produce a large calcium entry. Why does the tension decline with time? Because internal sodium declines in a sodium-free medium, and as internal sodium declines the coupled calcium entry declines; calcium entry finally reaches the point where it is slower than the ability of the SR to remove calcium from the myoplasm; hence tension declines as this calcium is taken by internal stores. Trace (b) shows what happens during the brief sodium-free pulse. Tension rises rapidly as before, and it declines just as rapidly when sodium is replaced. This suggests that if internal sodium is maintained, tension declines because calcium is pumped out not by internal stores but by the sodium-calcium exchanger. A long, sodium-free treatment does nothing to the SR, because, as trace (c) shows [the same kind of experiment as that shown in trace (a)], after tension has declined to nothing a pulse of

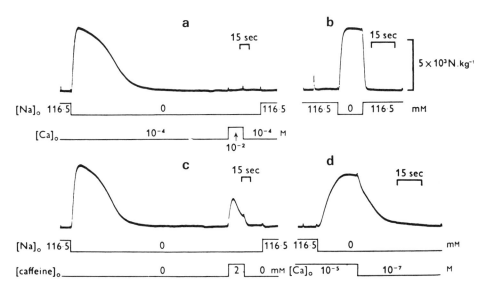

FIG. 10. This is a plot of tension versus time for frog cardiac cells as measured by Chapman (4). **(a)** The preparation is subjected to low Na_o for a substantial period of time. Note that after an increase, tension declines slowly. **(b)** The same sequence of Na-free treatment is repeated, but external Na is replaced before $[Na]_i$ declines. **(c)** The protocol is the same as in **(a)**, but, after the decline in tension, an externally applied pulse of caffeine is made. **(d)** The removal of Na_o (with $Ca_o = 10^{-5}$) leads to the development of tension; a change to $[Ca]_o$ of 10^{-7} leads to a decline in tension.

caffeine in the external solution will produce an internal release of calcium that is reflected in tension. Chapman's experiments do not tell whether the sodium-calcium exchange is electrogenic but only that sodium-calcium exchange clearly exists and depends on internal sodium for its operation.

The most important and persuasive data about a role for sodium-calcium exchange in controlling tension developed in cardiac cells have been obtained recently by a number of laboratories that have either conducted analytical studies or have measured internal sodium with electrodes and have varied it often by reducing external potassium. These data show that tension measured upon depolarization is critically dependent on the internal sodium concentration.

CONTROL OF CONTRACTILITY BY $[Na]_i$

Recent studies on the Na/K pump of cardiac cells are described in detail by Drs. Lieberman, Eisner, and Gadsby in this volume, so I shall not refer further to this mechanism which, by all the tests applied thus far, appears identical to the one in red blood cells, skeletal muscle fibers, and giant axons. It does, however, effectively control $[Na]_i$ and, as we shall see, therefore, contractility. It has become clear that tension development in cardiac fibers depends on internal Na concentration (7,9).

The simplest explanation for a dependence of tension on $[Na]_i$ is that Na/Ca exchange is the principal process in setting the level of $[Ca]_i$. Other explanations are, of course, not ruled out, but in spite of some substantial effort it has not been possible to show that changes in $[Na]_i$ affect the contractile machinery directly. We know that changes in $[Ca]_i$ affect pH_i (15), so a modulation of Ca effects is possible if contractile sensitivity to Ca is pH dependent.

The experiments of Eisner et al. (7) as cited above, can be analyzed to yield useful information as to what occurs when the Na pump is slowed by removing Rb (a substitute for K^+ in activating the Na pump) from the perfusion medium. An experiment is shown in Fig. 11 for the condition where the holding potential is -70 mV, and 0.5-sec pulses to -33 mV are applied under voltage clamp at a frequency of 0.1 Hz. Note that before Rb-free conditions are applied there is 0.1 mg of tension during a depolarization. A conclusion is that the level of $[Ca]_i$ is so low at -70 mV that there is negligible tension, and, furthermore, with depolarization there is insufficient Ca introduced from outside or released from internal stores to produce any real tension. It is also clear that at -70 mV the fiber is already depolarized by 15 mV, so that at normal resting potential of -85 mV, $[Ca]_i$ would be expected to be even lower, if Ca_i were dependent on the Na electrochemical gradient.

The assumptions to be made in this analysis are: (a) For the depolarizations applied there is never a contribution of Ca from Ca channels (3); (b) absence of a steady tension (in excess of the tension normally used in mounting the preparation) at -70 mV indicates that Na/Ca exchange or other Ca extrusion mechanisms (including the SR) hold $[Ca]_i$ below a level of 100 nM (the contractile

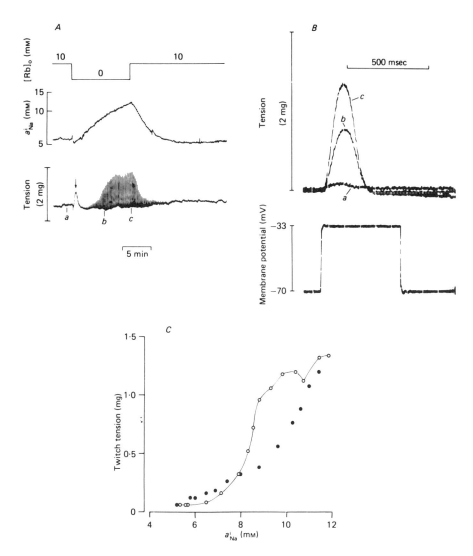

FIG. 11. A: The response to a slowing of the Na/K pump in cardiac Purkinje fibers produced by Rb_o-free solution. The preparation is clamped at −70 mV and pulsed to −33 mV at 0.1 Hz. Note that $[Na]_i$ approximately doubles and twitch tension rises steeply. **B:** Tension and clamp potential (a,b,c) correspond to labels in **A**. **C:** *Open circles* show the rise in tension with rise in a^i_{Na}; *filled circles* show the fall in tension as internal Na falls. [From Eisner et al. (6), with permission.]

threshold); and (c) absence of steady tension at −33 mV indicates that neither the SR nor Na/Ca exchange delivers sufficient Ca to the myoplasm to bring about an increase in tension development. Contractile threshold is poorly defined, but for present purposes it might be considered as the [Ca] that yields 1% of maximal tension. Relevant data are reviewed by Chapman (5), who concludes that threshold may vary from 100 nM (cited above) to 316 nM. It seems simpler to consider the lower of these values as valid. With these assumptions, we then ask what the value of $[Ca]_i$ must be such that with $[Na]_i = 8$ mM ($a_{Na} = 6$ and an activity coefficient 0.75) a depolarization of 37 mV (-70 to -33) does not result in a $[Ca]_i$ greater than 100 nM. There are two processes (Na/Ca exchange and SR Ca release) that could deliver Ca. First, the Na/Ca exchange that might be coupled with 3 or 4 Na/Ca in an electrogenic mode. At equilibrium this could increase $[Ca]_i$ by $\exp -EF/RT$ or $\exp -2EF/RT$, which for 37 mV of depolarization is 4.4- or 19.3-fold. Since there is no tension, $[Ca]_i$ before depolarization must be less than 100 nM/4.4 = 23 nM or 100/19.3 = 5 nM. The experimental change to Rb-free solution doubles $[Na]_i$, and this should, according to Na/Ca exchange thermodynamics, increase the equilibrium level of $[Ca]_i$ by 2^3 or 2^4, depending on whether $r = 3$ or 4. Again, since there is no change in steady tension, $[Ca]_i$ must be less than 100/8 or 100/16 or 12.5 or 6.25 nM. These figures are not really different from those given above. A conclusion from the analysis thus far is that Na/Ca exchange acting with a coupling ratio of either 3 or 4 should have produced some steady tension at -70 mV when $[Na]_i$ was 16 mM, unless the value of $[Ca]_i$ at $[Na]_i = 8$ mM was ≤ 13 nM.

Is it possible that an internal Ca-complexing system such as the SR holds $[Ca]_i$ constantly below the equilibrium levels as calculated above? This seems impossible since a test depolarization is applied only once in 10 sec, more than ample time for all Ca-transporting systems to come to a steady state, and it is applied for 500 msec, also time to allow $[Ca]_i$ to reach a steady state. If the SR were accumulating Ca for 10 sec, it should have some Ca to release during the test depolarization when $[Na]_i = 8$ mM, but since tension is barely detectable it clearly has little Ca to release. On the other hand, when $[Na]_i = 16$ mM, the SR does have Ca to release, and tension is in fact produced. Measurement of part B of Fig. 11 illustrates 15-fold greater tension when $[Na]_i = 16$ as opposed to 8 mM, suggesting that $[Ca]_i$ is much lower when $[Na]_i$ is 8 mM, and that $[Ca]_{SR}$ simply reflects $[Ca]_i$ multiplied by an accumulation factor. Again, the conclusion seems inescapable that, however it is produced, $[Ca]_i$ is *very* much lower than 100 nM.

The response of Purkinje fibers is not always as shown in Fig. 11, since in other fibers (Fig. 9 of ref. 7), there is a rise of steady tension at -70 mV with a rise in $[Na]_i$, as well as a rise in tension with depolarization. Indeed, tension of these two sorts rises in parallel with increases in $[Na]_i$. This suggests that we are now on a different part of the $[Ca]_i$–$[Na]_i$ curve where, over a limited range of $[Na]_i$, tension rises approximately linearly with this ion concentration. Why is there a difference between fibers in their response? The most likely explanation is that the inward leak of Ca is another variable and one that can displace Na/Ca exchange from equilibrium.

It is difficult, therefore, not to conclude that the results obtained from these studies can only be encompassed by assuming that $[Ca]_i$ in cardiac cells is in the range of 15 to 25 nM when $[Na]_i$ is at the level of 8 mM (concentration) and that at this level, on depolarization, internal Ca stores are incapable of supplying enough Ca to produce contraction. One must then ask two questions: First, why do Ca electrode studies show $[Ca]_i$ to be approximately 200 nM at the resting potential of Purkinje fibers and second, if the former value is incorrect, how is it possible to have values as low as 15 to 25 nM? To answer the first question, one can note that most measurements with intracellular Ca electrodes show $[Ca]_i$ to be at the limit of detection sensitivity of the electrode. Obviously, if the detector cannot measure lower levels, then one settles for higher values. It is not reassuring to note that Na-free conditions that produce a contraction in cardiac cells do not show an impressive increase in $[Ca]_i$ measured with a Ca electrode, which should be associated with such a concentration. In short, there are reasons for doubting that Ca-electrode measurements are in fact accurate.

To address the question of how very low values of $[Ca]_i$ can be approached, one must either assume that substrate-mediated pumps reduce $[Ca]_i$ from values that, as indicated earlier, are thermodynamically appropriate to a 3 Na/Ca exchange (100 nM) or that the ratio is 4 Na/Ca, in which case extremely low values (<1 nM) of $[Ca]_i$ are possible.

ELECTROGENIC VS. VOLTAGE SENSITIVE TRANSPORT

Thus far, I have been considering mechanisms for Na/Ca exchange that, although they *might* be electrogenic, could also work if they were only voltage sensitive. How might voltage sensitivity be brought about? If one imagines a carrier that can move either Na or Ca by countertransport, and if one makes the further stipulation that this carrier has a negative charge so that at normal membrane polarization most of the carrier is at the outer face of the membrane, then depolarization will release such a carrier from its immobilized state and allow it to transfer Ca inward in exchange for Na_i. Similarly, if the membrane polarization is reversed by a voltage clamp so that the cell interior is made positive, then most of the carrier will be immobilized at the inside surface of the membrane and Ca influx will again fall. Such a carrier would indeed reproduce the main findings with respect to Ca influx with depolarization which are: (a) influx increases as depolarization approaches a membrane potential level of zero and then declines to very low values; and (b) influx would be dependent on Na_i as well (assuming that the unloaded carrier cannot move). What would *not* be expected from such a system is that Ca efflux would decrease with depolarization unless: (a) the efflux reaction was more electrogenic (in the sense of charge separation) than the fixed charge of the carrier, or (b) the Ca efflux Na/Ca exchange is a system quite separate from the influx reaction.

Let me consider first the idea that Ca efflux is produced by a separate system from that bringing Ca into the cell. One proposal is that Ca efflux is brought about by an uncoupled Ca pump driven by ATP as a substrate. What is the expectation if the cell is hyperpolarized and the Ca efflux is brought about by an ATP-driven pump? The electrochemical gradient for Ca is increased by hyperpolarization; hence, the work required is increased so that either the pump is unaffected or it decreases its rate of operation. Experimentation shows, however, that at physiological values of $[Ca]_i$, the efflux of Ca is *increased* by hyperpolarization (11), so that an answer based on an ATP-driven pump simply cannot explain experimental results. It might be noted parenthetically that in the context of the cardiac cell an ATP-driven pump in the surface membrane is a liability, since it must be expected to run at a rate proportional to $[Ca]_i$ and to be membrane-potential insensitive. It therefore prevents the rapid rise of $[Ca]_i$ on depolarization, which is an essential feature involved in contraction development in cardiac cells. The SR, of course, does much the same thing but has as a redeeming feature the fact that it first produces a gated release of Ca.

Having rejected an ATP-driven Ca efflux pump as being in conflict with experimental results (insofar as these relate to a membrane-potential-sensitive efflux pump), it is possible that some sort of Na/Ca exchange is involved but a system separate from that for Ca influx. Here, the results from squid axons argue strongly against such an arrangement. Treatments that increase Ca efflux (increase Na_o, decrease Na_i) decrease Ca influx, and the converse is true. Changes in membrane potential that increase Ca efflux, decrease Ca influx, and the converse is also true. All of this experimental information argues for a high degree of coupling between the system moving Ca inward and outward.

What then can be concluded about the relationship between voltage sensitivity of the Na/Ca exchange and electrogenic action? Since Ca efflux requires energy input to bring about this movement, I can see no alternative to the idea that its voltage sensitivity is the direct result of the electrogenicity of the carrier reaction that brings about the flux. Stated differently, an increase in Ca efflux as a result of an increase in membrane potential must mean that the energy to drive the efflux has been increased. This can only occur if 3 or more Na ions are involved in driving Ca from the cell.

The conclusion arrived at above need not mean that the Na/Ca carrier is uncharged; it is possible to combine the idea of a charged, unloaded carrier with that of an electrogenic Na/Ca exchange to reproduce all the effects that have been discussed above. Thus, such a carrier would have a maximum for the inward movement of Ca at zero membrane potential and still transport more Na charges in one direction than Ca charges in the opposite direction.

In conclusion, I would like to emphasize that new channels are being found at a truly unbelievable rate. Starting with the Ca_i activated K channel of Meech (10), which serves to effect repolarization after a certain low threshold level of Ca_i has been reached, we have the relatively ion nonspecific Ca-activated channel that acts as a short circuit on the former channel. I believe it is highly unlikely that a Ca channel will be found that is activated by $[Na]_i$ (this would reproduce some of the results I have discussed), mainly because of the close coupling that appears between Ca entry and exit. There is no channel that, when activated, could produce a net efflux of Ca.

ACKNOWLEDGMENTS

The experimental work reported on was aided by grants BNS 76–19728 A01 from the National Science Foundation and NS 13402 from the National Institutes of Health.

REFERENCES

1. Baker, P. F., Hodgkin, A. L., and Ridgway, E. B. (1971): Depolarization and calcium entry in squid axons. *J. Physiol. (Lond.)*, 218:709–755.
2. Baker, P., Meves, H., and Ridgway, E. B. (1973): Calcium entry in response to maintained depolarization of squid axons. *J. Physiol. (Lond.)*, 231:527–548.
3. Beeler, G. W., and Reuter, H. (1970): The relation between membrane potential membrane currents and activation of contraction in ventricular myocardial fibres. *J. Physiol. (Lond.)*, 207:211–229.
4. Chapman, R. A. (1974): A study of the contractures induced in frog atrial trabeculae by a reduction of the bathing sodium concentration. *J. Physiol. (Lond.)*. 237:295–313.
5. Chapman, R. A. (1979): Excitation-contraction coupling in cardiac muscle. *Prog. Biophys. Mol. Biol.*, 35:1–52.
6. DiPolo, R., Requena, J., Brinley, F. J., Jr., Mullins, L. J., Scarpa, A., and Tiffert, T. (1976): Ionized calcium concentration in squid axons. *J. Gen. Physiol.*, 67:433–467.
7. Eisner, D. A., Lederer, W. J., and Vaughan-Jones, R. D. (1981): The dependence of sodium pumping and tension on intracellular sodium activity in voltage-clamped sheep purkinje fibres. *J. Physiol. (Lond.)*, 317:163–187.
8. Hodgkin, A. L., and Keynes, R. D. (1957): Movements of labelled calcium in squid giant axons. *J. Physiol. (Lond.)*, 138:253–281.
9. Lee, C. O., Kang, D. H., Sokol, J. H., and Lee, K. S. (1980): Relation between intracellular Na ion activity and tension of sheep cardiac Purkinje fibres exposed to dihydro-ouabain. *Biophys. J.*, 29:315–330.
10. Meech, R. W. (1972): Intracellular calcium injection causes increased potassium conductance in *Aplysia* nerve cells. *Comp. Biochem. Physiol.*, A42:493–499.
11. Mullins, L. J., and Brinley, F. J., Jr. (1975): Sensitivity of calcium efflux from squid axons to changes in membrane potential. *J. Gen. Physiol.*, 65:135–152.
12. Mullins, L. J., and Requena, J. (1981): The "late" Ca channel in squid axons. *J. Gen. Physiol.*, 78:683–700.
13. Mullins, L. J., Tiffert, T., Vassort, G., and Whittembury, J. (1983): Effects of internal sodium and hydrogen ions and of external calcium ions and membrane potential on calcium entry in squid axons. *J. Physiol. (Lond.)*, 338:295–319.
14. Requena, J., DiPolo, R., Brinley, F. J., Jr., and Mullins, L. J. (1977): The control of ionized calcium in squid axons. *J. Gen. Physiol.*, 70:329–353.
15. Vassort, G., Tiffert, T., Whittembury, J., and Mullins, L. J. (1982): The effects of internal Na^+ and H^+, external Ca^{++} and of membrane potential on Ca entry in squid axons. *Biophys. J.*, 37:68a.

Electrogenic Transport: Fundamental Principles and Physiological Implications,
edited by Mordecai P. Blaustein and Melvyn Lieberman. Raven Press, New York © 1984.

Physiologic Criteria for Electrogenic Transport in Tissue-Cultured Heart Cells

Melvyn Lieberman, C. Russell Horres, Ron Jacob, Elizabeth Murphy, David Piwnica-Worms, and *David M. Wheeler

Department of Physiology, Duke University Medical Center, Durham, North Carolina 27710

Heart cells in tissue culture have served as model preparations for investigators whose interests focus on membrane processes fundamental to cardiac physiology. In particular, the preparations have provided considerable insight into the electrophysiological properties (25,26,34,40) and underlying ion-transport phenomena (15,16,22,26,28,37,43) of the cardiac membrane. Although evidence from voltage-clamp studies convincingly demonstrated that the rapid depolarization phase of the action potential of cultured heart cells results from a transient increase in sodium conductance (10), the membrane currents responsible for the repolarization phase are unresolved. Identification of the mechanisms contributing to this phase of the cardiac action potential, whether from naturally occurring preparations or those in culture has been complicated by changes in membrane permeability and other electrogenic transport mechanisms (8, 20,29).

A further complication in describing the ionic currents of the prolonged repolarization phase is presented by the morphological complexity of naturally occurring preparations of cardiac muscle. In particular, the presence of an endocardial sheath, collagenous matrix, and intercellular clefts promotes the accumulation and depletion of ions within extracellular space during experimental manipulations (3,7,27). Heart cells grown in tissue culture reduce these morphological complexities and are well suited to study the contribution of electrogenic transport to the electrophysiological properties of cardiac cells.

Electrogenic mechanisms can be identified from studies that correlate ion transport with electrophysiological phenomena. Ion transport across the cardiac cell membrane can be broadly categorized as electrodiffusive, electrogenic, or electroneutral. Examples of such mechanisms, for which evidence has been obtained in our laboratory, are summarized in Fig. 1. As indicated, the transport processes include those solely dependent on electrochemical gradients (e.g., K, Na, and, to a small extent, Cl) and those passively coupled to electrochemical gradients (e.g., Na-Ca, Na-H, Cl-HCO_3) as well as active processes that operate

---

* Present address: Department of Anesthesiology, University of Alabama Hospitals, Birmingham, Alabama 35233.

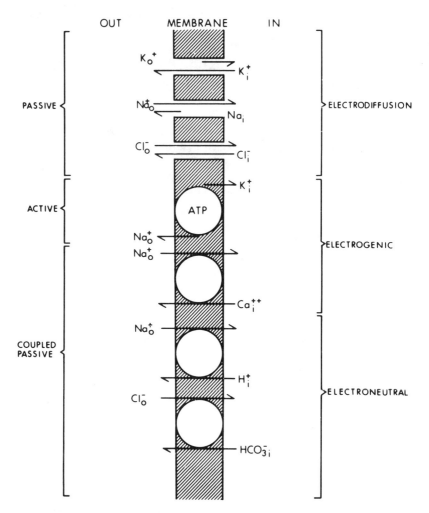

FIG. 1. Schematic representation of active and passive membrane-transport processes identified in growth-oriented heart cells in tissue culture.

against electrochemical gradients (e.g., Na-K). Clearly, the existence of parallel mechanisms, as in the case of Na transport, can complicate attempts to quantify electrogenic transport. Other factors to be considered when examining electrogenic mechanisms include alterations in electrochemical gradients, and the possible influence of the ionic milieu on both passive membrane permeabilities, e.g., Ca-activated K conductance, and intercellular coupling (12,16,21).

PREPARATION

Experimental maneuvers designed to study electrogenic transport often involve either the rapid change of the extracellular ionic composition or the addition of specific inhibitors within a time course during which intracellular ionic composition and membrane conductances are minimally altered. For such studies, we have chosen a growth-oriented preparation of heart cells (17), which was developed in our laboratory (see Fig. 2). This preparation, also referred to as the polystrand, consists of heart cells grown as a thin, cylindrical annulus around a substrate of nylon monofilament in a manner that limits the maximum extracellular diffusional distance to < 50 μm. The geometric arrangement of the cells renders the preparation well-suited for radiotracer kinetic analyses and ion content measurements, whereas the growth orientation enhances the mechanical stability of the preparation for electrophysiological experiments. Data from such analyses are presented in the following section to provide pertinent background information needed to establish electrogenic transport in cultured heart cells.

ELECTROCHEMICAL DATA

The dependence of the membrane potential on extracellular K, as shown in Fig. 3, is characteristic of a membrane primarily permeable to K with a finite permeability to Na. Representative electrochemical data obtained under steady-state conditions are summarized in Table 1. The electrochemical gradients for Na, K, Ca, and Cl are consistent with those reported in the literature for the adult vertebrate myocardium (39,41).

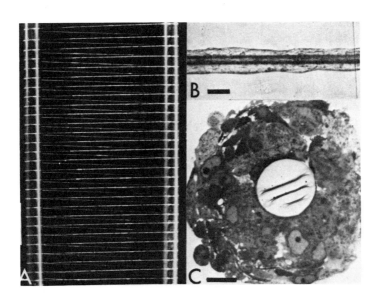

FIG. 2. Composite photomicrographs of the growth-oriented preparations of heart cells. **A:** Polystrand preparation in which cellular growth is limited to the central portion of the nylon substrate. *Bar:* 3 mm. **B:** Segment of a contractile preparation viewed longitudinally. *Bar:* 50 μm. **C:** Cross-section of a contractile preparation. *Bar:* 10 μm. For details of the preparation, see Horres et al. (19).

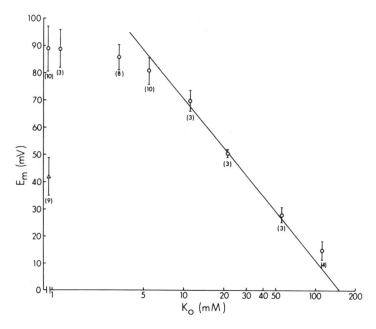

FIG. 3. Transmembrane resting potential of the cultured heart cell as a function of external potassium. *Open circles:* Potentials recorded immediately after a rapid solution change for a single impalement. *Open triangle:* Potential recorded after several minutes in K-free solution. *Straight line* has a slope of 60 mV/decade. Number of experiments is indicated in parentheses; values are mean ±SD. [From Lieberman et al. (24), with permission.]

Steady-state flux measurements obtained in cultured heart cells differ significantly from most measurements obtained from naturally occurring preparations of cardiac muscle, chiefly because of the anatomic complexities of the intact embryonic and adult heart tissue (see ref. 15, for discussion). In the polystrand preparation, the steady-state K flux value of 16 pmol cm^{-2} sec^{-1} is sufficient to account for most of the resting membrane conductance and varies with exter-

TABLE 1. *Electrochemical data from polystrands of cultured embryonic heart cells*

	K[a]	Na[b]	Cl[c]	Ca
Extracellular concentration (mM)	5.4	144	128	2.7
Intracellular content (nmol mg dry wt^{-1})	570	69	107	5.3
Intracellular concentration (mM)	133	16	25	2.3×10^{-4}[d]
Equilibrium potential (mV)	−83	+57	−42	+122
Steady-state flux (pmol cm^{-2} sec^{-1})	16	98	30	—

[a] Horres and Lieberman (15).
[b] Wheeler et al. (43).
[c] Piwnica-Worms et al. (37).
[d] Murphy et al. (31). Isolated heart cells using the null-point method.

nal K (1 to 20 mM) in a manner consistent with a constant K permeability (16). The steady-state Na flux of 98 pmol cm^{-2} sec^{-1} is substantially larger than almost all Na fluxes reported for cardiac muscle (43). The implications of the finding that steady-state Na flux is 6 times greater than the K flux will be discussed in the following section. The Cl flux of 30 pmol cm^{-2} sec^{-1} is one order of magnitude greater than most published values for intact cardiac muscle (37) and nearly twofold greater than the K flux in the polystrand. These findings suggest that a major component of chloride transport does not contribute to membrane conductance in the polystrand and, hence, is electrically silent.

These rapid fluxes in cultured heart cells imply the presence of a significant energy-dependent transport mechanism to maintain the steep electrochemical gradients. Indeed, when spontaneously beating preparations of heart cells were exposed to a K-free solution, the electrochemical gradients for K and Na were rapidly dissipated (16) and the membrane potential declined (see Fig. 3). This response to the absence of K is a well-known characteristic of the Na-K ATPase pump mechanism (35), which is often considered to be electrogenic and has been the subject of a continuing investigation in our laboratory.

ELECTROGENIC Na-K TRANSPORT

The effects of the cardiac glycoside, ouabain, on the Na-K pump are often used to estimate the contribution of this mechanism to the membrane potential in steady state (9,19). As shown in Fig. 4, ouabain (10^{-4}M) caused the membrane to depolarize slightly (about 4 mV) before a significant change could occur in the electrochemical gradients. Within 3 min after the introduction of ouabain, the heart cells depolarized to approximately −40 mV, a value comparable to the potential obtained in K-free solution. This level of depolarization cannot be directly attributed to inhibition of an electrogenic active transport mechanism because of the combined problems of a nonlinear current-voltage relationship, changes in ion gradients, and possible ouabain-induced changes in membrane

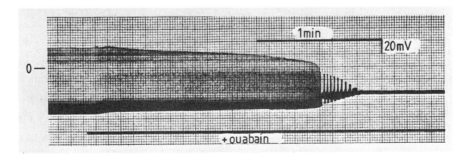

FIG. 4. Continuous recording of transmembrane potentials from a spontaneously active polystrand preparation exposed to control solution (5 mM K) containing ouabain (10^{-4}M). Note the immediate depolarization that occurs in response to ouabain.

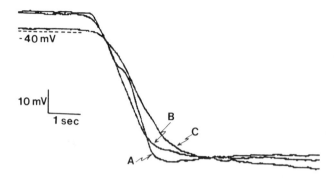

FIG. 5. Transmembrane potentials recorded from polystrands during reactivation of the Na-K pump by switching from a K-free solution to one containing 5.4 mM K. Tracings marked **A**, **B**, and **C** represent the response of preparations with overall diameters of 100 μm, 200 μm, and 700 μm, respectively.

permeability. Convincing evidence for electrogenic transport in cultured heart cells was obtained when, in the presence of elevated Na_i, K reactivation of the Na-K pump caused the transmembrane potential to exceed the measured K equilibrium potential (18). However, the response to Na-K pump reactivation can be affected by the complexity of the preparation. As shown in Fig. 5, when polystrands are intentionally grown to different diameters, thick preparations slowed the hyperpolarizing responses. This finding emphasizes the need for thin preparations with minimal diffusion distances when investigating the kinetics of the Na-K pump.

Attempts to determine the stoichiometry of active Na-K transport from the measurements of steady-state flux for Na and K are complicated by a significant ouabain-insensitive Na efflux (Table 2). The ouabain-sensitive component of Na efflux, when compared with the ouabain-sensitive K influx in the cultured heart cell, suggests a 3:2 pump stoichiometry. However, the sodium component is derived from the difference between two large fluxes and represents only a small component of the total Na efflux.

TABLE 2. *Potassium and sodium fluxes from polystrands of cultured embryonic heart cells*

Flux	K (pmol cm^{-2} sec^{-1})	Na (pmol cm^{-2} sec^{-1})
Efflux (steady state)	16[a]	98[c]
Influx (ouabain sensitive)	11[b]	—
Efflux (ouabain sensitive)	—	16[c]

[a] Horres and Lieberman (15).
[b] Horres (14).
[c] Wheeler et al. (43).

Even though the K flux is much less than the Na flux, the membrane behaves as a K electrode (see Fig. 3) implying the presence of a considerable electroneutral Na exchange. This exchange, as measured by the ^{24}Na efflux rate constant, is inhibited by 40% if either Na or Ca is removed from the extracellular medium (42). These results and those from other laboratories (1,2,6,6a,11,13,23), suggest the presence of a Na-Na and Na-Ca exchange mechanism in cultured heart cells.

ELECTROGENIC Na-Ca TRANSPORT

Evidence for Na-Ca transport has been obtained from experiments in which a net transport of these cations is produced by manipulating the ionic gradients across the membrane (5,30,38). For example, incubation of cultured heart cells in low K medium (6a,32) or ouabain (1,2,4,11,13) dissipates the Na gradient and promotes a substantial increase in intracellular calcium. Likewise, Fig. 6 shows that incubation of preparations in the absence of sodium causes a rapid increase of intracellular calcium concomitant with a decrease in intracellular sodium (see also 2,11,13,23). Additional evidence for Na-Ca exchange was obtained by investigating calcium uptake in response to changes in extracellular Na in the range of 0 to 144 mM (33). At concentrations of $Na_o \leq 20$ mM,

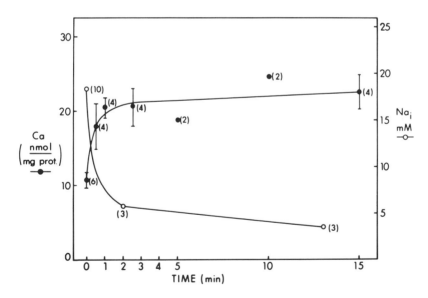

FIG. 6. Changes in intracellular calcium and sodium content of cultured heart cells in response to removal of external sodium as measured by atomic absorption spectroscopy. *Solid circles* represent Ca content obtained from monolayer cultures. *Open circles* represent Na concentration obtained from polystrands.

calcium content increased, whereas calcium uptake fell off sharply at values of $Na_o > 20$ mM. Furthermore, quinidine (65 μM) decreased the uptake of calcium by approximately 70%, a finding consistent with its effects on the kinetics of ^{24}Na efflux (42). Thus, it appears that a Na-Ca exchange mechanism, functionally dependent on the sodium gradient, is maintained by the Na-K pump and may contribute to the regulation of intracellular calcium.

To determine whether Na-Ca exchange is electrogenic, cultured heart cells were perfused with K-free solution plus ouabain to inhibit the electrogenic Na-K pump and to enhance intracellular Na concentration. The Na gradient was then reversed by perfusion with a K-free solution containing 30 mM Na, and as indicated in Fig. 7, the membrane potential hyperpolarized by 45 mV within a few seconds after the onset of this maneuver. In addition, the hyperpolarization could be reduced by approximately 50% in the presence of quinidine. These findings are not uniquely attributable to an electrogenic transport process because the data is also consistent with an electroneutral Na-Ca exchange coupled to a calcium-sensitive increase in K permeability (8), which would bring the membrane potential closer to the potassium equilibrium potential. Support for this mechanism was obtained from the observation that the hyperpolarization caused by switching from 0 mM K to 0 mM K, 30 mM Na was reduced when the K gradient was reduced by raising extracellular K to 5.4 mM. The interpretation of experiments involving the removal of extracellular sodium are further complicated by the presence of a significant Na-H exchange in the cultured heart cell (36). Such a mechanism, although not necessarily electrogenic, can influence, either directly or indirectly, other electrogenic and electrodiffusive mechanisms in the cardiac membrane. Therefore, the above data does not permit confirmation of the electrogenicity of Na-Ca transport in cultured heart cells at this time.

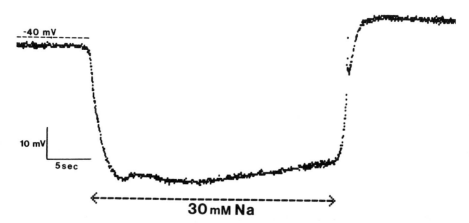

FIG. 7. Transmembrane potential recorded from a polystrand depolarized by perfusion with a K-free solution containing 10^{-4}M ouabain and then with the same solution but with 30 mM Na (tetramethyl ammonium substitute).

SUMMARY

A description of the physiologic criteria for electrogenic transport in cultured heart cells must consider the direct as well as the indirect contributions of the electrodiffusive and electroneutral transport mechanisms of the kind described in this chapter. It is important to recognize that fundamental transport processes whether active, passive, or coupled-passive, do not operate in isolation but function rather as an interrelated network represented by the example in Fig. 8. The results of our studies point to the importance of providing an accurate, quantitative description of the transmembrane movement of sodium, because this ion is intricately involved in the overall regulation of several transport mechanisms in cardiac muscle. Cultured heart cell preparations enable the rapid exchange of extracellular space essential to the study of complex transport mechanisms. These preparations have thus enhanced our understanding of the magnitude and interplay of exchange fluxes that may eventually be implicated in the repolarization phase of the cardiac action potential and genesis of pacemaker potentials.

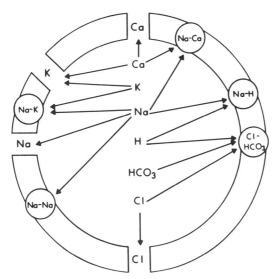

FIG. 8. Schematic representation of the cultured heart cell demonstrating the interaction between electrodiffusive, electrogenic, and electroneutral transport processes.

ACKNOWLEDGMENTS

We thank Owen Oakeley for expert technical advice and Chuck Henry and Dora Lyons for their efforts in maintaining the tissue-culture facility. This research was supported in part by grants from the National Institutes of Health (HL07101, HL27105, GM07171, HL17670), North Carolina Heart Association and the Insurance Medical Scientist Scholarship Fund (DMW).

REFERENCES

1. Barry, W. H., Biedert, S., Miura, D. S., and Smith, T. W. (1981): Changes in cellular Na^+, K^+, and Ca^{2+} contents, monovalent cation transport rate, and contractile state during washout of cardiac glycosides from cultured chick heart cells. *Circ. Res.*, 49:141–149.
2. Barry, W. H., and Smith, T. W. (1982): Mechanisms of transmembrane calcium movement in cultured chick embryo ventricular cells. *J. Physiol. (Lond.)*, 325:243–260.
3. Baumgarten, C. M., and Isenberg, G. (1977): Depletion and accumulation of potassium in the extracellular clefts of cardiac Purkinje fibers during voltage clamp hyperpolarization and depolarization. *Pfluegers Arch.*, 368:19–31.
4. Biedert, S., Barry, W. H., and Smith, T. W. (1979): Inotropic effects and changes in sodium and calcium contents associated with inhibition of monovalent cation active transport by ouabain in cultured myocardial cells. *J. Gen. Physiol.*, 74:479–494.
5. Blaustein, M. P. (1974): The interrelationship between sodium and calcium fluxes across cell membranes. *Rev. Physiol. Biochem. Pharmacol.*, 70:34–82.
6. Burt, J. M. (1982): Electrical and contractile consequences of Na^+ or Ca^{2+} gradient reduction in cultured heart cells. *J. Mol. Cell. Cardiol.*, 14:99–110.
6a. Burt, J. M., and Langer, G. A. (1982): Ca^{++} distribution after Na^+ pump inhibition in cultured neonatal rat myocardial cells. *Circ. Res.*, 51:543–550.
7. Cohen, I., and Kline, R. (1982): K^+ fluctuations in the extracellular spaces of cardiac muscle. *Circ. Res.*, 50:1–16.
8. Coraboeuf, E. (1978): Ionic basis of electrical activity in cardiac tissues. *Am. J. Physiol.* 234:H101–H116.
9. Daut, J., and Rüdel, R. (1982): The electrogenic sodium pump in guinea-pig ventricular muscle: Inhibition of pump current by cardiac glycosides. *J. Physiol. (Lond.)*, 330:243–264.
10. Ebihara, L., Shigeto, N., Lieberman, M., and Johnson, E. A. (1980): The initial inward current in spherical clusters of chick embryonic heart cells. *J. Gen. Physiol.*, 75:437–456.
11. Fosset, M., deBarry, J., Lenoir, M. C., and Lazdunski, M. (1977): Analysis of molecular aspects of Na^+ and Ca^{2+} uptakes by embryonic cardiac cells in culture. *J. Biol. Chem.*, 252:6112–6117.
12. Glitsch, H. G. (1979): Characteristics of active Na transport in intact cardiac cells. *Am. J. Physiol.*, 236:H189–H199.
13. Goshima, K., and Wakabayashi, S. (1981): Involvement of an Na^+-Ca^{2+} exchange system in genesis of ouabain-induced arrhythmias of cultured myocardial cells. *J. Mol. Cell. Cardiol.*, 13:489–509.
14. Horres, C. R. (1975): *Potassium Tracer Kinetics of Growth-Oriented Heart Cells in Tissue Culture* (PhD Thesis). Duke University, Durham, North Carolina.
15. Horres, C. R., and Lieberman, M. (1977): Compartmental analysis of potassium efflux from growth-oriented heart cells. *J. Membr. Biol.* 34:331–350.
16. Horres, C. R., Aiton, J. F., and Lieberman, M. (1979): Potassium permeability of embryonic avian heart cells in tissue culture. *Am. J. Physiol.*, 236:C163–C170.
17. Horres, C. R., Lieberman, M., and Purdy, J. E. (1977): Growth orientation of heart cells on nylon monofilament: Determination of the volume-to-surface area ratio and intracellular potassium concentration. *J. Membr. Biol.* 34:313–329.
18. Horres, C. R., Aiton, J. F., Lieberman, M., and Johnson, A. E. (1979): Electrogenic transport in tissue cultured heart cells. *J. Mol. Cell. Cardiol.*, 11:1201–1205.
19. Isenberg, G., and Trautwein, W. (1974): The effects of dihydro-ouabain and lithium ions on the outward current in cardiac Purkinje fibers. Evidence for electrogenicity of active transport. *Pfluegers Arch.*, 350:41–54.
20. Johnson, E. A. (1982): The origin of slow currents in cardiac muscle. In: *Normal and Abnormal Conduction in the Heart*, edited by A. Paes de Carvalho, B. F. Hoffman, and M. Lieberman, pp. 249–276. Futura Publishing Co., Mt. Kisco, New York.
21. Kootsey, J. M., Johnson, E. A., and Chapman, J. B. (1981): Electrochemical inhomogeneity in ungulate Purkinje fibers: Model of electrogenic transport and electrodiffusion in clefts. In: *Cardiovascular Physiology. Heart, Peripheral Circulation and Methodology*, edited by A. G. B. Kovách, E. Monos, and G. Rubány, pp. 83–92. Pergamon Press, New York.
22. Langer, G. A. (1981): Calcium exchange in myocardial tissue culture. In: *Excitable Cells in Tissue Culture*, edited by P. G. Nelson and M. Lieberman, pp. 359–377. Plenum Press, New York.

23. Langer, G. A., Nudd, L. M., and Ricchiuti, N. V. (1976): The effect of sodium deficient perfusion on calcium exchange in cardiac tissue culture. *J. Mol. Cell. Cardiol.*, 8:321–328.
24. Lieberman, M., Horres, C. R., Aiton, J. F., and Shigeto, N. (1982): Developmental aspects of cardiac excitation: Active transport. In: *Normal and Abnormal Conduction in the Heart*, edited by A. Paes de Carvalho, B. F. Hoffman, and M. Lieberman, pp. 313–326. Futura Publishing Co., Mt. Kisco, New York.
25. Lieberman, M., Sawanobori, T., Kootsey, J. M., and Johnson, E. A. (1975): A synthetic strand of cardiac muscle: Its passive electrical properties. *J. Gen. Physiol.*, 65:527–550.
26. Lieberman, M., Horres, C. R., Shigeto, N., Ebihara, L., Aiton, J. F., and Johnson, E. A. (1981): Cardiac muscle with controlled geometry. Application to electrophysiological and ion transport studies. In: *Excitable Cells in Tissue Culture*, edited by P. G. Nelson and M. Lieberman, pp. 379–408. Plenum Press, New York.
27. Martin, G., and Morad, M. (1982): Activity-induced potassium accumulation and its uptake in frog ventricular muscle. *J. Physiol. (Lond.)*, 328:205–227.
28. McCall, D. (1979): Cation exchange and glycoside binding in cultured rat heart cells. *Am. J. Physiol.*, 236:C87–C95.
29. Morad, M., and Tung, L. (1982): Ionic events responsible for the cardiac resting and action potential. *Am. J. Cardiol.*, 49:584–594.
30. Mullins, L. J. (1979): The generation of electric currents in cardiac fibers by Na/Ca exchange. *Am. J. Physiol.*, 236:C103–C110.
31. Murphy, E., Henry, S. C., and Lieberman, M. (1983): Cytosolic free Ca and myocardial cell injury. *Fed. Proc.*, 42(3): 722(Abstr.).
32. Murphy, E., Aiton, J. F., Horres, C. R., and Lieberman, M. (1981): Na-K pump inhibition and calcium accumulation in cultured heart cells. *Fed. Proc.*, 40:617(Abstr.).
33. Murphy, E., Wheeler, D. M., Anderson, L., Horres, C. R., and Lieberman, M. (1981): Sodium-calcium exchange in cultured chick heart cells. *J. Gen. Physiol.*, 80:16a–17a.
34. Nathan, R. D., and DeHaan, R. L. (1979): Voltage clamp analysis of embryonic heart cell aggregates. *J. Gen. Physiol.*, 73:175–198.
35. Page, E., Goerke, R. J., and Storm, S. R. (1964): Cat heart muscle *in vitro*. IV. Inhibition of transport in quiescent muscles. *J. Gen. Physiol.*, 47:531–543.
36. Piwnica-Worms, D., Jacob, R., Horres, C. R., and Lieberman, M. (1982): Na-H exchange in cultured chick heart cells coupled to electrogenic Na-K transport. *J. Gen. Physiol.*, 80:20a–21a.
37. Piwnica-Worms, D., Jacob, R., Horres, C. R., and Lieberman, M. (1983): Transmembrane chloride flux of tissue-cultured chick heart cells. *J. Gen. Physiol.*, 81:731–748.
38. Reuter, H. (1974): Exchange of calcium ions in the mammalian myocardium. *Circ. Res.*, 34:599–605.
39. Sperelakis, N. (1980): Origin of the cardiac resting potential. In: *Handbook of Physiology—The Cardiovascular System, I*, edited by R. M. Berne, N. Sperelakis, pp. 187–267. Williams & Wilkins, Baltimore.
40. Sperelakis, N. (1982): Cultured heart cell reaggregate model for studying problems in cardiac toxicology. In: *Cardiovascular Toxicology*, edited by E. W. Van Stee, pp. 57–108. Raven Press, New York.
41. Vaughan-Jones, R. D. (1979): Non-passive chloride distribution in mammalian heart muscle: Microelectrode measurement of the intracellular chloride activity. *J. Physiol. (Lond.)*, 295:83–109.
42. Wheeler, D. M. (1981): *Sodium Tracer Kinetics and Transmembrane Flux in Tissue-Cultured Chick Heart Cells*. (PhD Thesis). Duke University, Durham, North Carolina.
43. Wheeler, D. M., Horres, C. R., and Lieberman, M. (1982): Sodium tracer kinetics and transmembrane flux in tissue-cultured heart cells. *Am. J. Physiol.*, 243:C169–C176.

*Electrogenic Transport: Fundamental
Principles and Physiological Implications,*
edited by Mordecai P. Blaustein and Melvyn
Lieberman. Raven Press, New York © 1984.

The Electrogenic Na Pump in Mammalian Cardiac Muscle

*D. A. Eisner, **W. J. Lederer and †R. D. Vaughan-Jones

*Department of Physiology, University College London, London WC1E 6BT, England;
**Department of Physiology, University of Maryland School of Medicine, Baltimore,
Maryland 21201; †Department of Physiology, South Parks Rd,
Oxford OX1 3QT, England*

The sodium-potassium pump is generally believed to be electrogenic. This electrogenicity is a consequence of the fact that the Na pump expels more Na ions from the cell than it brings in K ions (32). The best estimates in the red cell suggest that each cycle of the Na-K pump hydrolyzes one molecule of adenosine triphosphate (ATP), extrudes three Na ions, and pumps two K ions into the cell (20). A similar ratio of Na:K transport has also been measured in the squid axon (2). Only in these simple preparations can the stoichiometry of the sodium pump be measured from isotopic flux experiments. Because of experimental complications, flux experiments have not contributed much information about the stoichiometry of the Na pump in cardiac muscle. For this reason a favorite method to study the pump has been to look for the electric current that such an electrogenic pump should produce. In small nerve fibers, if the intracellular Na is elevated by rapid stimulation, a transient hyperpolarization is seen on cessation of stimulation (33). Similar hyperpolarizations can be seen after elevating internal Na either by exposure to a K-free solution or by cooling (see ref. 37 for a review). These hyperpolarizations can be abolished by ouabain and other cardioactive steroids and have therefore been attributed to the operation of a Na-K pump. It is necessary, however, to establish whether the ouabain-sensitive hyperpolarization is a result of (a) an electrogenic Na-K pump or (b) an electroneutral pump which, when stimulated by an elevation of Na_i, could decrease the K concentration in the region adjacent to the outside of the cell and thereby produce hyperpolarization. That an electrogenic pump must produce at least part of the hyperpolarization was shown by the fact that the hyperpolarization following cessation of stimulation reached a less negative level when external K was reduced, presumably because the Na-K pump was less activated (33).

Thomas (36) injected Na into a snail neuron and recorded an ouabain-sensitive hyperpolarization attributed to activation of the Na-K pump. The current underlying this hyperpolarization was measured under voltage-clamp conditions. The results showed that 1/4 to 1/3 of the Na injected appeared as net charge movement across the cell membrane. This is consistent with a Na/K transport ratio of 3:2 or 4:3.

As far as cardiac muscle is concerned, the existence of an electrogenic Na-K pump was, until recently, less clear. Délèze (8) first suggested that the pump was electrogenic. This was supported by the observation that a transient hyperpolarization appeared on rewarming, after Na loading by cooling (21,25,31). This hyperpolarization is abolished by ouabain or by application of K-free solutions. A hyperpolarization that is abolished by metabolic inhibition or Na removal is also seen following the cessation of a period of Na loading produced by rapid stimulation (39).

These experiments suggest that changes of Na-pump activity in cardiac muscle can produce changes of membrane potential. The question of immediate concern is whether these membrane potential changes are a consequence of an electrogenic Na-K pump or whether they reflect an extracellular K depletion produced by an electroneutral Na pump. Some of the evidence reviewed above suggests that an electrogenic mechanism must be involved. Furthermore, evidence suggests that the changes of membrane potential produced by reactivating the Na pump cannot be entirely accounted for by extracellular K depletion (16). However, no direct measurements of this putative electrogenic pump current were available in cardiac muscle. In this chapter we therefore consider four related questions. (i) If the Na pump is electrogenic can we measure the current it produces without contamination from secondary problems of ionic redistribution? (ii) How does the activity of the Na pump depend on the activation by cations at intracellular and extracellular sites? (iii) How much of the sodium extruded from the cell is electrogenic rather than being in exchange for the entry of K ions? (iv) Does the electrogenic Na-pump current make a significant contribution to the resting potential of cardiac muscle? A related question concerning the importance of the electrogenic-pump current on electrical activity is dealt with in the chapters by Drs. Lieberman et al. and Gadsby.

METHODS

The experimental methods have been described in detail elsewhere (13). Sheep cardiac Purkinje fibers were shortened to 1 to 2 mm in length. Voltage-clamp control was imposed with a two-microelectrode clamp and intracellular Na activity measured with a recessed-tip, Na-selective glass microelectrode (38). The Na-selective microelectrodes had sufficiently fast responses (half-complete in less than 6 sec) to record the changes of Na in the present work. Intracellular Na measurements have been expressed in terms of the activity (a^i_{Na}) assuming the activity coefficient of the external calibrating solution to be 0.75 (see ref. 13 for further details). Solutions contained: 145 mM NaCl; 1 mM MgCl$_2$; 10 mM Tris-HCl (pH 7.4 at 37°C); 10 mM glucose. Various amounts of RbCl or

KCl were added as described; $CaCl_2$ was 5 mM in the experiments of Figs. 1 to 5 and 2 mM in the experiments of Figs. 6 to 13.

RESULTS

The Identification of a Na-K-Pump-Dependent Current

In the present experiments the Purkinje fiber was exposed to a K-free solution to inhibit the Na pump and load the cells with sodium. The Na pump was then reactivated by adding back either K or Rb ions to the superfusing solution. Under these conditions the sodium pump should be stimulated above its steady-state level by the raised internal Na; therefore, any current that depends on sodium-pump activity should be increased. It will be shown later that Rb is a more satisfactory ion than K to use for this purpose and, since K and Rb have similar effects on the Na pump in the sheep Purkinje fiber (12,13,22), the Rb ion is used in the experiment of Fig. 1. Here 20 mM Rb was added after the preparation had been exposed to a K-free, Rb-free solution for 20 min. The crosses show that the membrane potential rapidly hyperpolarized after the addition of Rb. This hyperpolarization then decayed over a period of several minutes. In order to investigate the changes of membrane current that produce this hyperpolarization, the voltage clamp was applied at regular

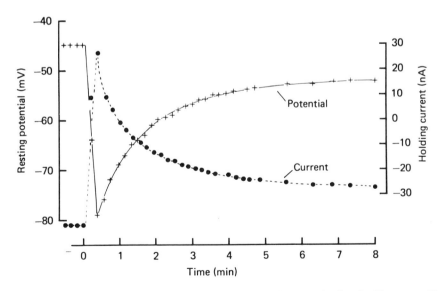

FIG. 1. Membrane potential and current changes produced by reactivating the Na pump with 20 mM Rb_o. The preparation was exposed to a Rb-free solution for 10 min, and then 20 mM Rb_o was added at time zero. The solution was always K-free. Symbols: (+), resting potential; (●), membrane current at −68 mV. In order to measure current the voltage clamp was applied for 4 sec at 0.1 Hz. [From Eisner and Lederer (11), with permission.]

intervals. The current record shows a transient outward overshoot of current, which then decays with a time-course similar to that of membrane potential.

The Effects of Strophanthidin

If the overshoot of current found in Fig. 1 is produced by the activation of an electrogenic Na pump it should be inhibited by blockers of this pump such as the cardioactive steroid, strophanthidin. In the experiment of Fig. 2 the preparation was exposed to a K-free, Rb-free solution for 15 min to elevate Na_i before adding 10 mM Rb to reactivate the Na pump. The membrane potential was clamped throughout so that changes of membrane current could be recorded. The control record (a) shows a large overshoot of current in the absence of

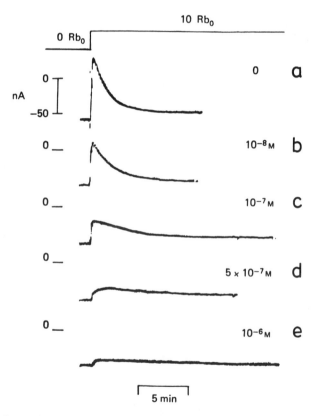

FIG. 2. The effects of strophanthidin on the overshoot of outward current produced by Na-pump reactivation. The preparation was initially in the control solution (10 mM Rb_o). It was then exposed for 15 min to a Rb-free solution containing the test concentration of strophanthidin; 10 mM Rb was then added to this solution at the time indicated. After the current had reached a steady state the preparation was returned to the control solution (strophanthidin-free). The entire cycle was then repeated at other strophanthidin concentrations. The strophanthidin concentration is indicated by each record (M). The solutions were K-free throughout. [From Eisner and Lederer (10), with permission.]

strophanthidin. The other records (b–e) demonstrate that increasing strophanthidin concentration produces a graded reduction of the magnitude of the overshoot of outward current. High concentrations of cardioactive steroids have also been shown to abolish the outward current transient in canine Purkinje fibers (17).

Is the Outward Current Transient a Pure Electrogenic Na-Pump Current?

The above results demonstrate that the outward current transient produced by adding back Rb ions is related to activation of the sodium pump. It remains to be established, however, how much of this current is the electrogenic-pump current as opposed to being due to secondary changes of intra- or extracellular ionic concentrations produced by the increase of pump activity. In order to examine this point, we have separated the various components by measuring the current overshoot as a function of membrane potential. The experiments have been performed using either K or Rb. An experiment using potassium is illustrated in Fig. 3. The experiment is similar to that of Fig. 1 except that,

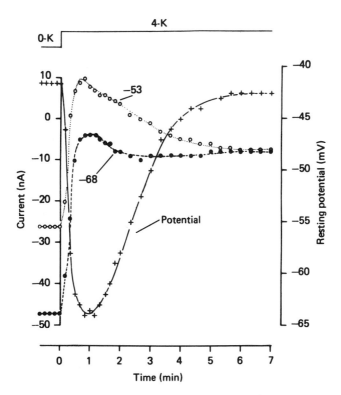

FIG. 3. Potential dependence of membrane currents during the transient hyperpolarization produced by Na-pump reactivation with 4 mM K_o. The preparation had been exposed to a K-free solution for 10 min and, at the time shown, 4 mM K_o was added. The symbols indicate: (+), resting potential; (○), current at −53 mV; (●), current at −68 mV. [From Eisner and Lederer (11), with permission.]

after a 10-min exposure to K-free solution, 4 mM K was added to reactivate the Na pump. This produces a large hyperpolarization (crosses), which decays over a period of several minutes. The voltage clamp was applied at regular intervals to measure the current required to hold the membrane potential at −53 and −68 mV. Although a peak of outward current is seen at both potentials, it is much bigger at the less negative level. In other words, the apparent membrane conductance decreases with time after adding 4 mM K. This change of conductance cannot simply be explained by the activation of an electrogenic pump and suggests that other current changes may be occurring.

Figure 4 illustrates an experiment designed to investigate these current changes over a wider range of membrane potentials. The control current–voltage relationship (I–V) obtained after a prolonged exposure to 4 mM K (open circles) shows the typical N-shaped I–V of the Purkinje fiber and, in particular, a region of negative slope conductance. The Na pump was then inhibited by removing external K for 20 min, and then 4 mM K was added back to reactivate the Na pump. Another I–V relationship (closed circles) was obtained after 2.5 min in high K. This has no region of negative slope conductance and intersects the control relationship. Furthermore, this I–V has a less negative resting potential than the control (−72 compared to −73 mV). Figure 4B shows the result of subtracting the control relationship from that obtained 2.5 min after adding back K. The difference current has an apparent reversal potential at about −53 mV. This complicated behavior is unlikely to be produced by an electrogenic pump current, since such a current should always decline with time. The results at more negative potentials where there is *less* outward current during the period of Na-pump reactivation therefore require another explanation. The peculiar shape of the difference curve could be explained if stimulation of the sodium pump produced a depletion of extracellular potassium ions. This depletion could also explain the depolarization of the resting potential, since it is well established that *decreasing* external K can produce a depolarization of the Purkinje fiber membrane potential (40). To investigate this hypothesis we have compared the effects of reactivating the Na pump with those produced by lowering the bathing potassium concentration to mimic the effects of potassium depletion. The crosses in Fig. 4 show an I–V relationship obtained in 2 mM K. This has no region of negative slope conductance, and its shape is more similar to that of the 2.5-min pump reactivation curve than the steady-state, 4 mM K curve. The difference current (B) shows the result of subtracting the control (4 mM K) I–V from the 2 mM K curve. The general shape of this difference curve is similar to that of the Na-pump reactivation difference curve. The Na-pump reactivation curve is, however, lifted up on the current axis with respect to the 2 mM K curve. The I–V produced by Na-pump reactivation therefore appears to consist of two components: (a) distortion of its shape, which resembles that produced by lowering external K, and (b) an upward shift, which may possibly be due to an electrogenic Na-pump current. It is, however, obvious that no meaningful measurements of the pump current can be made in these K-contain-

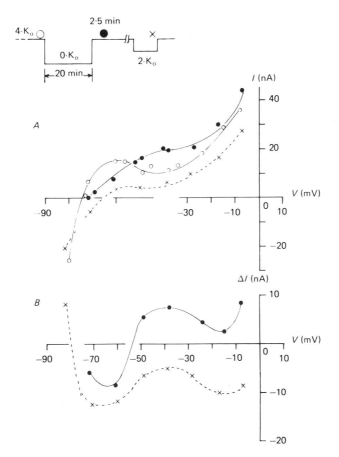

FIG. 4. Effects on the *I–V* relationship of (a) Na-pump reactivation with 4 mM K_o and (b) K depletion per se. **A:** *I–V* relationships. Steady-state *I–V* relationships were obtained by increasing the membrane potential in approximately 10-mV increments. To obtain steady state currents required up to 6 to 8 sec depending on the potential range. The control *I–V* (○) was obtained in 4 mM K_o and, following a 20-min exposure to K-free solution, K_o was increased to 4 mM. Another *I–V* was obtained after 2.5 min in 4 mM K_o (●). After the *I–V* had returned to control, K_o was decreased to 2 mM and another *I–V* obtained (×). These solution changes are shown in the *inset*. **B:** *I–V* difference relationships. The curves show the result of subtracting the control (○) *I–V* from each of the other two. The symbols show the result of subtracting the control *I–V* from: (●), the 2.5-min Na-pump reactivation curve; (×), the 2 mM K_o curve. [From Eisner and Lederer (11), with permission.]

ing solutions. We therefore decided to examine the effects of the Rb ion which, although it substitutes for K on the Na pump, has very different properties on membrane channels (1) and might therefore be expected to produce different extracellular depletion effects.

In the experiment of Fig. 5 the preparation was exposed to a K-free, Rb-free solution for 20 min before adding 10 mM Rb to reactivate the Na pump.

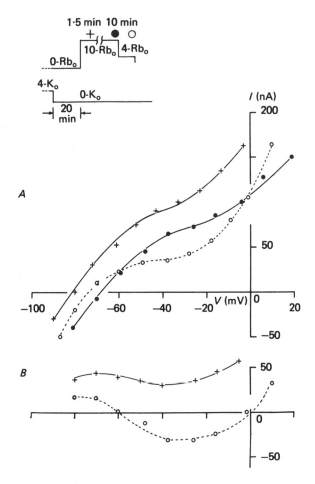

FIG. 5. Effects on the *I–V* relationship of (a) Na pump reactivation with 10 mM Rb_o or (b) Rb depletion per se. **A:** *I–V* relationships. Steady-state currents were obtained as described for Fig. 4. The preparation was initially in a Rb-free, 4 mM K_o solution and was then exposed to a Rb-free, K-free solution for 20 min. Rb_o was increased to 10 mM to reactivate the Na pump and an *I–V* obtained after 1.5 min (+). After a 10-min exposure to 10 mM Rb_o the current had reached a steady level and another *I–V* was obtained (●). Rb_o was then decreased to 4 mM and another *I–V* obtained (○). The solution protocol is illustrated in the *inset*. **B:** *I–V* difference relationships. These curves were calculated by subtracting the steady-state 10 mM Rb_o curve from each of the other records. The symbols show which of the relationships in **A** has been used as the minuend in the calculation. [From Eisner and Lederer (11), with permission.]

I–V relationships were obtained both 1.5 min after adding Rb (during the hyperpolarization) and in the steady state when the membrane potential had relaxed fully (after 25 min). It is clear that, at every potential, there is more outward current at 1.5 than at 25 min. This is to be expected if the current changes are due to the reactivation of an electrogenic Na-K pump but not if they are produced by extracellular depletion. This distinction is illustrated by the third $I-V$ of Fig. 5, which shows the effects of decreasing the bathing Rb concentration from 10 to 4 mM. The 4 mM Rb curve intersects the 10 mM curve such that, although the resting potential is more negative in 4 than in 10 mM Rb, over a wide range of potentials there is more outward current at 10 than at 4 mM Rb. These results therefore show that the transient overshoot of current produced by reactivation of the Na pump with Rb is not due to extracellular depletion but to the electrogenic pump current. This result is emphasized in Fig. 5B, which shows $I-V$ differences obtained by subtracting the steady-state (25 min) $I-V$ from both the 1.5-min 10 mM Rb and also the 4 mM Rb curves. It is clear that the 4 mM Rb difference curve intersects the voltage axis, whereas the pump reactivation curve does not. It should be noted that the pump reactivation curve is not completely flat. The slight bending may be attributed to some depletion effects.

The above results show that, although the electrogenic Na-pump current can be measured in Rb-containing solutions, current changes due to depletion of extracellular K obscure it in K-containing solutions. Since Rb and K are pumped at similar rates (12,13,22) one would expect the amount of depletion of Rb to be similar to that of K. This suggests that Rb depletion has a smaller effect on membrane current than does K depletion. This conclusion is supported by the fact that a change of external K concentration has a greater effect on the efflux of potassium from the Purkinje fiber than does the same change of external Rb (28). Further experiments (not shown) have demonstrated that the electrogenic Na-pump current can also be measured when the Na pump is reactivated by either Cs or Li ions (11). The results are qualitatively similar to those in Rb solutions except that Cs is a less potent activator of the Na pump, and Li is even less effective.

The Dependence of the Electrogenic Na Pump on a_{Na}^i

The above results suggest that, at least in Rb-containing solutions, it is possible to measure the current produced by the electrogenic Na pump. We will now consider the dependence of this current on the intracellular Na activity, a_{Na}^i. Figure 6 shows data from an experiment in which the preparation was exposed to a K-free, Rb-free solution for 6 min, and then 4 mM Rb was added to reactivate the Na pump. This produces the expected fall of a_{Na}^i and transient overshoot of the electrogenic Na-pump current. The same data are plotted on a semilogarithmic scale in Fig. 7A. It is obvious that both current and a_{Na}^i decay exponentially. Furthermore, the rate constant of the decay of current is similar to that of a_{Na}^i. This identity of the rate constants of a_{Na}^i and current is a consistent finding (13). Figure 7A also shows that 4 mM K and 4 mM Rb produce similar rates of fall of a_{Na}^i, confirming that they activate the Na pump by a similar amount. The level of current above its final steady state has been plotted as a function of a_{Na}^i throughout the period of Na-pump reactivation

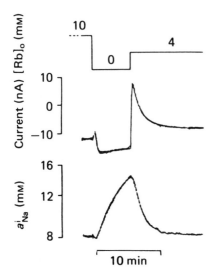

FIG. 6. Measurement of a_{Na}^i and membrane current during Na-pump reactivation with 10 mM Rb_o. *Top trace* shows the membrane current and the *lower trace* a_{Na}^i. The solution protocol is illustrated above the record. All solutions were K-free. Apart from brief depolarizing pulses, the membrane potential was held at −68 mV throughout. From a control solution of 10 mM Rb_o, Rb_o was lowered to zero for 6 min and then increased to 4 mM. [From Eisner et al. (13), with permission.]

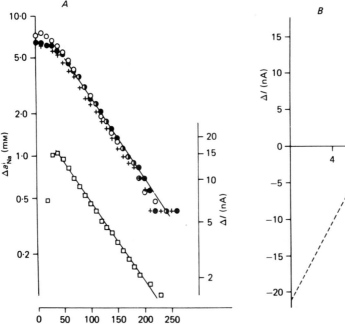
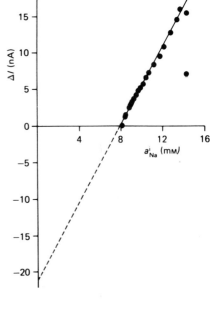

FIG. 7. The relationship between a_{Na}^i and current following Na-pump reactivation. **A:** Semilogarithmic plot of the time course of decay of a_{Na}^i and current following Na-pump reactivation. Data taken from the experiment of Fig. 6. The levels of current and a_{Na}^i are plotted as the magnitude above their steady-state levels (ΔI and Δa_{Na}^i, respectively) as a function of time after adding either 4 mM Rb_o (□, current; ●, a_{Na}^i) or 4 mM K_o (○, a_{Na}^i). *Crosses* show the a_{Na}^i record in 4 mM Rb_o corrected for the response time of the Na^+ electrode. The lines have been fitted by linear regression to all points after the first 30 sec. **B:** ΔI as a function of a_{Na}^i. Data are taken from the reactivation with 4 mM Rb_o. [From Eisner et al. (13), with permission.]

(Fig. 7B). The magnitude of the electrogenic Na-pump current is a linear function of a^i_{Na}. Such a linear dependence of pump activity on a^i_{Na} has been suggested previously (7). It is possible, however, that the relationship deviates from linearity at either lower or higher a^i_{Na} (23).

The Coupling Ratio of the Sodium Pump

On a simple analysis, the amount of Na extruded electrogenically during Na-pump reactivation is given by the integral of the current record above its steady-state level. On the other hand, the fall of a^i_{Na} gives a measure of the total amount of Na pumped out of the cell. If the Na pump is the only mechanism producing the Na extrusion then, by comparing these measurements, it should be possible to estimate how much of the sodium leaves the cell as net charge transfer rather than in exchange for K (Rb) ions entering. This has been done for the experiment of Fig. 8. Here the preparation was exposed to Rb-free solutions for various periods. The longer the duration in Rb-free solution, the greater the increase of a^i_{Na}. Furthermore, the size of the electrogenic Na-pump current and also its integral increase with prolonged exposure to Rb-free solution. This integral has been measured as the area under the electrogenic Na-pump

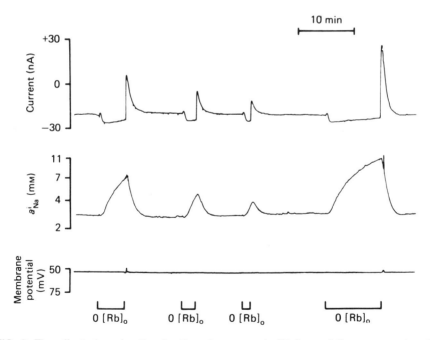

FIG. 8. The effect of varying the duration of exposure to Rb-free solution on current and a^i_{Na}. **Top:** Current; **middle:** a^i_{Na}; **bottom:** membrane potential. The membrane potential was held at -54 mV throughout the experiment. The control solution was 10 mM Rb_o. As shown by the bars, Rb_o was lowered to zero for periods of 5, 2.5, 1.5, and 10 min. Following these exposures Rb_o was increased back to 10 mM. [From Eisner et al. (13), with permission.]

current transient and the magnitudes of these areas are plotted as a function of the fall of a_{Na}^i (Δa_{Na}^i). The area is proportional to Δa_{Na}^i, suggesting that a constant fraction of the sodium extruded from the cell during pump reactivation leaves as net charge transfer rather than in exchange for K (Rb) ions entering. Therefore, the electrogenic fraction of Na extrusion is independent of a_{Na}^i, at least over the measured range. Further experiments (14) have shown that the coupling ratio of the Na pump is also independent of Rb_o.

It is more difficult to calculate the absolute value of this electrogenic fraction of Na extrusion. It can be shown (13) that the total amount of charge (Q) extruded during Na-pump reactivation is given by:

$$Q = F r V_c \Delta a_{Na}^i / \gamma \tag{1}$$

where γ is the activity coefficient for Na ions, V_c the intracellular volume of the preparation, r the fraction of Na which leaves the cell electrogenically, and F is the Faraday. Therefore, the gradient of the line in Fig. 9 should be $F r V_c / \gamma$. A knowledge of the values of γ and V_c would allow r to be calculated. Unfortunately, uncertainty in these values makes it difficult to obtain accurate values for r. Despite these problems we have obtained a value for r of 0.26 ± 0.06 (mean ± SEM). This value is similar to those obtained for the snail neuron and the erythrocyte of 0.25 to 0.33 (32,36). As well as the errors produced by the above uncertainties, it should be noted that mechanisms other than the Na pump (such as Na-Ca exchange) may extrude Na ions from the cell (6). If these mechanisms are not electrogenic, then r will be underestimated. Nevertheless, it appears that somewhere between one-third and one-quarter of the Na extruded by the Na pump leaves the cell as net charge transfer. This would be consistent with a Na pump, which exchanges either 3Na/2K or 4Na/3K.

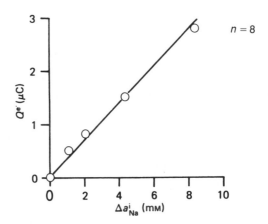

FIG. 9. The relationship between the charge extruded during the electrogenic Na-pump current transient and the decrease of a_{Na}^i. Data taken from Fig. 8. The *ordinate* is the area under the electrogenic Na-pump current transient (μC), and the *abscissa* shows the decrease of a_{Na}^i (Δa_{Na}^i) produced by Na-pump reactivation. [From Eisner et al. (13), with permission.]

The Dependence of the Na Pump on External K and Rb

The effects of various concentrations of Rb on the electrogenic Na pump were investigated in Fig. 10. The control solution was K-free and contained 10 mM Rb. External Rb was then removed for 5-min periods before adding a test concentration of Rb to reactivate the Na pump. The cycle was then completed

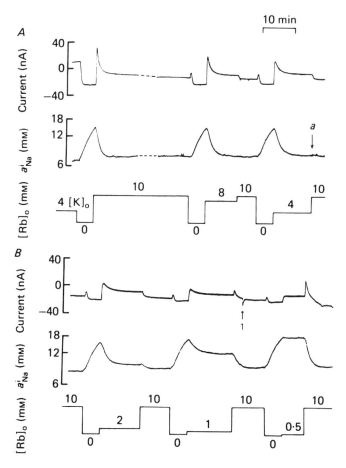

FIG. 10. The effects of various concentrations of Rb on a_{Na}^i and current. In each panel the *top* trace shows current and the *middle* a_{Na}^i. Changes of Rb_o are indicated in the *lower* trace. The bottom panel (**B**) is a continuation of **A**. Apart from a few minutes at the beginning of the record (as indicated), K_o was zero throughout. The control solution contained 10 mM Rb. The membrane potential was held at −70 mV and Rb_o reduced to zero for 5-min periods. Following this, a test concentration of Rb was added to reactivate the Na pump. The cycle was completed by returning to the control solution, and this protocol was repeated with various test concentrations of Rb. *Arrow 1* denotes the effects of perturbations in the flow of superfusing solution. The record has been interrupted for the period indicated by the *dotted line* after the first reactivation with 10 mM Rb_o. [From Eisner et al. (14), with permission.]

by returning to 10 mM Rb. The results show that the greater the concentration of external Rb, the larger the electrogenic Na-pump current and the lower the steady-state level of a_{Na}^i. These data are replotted in Fig. 11 on a semilogarithmic axis and show that both a_{Na}^i (open symbols) and current (solid symbols) decay monoexponentially. It is also obvious that the higher the Rb_o, the faster the decay of both current and a_{Na}^i. Furthermore, at all Rb_o, the decay of current is roughly parallel to that of a_{Na}^i.

The observation that current and a_{Na}^i decay faster at higher Rb_o shows that higher concentrations of Rb produce greater activations of the Na pump. Another estimate of the degree of activation of the pump can be obtained from the steady-state level of a_{Na}^i. If it is assumed that the net passive Na influx is constant, then a lower level of a_{Na}^i would indicate a greater activation of the Na pump by external Rb. However, the two methods produce different estimates for the sensitivity of the Na pump to Rb_o. This is illustrated by comparing the effects of 4 and 10 mM Rb in Fig. 10. The fact that 10 mM Rb produces a much larger and faster-decaying current than does 4 mM Rb suggests that the Na

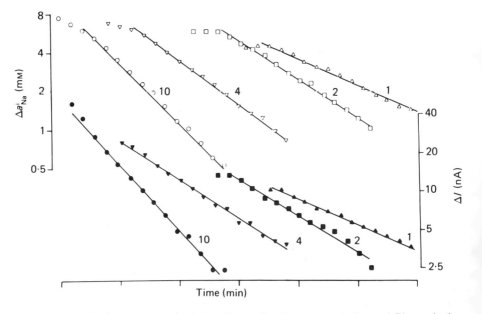

FIG. 11. Semilogarithmic analysis of the effects of various concentrations of Rb_o on both the decay of a_{Na}^i and the electrogenic Na-pump current transient. Data taken from Fig. 10 for a_{Na}^i and current are plotted as their levels above the steady-state value (a_{Na}^i, *open symbols;* I, *closed symbols,* respectively). Data are plotted on a logarithmic scale as a function of time. Different symbols correspond to different Rb_o: 10 mM (○,●); 4 mM (▽,▼); 2 mM (□,■); 1 mM (△,▲). The records for the various Rb_o have been arbitrarily shifted along the time axis to avoid overlap. The lines have been fitted by a least-squares method to all the points except those in the first 15 sec after the peak of the electrogenic pump current transient. [From Eisner et al. (14), with permission.]

pump is incompletely activated at 4 mM Rb. However, increasing Rb_o from 4 to 10 mM (arrow a) has no significant effect on a_{Na}^i. This result suggests that the Na pump is fully saturated with respect to external Rb at 4 mM. This discrepancy is reinforced by Fig. 12, which shows these two measures of pump activation plotted as a function of Rb_o. The steady-state a_{Na}^i data can be described by a K_m for external Rb of < 1 mM, whereas the data obtained during the decay of the electrogenic pump current are better fitted by a much lower affinity (higher K_m of about 4 mM). Possible reasons for this discrepancy will be addressed in the discussion.

The Magnitude of the Steady-State Na-Pump Current

In the experiments described above, the electrogenic Na-pump current has been measured after stimulating it by an increase of a_{Na}^i. It is of obvious interest to measure the magnitude of this current under normal conditions. A crude estimate can be obtained from experiments such as that illustrated in Fig. 7B.

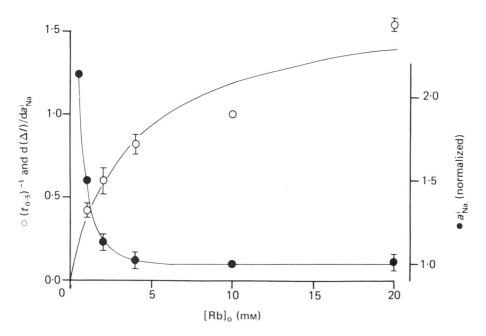

FIG. 12. The dependence on Rb_o of various measures of Na-pump activity (collected data from four experiments). *Solid circles* show the steady-state level of a_{Na}^i, which has been normalized by dividing by the value in 10 mM Rb_o in that particular experiment. *Open circles* show the average of two kinetic measures of pump activity. One is $(t_{0.5})^{-1}$ for the rate of fall of a_{Na}^i and the other $(d(\Delta I)/da_{Na}^i)$ is the slope of the relationship between the electrogenic pump current and a_{Na}^i (i.e., the slope of the line in experiments such as that of Fig. 7B), which is a measure of the activation of the Na pump. These kinetic values have also been normalized to the value in 10 mM Rb_o. [From Eisner et al. (14), with permission.]

This shows that there is a linear dependence of the electrogenic Na-pump current transient on a_{Na}^i. This relationship can be extrapolated back to zero a_{Na}^i as shown by the dotted line. The level of this intercept then gives the value of the extrapolated resting electrogenic pump current, in this case about 20 nA. This method suffers from the questionable validity of extrapolating the line back to zero a_{Na}^i. A simpler method is to block the Na pump with strophanthidin and then measure the inward shift of current produced by the abolition of the electrogenic Na-pump current. If strophanthidin had no other immediate effects on current, this should give a good estimate of the resting pump current. Such an experiment has been performed in both sheep (3,26) and canine (17) cardiac Purkinje fibers as well as in ventricular muscle (5). Isenberg and Trautwein (26) found that the magnitude of the inward shift of current was essentially independent of membrane potential. However, Cohen et al. (3) found a marked dependence on membrane potential, which was attributed to the effects of extracellular K accumulation, which is known to result from Na-pump inhibition (27). We have reexamined this point in the experiment of Fig. 13, which was performed in a K-free, 10 mM Rb solution to decrease effects of external K accumulation on membrane currents. In order to measure current, the membrane potential was stepped repeatedly to the 5 levels shown. Strophanthidin (10^{-5} M) was then added producing an inward shift of holding current as shown by the high-gain current trace (Fig. 13A). The low-gain current trace shows that the membrane conductance increases with the duration of exposure to strophanthidin along with the increase of a_{Na}^i. I–V relationships obtained at various times are shown in Fig. 13B. It is clear that the initial effect is a downward displacement of the I–V followed by a slower increase of conductance. This secondary increase of conductance occurs slowly and is therefore unlikely to be caused by extracellular K or Rb accumulation. A more likely explanation would be that the increase of a_{Na}^i or the consequent increase of intracellular calcium produces the increased membrane conductance (cf. ref. 11, Fig. 10). Isenberg and Trautwein (26) similarly found that, following an initial downward shift, the shape of the I–V relationship was distorted by cardioactive steroid application.

It seems likely that the initial, potential independent change of current reflects the contribution of the steady-state electrogenic pump current. In the present experiments the magnitude of this initial shift of current was typically 10 to 20 nA which, given the typical conductance of the fibers (0.5 to 1 μS), would contribute about 10 mV to the membrane potential.

DISCUSSION

Measurement of the Electrogenic Na-Pump Current

As noted in the introduction, there is little doubt that, in any tissue in which it has been studied, the Na pump is electrogenic. The alternative hypothesis that this apparent electrogenicity is entirely due to external potassium depletion

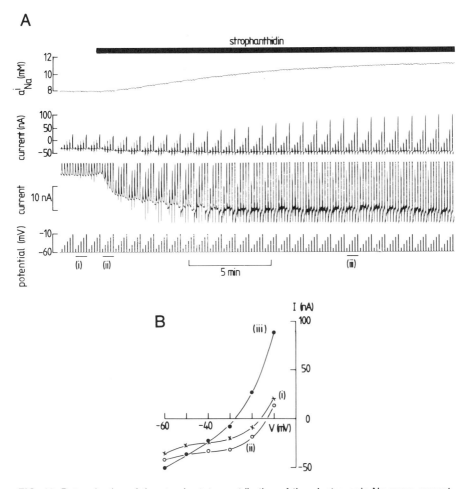

FIG. 13. Determination of the steady-state contribution of the electrogenic Na-pump current. **A:** Original records. Traces show (from *top* to *bottom*): a^i_{Na}, low-gain current, high-gain current, membrane potential. The membrane potential was held at -60 mV and 2-sec duration depolarizing voltage-clamp pulses were applied at 0.1 Hz. The pulses were applied in turn to -50, -40, -30, -20, and -10 mV, and then the cycle was repeated. The high-gain current trace shows only the current at -60 mV. The currents at other potentials are off-scale. Strophanthidin (10^{-5} M) was applied for the period shown. **B:** *I–V* relationships obtained at the times shown. The level of current indicated is that reached after the end of the 2-sec depolarizing pulse.

produced by an electroneutral sodium pump has had to be rejected (37). Nevertheless, little attention has been given to the related questions of (a) whether extracellular K depletion occurs and (b) whether this depletion produces significant changes of membrane current and potential. This problem has been studied in skeletal muscle, where it appears that temperature-dependent changes of Na-pump activity produce effects on the membrane potential attributable to both an electrogenic current and changes of extracellular K concentration (18).

The present results show that when the Na pump in cardiac muscle is reactivated by adding back K ions a large part of the resulting transient changes of membrane potential and current can be attributed to extracellular K depletion. It would be interesting to know whether such significant depletion currents are present in other tissues.

In contrast, when Rb is used to reactivate the Na pump, most of the change of membrane current appears to be produced by the electrogenic-pump current rather than by depletion effects. There is evidence to support this contention: (i) The current change produced by reactivating the Na pump with Rb is, more or less, independent of membrane potential. This is in contrast to the effects of external Rb depletion (Fig. 5). (ii) The value for the electrogenic fraction of Na pumping obtained here is similar to that found in other tissues. If a large fraction of the current produced by pump reactivation was due to depletion, then a different coupling ratio would have been expected. (iii) The value of this electrogenic fraction is independent of the external Rb concentration. Again, if significant depletion currents were present, a constant electrogenic fraction would not be expected.

Properties of the Electrogenic Na Pump in Cardiac Muscle

The present measurements of the electrogenic Na-pump current and its dependence on a_{Na}^i have established the following characteristics. (i) The pump rate is a linear function of a_{Na}^i in the experimentally accessible range of a_{Na}^i (about 3 to 20 mM). This is similar to the situation in the squid giant axon (2) and the snail neuron (36) but differs from the red blood cell, where clearly sigmoidal activation curves for internal Na are seen (19). This need not reflect any fundamental difference in the properties of the Na-binding site in the various tissues, since, even if the Na site itself has a very low real affinity for Na, a high apparent affinity for Na will be observed if one of the other reactions in the mechanism is rate limiting (34). (ii) The electrogenic fraction of Na extrusion is independent of a_{Na}^i and Rb_o, implying a constant stoichiometry of Na to K (Rb) transport. This agrees with the conclusion of work in the red cell (32) but not with that in some other preparations, such as the squid axon (30). Although we cannot exclude the possibility that the stoichiometry can be affected by more extreme changes of concentration, the cardiac Na pump appears to operate with a fixed stoichiometry over the experimentally available range. (iii) The activation of the Na pump by external Rb or K appears to be complicated. We have obtained a high apparent affinity ($K_m < 1$ mM) for Rb_o from measurements of the steady-state a_{Na}^i and a lower affinity (K_m about 4 mM) from kinetic measurements of the rate of fall of a_{Na}^i and the electrogenic Na-pump current on reactivating the Na pump. These values should be compared with the more typical value for the K_m of 1 to 2 mM found in other mammalian preparations (24). It should, however, be noted that other workers find no difference between the steady-state and kinetic values for the K_m (22). A plausible explanation for the discrepant values of the K_m obtained from steady-state and kinetic measurements is that the large increase of pump rate during reactivation by external Rb produces a local extracellular depletion of Rb. It is therefore possible that when the external Rb concentration is 4 mM the local concentration may be only 1 mM. This extracellular depletion will become less significant as a_{Na}^i and the Na-pump rate fall during pump reactivation, and the steady-state relationship between a_{Na}^i and Rb_o will give a more accurate representation of the activation curve for Rb_o. The experimental evidence presented for extracellular depletion (Fig. 3) is consistent with this hypothesis. Furthermore, in the canine Purkinje fiber, where depletion problems may be less important (35), a K_m of 1 mM is obtained from kinetic measurements (15). However, if depletion of extracellular K (Rb) is to explain the low affinity obtained from the kinetic measurements, it is not clear why a linear relationship is obtained between pump current and a_{Na}^i (Fig. 7B). This relationship would be expected to become steeper as a_{Na}^i falls and the pump rate decreases. In conclusion, although it is tempting to explain the discrepancy of the K_m in terms of depletion, this hypothesis cannot explain all the observations.

The Steady-State Electrogenic Na-Pump Current

The present work has also analyzed the steady-state component of the electrogenic Na-pump current. In agreement with previous work (5,17,26), it appears that this current contributes significantly to the resting potential of cardiac muscle. It is, therefore, worth considering how this steady-state electrogenic Na-pump current will be affected by inhibitors of the Na pump such as cardioactive steroids. At first sight it might appear that an agent that inhibited a certain fraction of the Na pumps would decrease the steady-state electrogenic Na-pump current by the same fraction. This is, however, not the case, since, in the steady state, there must be no net movement of Na or K across the cell membrane and the active Na-extrusion systems must balance the influx of Na into the cell (cf. ref. 30). There are three cases to consider. (i) The simplest case is that in which the Na pump is the only Na-extrusion system and balances a constant passive influx of Na. In this case, inhibiting a fraction of the Na pumps will produce an increase of a_{Na}^i until the pump flux is again equal to the passive influx. There will be no steady-state change of pump rate. (ii) It is likely that the increase of Ca_i^{2+} resulting from Na-pump inhibition will increase the Na permeability (4) and will, therefore, further increase a_{Na}^i. In this case, the Na-pump activity will increase to above control levels, and the effect of inhibiting a fraction of the Na pumps will be to *increase* the total Na-pump rate above its control level. This stimulation of total pump activity depends on a rise of a_{Na}^i and should not be confused with the stimulation suggested to be produced by low concentrations of ouabain (3). (iii) The final case is one in which, when the Na pump is inhibited, other Na-extrusion systems such as Na/Ca exchange (6) take over and extrude Na from the cell. In this case the rise of a_{Na}^i will not compensate for the inhibition of the Na pump and the total steady-state electrogenic Na-pump current will fall. The actual result of applying concentrations of cardioactive steroids, which inhibit a fraction of the Na pumps, will be a mixture of the above effects. It is difficult, therefore, to predict the effect of moderate Na-pump inhibition on the steady-state electrogenic Na-pump current and, therefore, on the electrophysiology of the cardiac cell. It should, however, be noted that the change of pump current is probably much less than the current changes produced by secondary effects of pump inhibition (ref. 9 and Fig. 13).

ACKNOWLEDGMENTS

The work described in this chapter was supported by grants from the following organizations: British Heart Foundation and MRC (D.A.E.); March of Dimes Birth Defects Foundation, NIH (HL-25675), Burroughs Wellcome Fund, and the American Heart Association and its Maryland Affiliate (W.J.L.); M.R.C. (R.D.V.-J.).

REFERENCES

1. Adrian, R. H., and Slayman, C. L. (1966): Membrane potential and conductance during transport of sodium, potassium and rubidium in frog muscle. *J. Physiol. (Lond.)*, 184:970–1014.
2. Baker, R. F., Blaustein, M. P., Keynes, R. D., Manil, J., Shaw, T. I., and Steinhardt, R. A. (1969): The ouabain-sensitive fluxes of sodium and potassium in squid giant axons. *J Physiol. (Lond.)*, 200:459–496.
3. Cohen, I., Daut, J., and Noble, D. (1976): An analysis of the actions of low concentrations of ouabain on membrane currents in Purkinje fibres. *J. Physiol. (Lond.)*, 260:75–103.
4. Colquhoun, D., Neher, E., Reuter, H., and Stevens, C. F. (1981): Inward current channels activated by intracellular Ca in cultured cardiac cells. *Nature*, 294:752–754.
5. Daut, J., and Rüdel, R. (1981): Cardiac glycoside binding to the Na/K-ATPase in the intact myocardial cell: Electrophysiological measurement of chemical kinetics. *J. Mol. Cell. Cardiol.*, 13:777–782.
6. Deitmer, J. W., and Ellis, D. (1978): Changes in the intracellular sodium activity of cardiac Purkinje fibres produced by calcium and other divalent cations. *J. Physiol. (Lond.)*, 277:437–453.
7. Deitmer, J. W., and Ellis, D. (1978): The intracellular sodium activity of cardiac Purkinje fibres during inhibition and re-activation of the sodium-potassium pump. *J. Physiol. (Lond.)*, 284:241–259.
8. Délèze, J. (1960): Possible reasons for drop of resting potential of mammalian heart preparations during hypothermia. *Circ. Res.*, 8:553–557.
9. Eisner, D. A., and Lederer, W. J. (1979): The role of the sodium pump in the effects of potassium depleted solutions on mammalian cardiac muscle. *J. Physiol. (Lond.)*, 294:279–301.
10. Eisner, D. A., and Lederer, W. J. (1979): Does sodium pump inhibition produce the positive inotropic effects of strophanthidin in mammalian cardiac muscle? *J. Physiol. (Lond.)*, 296:75–76P.
11. Eisner, D. A., and Lederer, W. J. (1980): Characterization of the electrogenic sodium pump in cardiac Purkinje fibres. *J. Physiol. (Lond.)*, 303:441–474.
12. Eisner, D. A., and Lederer, W. J. (1980): The relationship between sodium pump activity and twitch tension in cardiac Purkinje fibres. *J. Physiol. (Lond.)*, 303:475–494.
13. Eisner, D. A., Lederer, W. J., and Vaughan-Jones, R. D. (1981): The dependence of sodium pumping and tension on intracellular sodium activity in voltage-clamped sheep Purkinje fibres. *J. Physiol. (Lond.)*, 317:163–187.

14. Eisner, D. A., Lederer, W. J., and Vaughan-Jones, R. D. (1981): The effects of rubidium ions and membrane potential on the intracellular sodium activity of sheep Purkinje fibres. *J. Physiol. (Lond.)*, 317:189–205.
15. Gadsby, D. C. (1980): Activation of electrogenic Na^+/K^+ exchange by extracellular K^+ in canine cardiac Purkinje fibers. *Proc. Natl. Acad. Sci. USA*, 76:1783–1787.
16. Gadsby, D. C., and Cranefield, P. F. (1979): Electrogenic sodium extrusion in cardiac Purkinje fibers. *J. Gen. Physiol.*, 73:819–837.
17. Gadsby, D. C., and Cranefield, P. F. (1979): Direct measurements of changes in sodium pump current in canine cardiac Purkinje fibers. *Proc. Natl. Acad. Sci. USA*, 76:1783–1787.
18. Gadsby, D. C., Niedergerke, R., and Ogden, D. C. (1977): The dual nature of the membrane potential increase associated with the activity of the sodium/potassium exchange pump in skeletal muscle fibres. *Proc. R. Soc. Lond. [Biol.]*, 198:463–472.
19. Garay, R. P., and Garrahan, P. J. (1973): The interaction of sodium and potassium with the sodium pump in red cells. *J. Physiol. (Lond.)*, 231:297–325.
20. Garrahan, P., and Glynn, I. M. (1967): The stoichiometry of the sodium pump. *J. Physiol. (Lond.)*, 192:217–235.
21. Glitsch, H. G., Grabowski, W., and Thielen, J. (1978): Activation of the electrogenic sodium pump in guinea-pig atria by external potassium ions. *J. Physiol. (Lond.)*, 276:515–524.
22. Glitsch, H. G., Kampmann, W., and Pusch, H. (1981): Activation of active Na transport in sheep Purkinje fibres by external K or Rb ion. *Pfluegers Arch.*, 391:28–34.
23. Glitsch, H. G., Pusch, H., and Venetz, K. (1976): Effects of Na and K ions on the active Na transport in guinea-pig auricles. *Pfluegers Arch.*, 365:29–36.
24. Glynn, I. M. (1956): Sodium and potassium movements in human red cells. *J. Physiol. (Lond.)*, 134:278–310.
25. Hiraoka, M., and Hecht, H. H. (1973): Recovery from hypothermia in cardiac Purkinje fibres: Considerations for an electrogenic mechanism. *Pfluegers Arch.*, 339:25–36.
26. Isenberg, G., and Trautwein, W. (1974): The effect of dihydro-ouabain and lithium ions on the outward current in cardiac Purkinje fibers. *Pfluegers Arch.*, 350:41–54.
27. Kunze, D. L. (1977): Rate-dependent changes in extracellular K^+ in the rabbit atrium. *Circ. Res.*, 41:122–127.
28. Müller, P. (1965): Potassium and rubidium exchange across the surface membrane of cardiac Purkinje fibres. *J. Physiol. (Lond.)*, 177:453–462.
29. Mullins, L. J., and Brinley, F. J. (1969): Potassium fluxes in dialysed squid axons. *J. Gen. Physiol. (Lond.)*, 53:704–740.
30. Mullins, L. J., and Noda, K. (1963): The influence of sodium-free solutions on the membrane potential of frog muscle fibers. *J. Gen. Physiol.*, 47:117–132.
31. Page, E., and Storm, S. (1965): Cat heart muscle *in vitro*. VIII. Active transport of sodium in papillary muscle. *J. Gen. Physiol.*, 48:957–972.
32. Post, R. L., and Jolly, P. C. (1957): The linkage of sodium potassium and ammonium active transport across the human erythrocyte membrane. *Biochim Biophys. Acta*, 25:118–128.
33. Rang, H. P., and Ritchie, J. M. (1968): On the electrogenic sodium pump in mammalian non-myelinated nerve fibres and its activation by various external cations. *J. Physiol. (Lond.)*, 196:183–221.
34. Sachs, J. R. (1977): Kinetic evaluation of the Na-K pump reaction mechanism. *J. Physiol. (Lond.)*, 273:489–514.
35. Sommer, J. R., and Johnson, E. A. (1968): Cardiac muscle. A comparative study in Purkinje fibres and ventricular fibres. *J. Cell. Biol.*, 36:497–526.
36. Thomas, R. C. (1969): Membrane current and intracellular sodium changes in a snail neurone during extrusion of injected sodium. *J. Physiol. (Lond.)*, 201:495–514.
37. Thomas, R. C. (1972): Electrogenic sodium pump in nerve and muscle cells. *Physiol. Rev.*, 52:563–594.
38. Thomas, R. C. (1978): *Ion-Sensitive Micro-Electrodes*. Academic Press, New York and London.
39. Vassalle, M. (1970): Electrogenic suppression of automaticity in sheep and dog Purkinje fibres. *Circ. Res.*, 27:361–377.
40. Weidmann, S. (1956): *Elektrophysiologie der Herzmuskelfaser*. Huber, Bern.

Electrogenic Transport: Fundamental Principles and Physiological Implications, edited by Mordecai P. Blaustein and Melvyn Lieberman. Raven Press, New York © 1984.

Influence of the Sodium Pump Current on Electrical Activity of Cardiac Cells

David C. Gadsby

Laboratory of Cardiac Physiology, The Rockefeller University, New York, New York 10021

The normal, rhythmic beating of the heart that enables it to function as a pump depends on the orderly propagation of action potentials throughout the cardiac syncytium. The membrane currents that give rise to these action potentials comprise predominantly passive fluxes of ions down their electrochemical potential gradients, so that if the electrical activity is to persist, the ionic gradients must be maintained. In the case of sodium and potassium ions, this is largely accomplished by the Na/K exchange pump. In order to keep the intracellular concentrations of Na ($[Na]_i$) and K ($[K]_i$) approximately constant, the active, pumped fluxes of Na and K must be similar in size, but opposite in direction, to the passive fluxes. This homeostatic function of the Na/K pump is well served by the known dependence of the rate of pumped Na efflux on the level of $[Na]_i$; the dependence is roughly linear over a limited, but physiological, range of concentrations (10,14,39). Thus, an imbalance between influx and efflux of sodium ions that tends to raise $[Na]_i$ (e.g., a sudden increase in Na influx) will thereby increase pumped efflux: $[Na]_i$ and Na efflux will both rise toward new steady-state levels at which the Na fluxes are again in balance. Similarly, an imbalance that lowers $[Na]_i$ will result in reduced Na efflux that will, in turn, tend to counteract the original imbalance. By maintaining the transmembrane concentration gradients for Na and K, the Na/K pump exerts an important, but *indirect*, influence on the electrical activity of cardiac cells.

In addition, the Na/K pump has a *direct* influence on electrical activity because it is electrogenic. It transports more Na ions out than K ions in across the cell membrane and so continuously generates an outward (hyperpolarizing) component of membrane current (for reviews, see 12,24,27). The stoichiometric ratio of pumped movements of Na and K appears to remain approximately constant under physiological conditions in spite of experimentally induced changes in pump rate of up to several fold (14,15,18,27) so that the pump current remains proportional to the pump rate; since the ratio of the pumped Na and K fluxes seems to be about 3 Na : 2 K (cf. 14,40), this means that in cardiac cells, the Na/K pump must generate, over any given period of time, a substantial current equivalent in size to one-third of the time average of the inward Na currents associated with the electrical activity of the heart. Of course, the Na current underlying the upstroke of the action potential in Purkinje or working myocardial cells flows for only 1 msec and so is much more intense than the pump current, which presumably flows steadily throughout the cardiac cycle (about 1 sec). Nevertheless, the pump current may still be expected to influence the electrical activity of those cells, because the action potential plateau lasts for about one-third of the cardiac cycle, and at plateau potentials the membrane slope conductance is very low, so that a small current can make a substantial contribution to the shape of the action potential (21,31). In addition to their effects on the cardiac action potential, changes in the size of the pump current are known to influence the level of the resting, or maximum diastolic, potential as well as the rate of firing of pacemaker cells (see 24 for references).

The influence of the pump current on the electrical activity of cardiac cells is most easily investigated by determining the electrical effects of changes in pump rate. Experimentally, the pump rate can be diminished by reducing the extracellular concentration of K ions, ($[K]_o$), or by applying a cardiac steroid, and it can be enhanced by causing an increase in $[Na]_i$. Since, as previously mentioned, a balance between passive Na influx and active Na efflux normally maintains $[Na]_i$, it can be made to rise either by augmenting Na entry, by increasing the firing rate (4,8,36,42), or by impairing Na extrusion, e.g., by lowering $[K]_o$ (10,14,16,25,35). Sudden return to the original firing rate, or restoration of the original $[K]_o$, then unmasks the effects of the elevated $[Na]_i$ and of the consequent increase in pump rate, the latter effects being distinguishable by their sensitivity to cardiac steroids. The resulting imbalance between Na influx and efflux leads, as described above, to a gradual fall in $[Na]_i$, and hence in pump rate, until $[Na]_i$ has returned to its steady-state level at which influx and efflux are again balanced. To ensure that the extracellular K concentration, and hence the pump rate, can be changed rapidly, we select small preparations of cardiac cells and study them in a fast-flow system based on that designed by Hodgkin and Horowicz (29).

MATERIALS AND METHODS

For most of the experiments to be described here we used unbranched bundles of Purkinje cells, usually ≤ 2 mm long and ≤ 200 μm in overall diameter. These were dissected from the right ventricles of dog hearts, because the sheath of connective tissue surrounding the bundles generally appears to be thinner in the right than in the left ventricle, so that insertion of microelectrodes is expected to be easier. An additional advantage of these small preparations (see Discussion) is that preliminary morphological investigation suggests that, despite considerable variability in the size, number, and distribution of the Purkinje cells in the core of the bundle, those cells are more loosely packed than the cells in sheep and calf Purkinje fiber bundles (13,38; T. D. Pham and D. C. Gadsby, *unpublished observations*). Experiments to investigate the effects of pump activity on bursts of triggered action potentials (see Figs. 12 to 14) were carried out on small strips of atrial muscle, 2 to 4 mm long and about 1 mm wide, dissected from the canine coronary sinus; the coronary sinus was first isolated from the heart, then cut open along its length and pinned flat (44,45).

Both kinds of preparations are suspended between two fine (100-μm diameter) insect pins in the center of the narrow channel of the fast-flow chamber (Fig. 1). The hook-shaped pins are pushed into the Sylgard (#184, Dow Corning Corp., Midland, Michigan) floor of the channel, and the preparations are positioned in midstream, near the tops of the pins, so that they are exposed to the continuously moving fluid on all sides, except where they rest lightly on a 100-μm wide support to facilitate microelectrode impalement (see Fig. 1). All

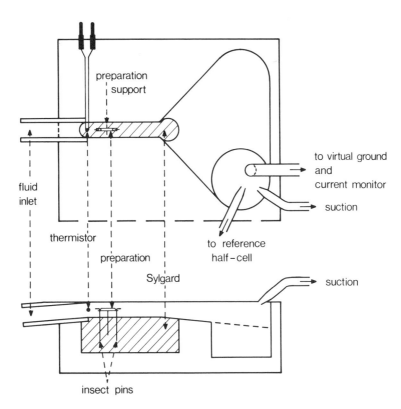

FIG. 1. Schematic diagram of fast-flow chamber. Superfusate continuously enters via the inlet at the left and flows along the narrow channel, which has a 3 mm square cross-section, before being removed by suction at the right. The fluid inlet is angled so that a cover slip can be slid over the channel to close it when necessary. Fibers are held in midstream at the tops of hook-shaped insect pins, pushed through connective tissue at each end of the fiber into the Sylgard floor of the channel. A third stainless steel insect pin, bent into a rectangular hoop, is positioned beneath the fibers to afford some support during microelectrode impalements. Downstream from the preparations, usually near the suction tube, a 3 M KCl-agar bridge connects the chamber, via a Ag/AgCl electrode, to an operational amplifier that holds the bath at virtual ground and monitors applied currents; nearby, a flowing 3 M KCl bridge connects the chamber to the reference half-cell which incorporates a sintered Ag/AgCl/Pt-black pellet. [From Gadsby and Cranefield (22), with permission.]

superfusion solutions are preheated and oxygenated before entering the channel, under gravity feed, via a system of valves that allows rapid selection of any one of 8 solutions to flow through the chamber: one output of a 2-position valve (#86410, Hamilton Co., Reno, Nevada) is attached to the channel inlet by a few millimeters of silicon rubber tubing, and the other output is connected to a drain. The two inputs to that valve are supplied by the outputs from two, 4-position, distribution valves (#86414, Hamilton) fitted with spring-loaded stops. At the normally used flow rate, 5 ml/min, solution changes at the middle of the channel occur with a half-time of about 0.5 sec, as determined from the time course of establishment of the liquid junction potential on switching between Cl-containing and Cl-free fluids. The junction potential was recorded between two broken microelectrodes, filled with 3 M KCl and Tyrode's solution, respectively, and placed in midstream with their tips close together.

Conventional glass microelectrodes filled with 3 M KCl are used to record membrane potentials and to inject current. The current-injecting microelectrode is inserted at the midpoint of the fiber and the voltage-recording microelectrode is inserted roughly one-third of the distance toward the fiber end. One operational amplifier is used to clamp the membrane potential to the chosen level and a second one to monitor the applied current; currents are filtered (time constant 100 msec) before being displayed on oscilloscopes and chart recorder.

The Cl-containing Tyrode's solution used during dissection and throughout some experiments contained: 137 mM NaCl; 4 mM KCl; 12 mM $NaHCO_3$; 1.8 mM NaH_2PO_4; 0.5 mM $MgCl_2$; 2.7 mM $CaCl_2$; 5.5 mM dextrose. This solution was bubbled with 95% O_2/5% CO_2 mixture while being preheated. The low Cl solution used in some experiments contained: 146 mM Na isethionate; 4 mM K methylsulfate; 5 mM Hepes (N-2-hydroxyethylpiperazine-N'-2-ethanesulfonic acid, pH 7.3); 0.5 mM $MgCl_2$; 2.7 mM Ca methanesulfonate; 5.5 mM dextrose. This solution was bubbled with oxygen during preheating. In both normal- and low-Cl solutions, K concentration was varied by substitution with Na. All solutions were filtered before use; experiments were carried out at $36.5 \pm 1°C$ (monitored with the small thermistor bead shown in Fig. 1). The cardiac steroid, 3-acetylstrophanthidin, was a gift from Eli Lilly & Co. (Indianapolis, Indiana) and was kept for several weeks as a refrigerated stock solution of 5 mM in ethanol. To induce delayed afterdepolarizations and bursts of triggered action potentials in coronary sinus preparations, L-norepinephrine bitartrate (Levophed®, Breon Laboratories, Sterling Drug Inc., New York) was added to superfusion solutions from a stock solution of concentration 1 mg/ml to produce the required uniform concentrations. To help prevent oxidation of the catecholamine, 10 μM Na_2EDTA was added to all superfusing solutions in experiments with coronary sinus preparations (17).

RESULTS

Figure 2a (middle trace) illustrates typical effects on the membrane potential of a small Purkinje fiber of a brief exposure to K-free solution (indicated by

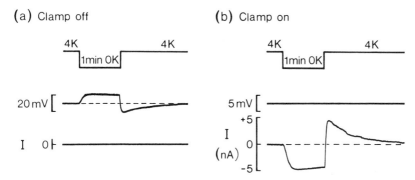

FIG. 2. Changes in membrane potential in the absence of applied current (a) and in net current (labeled I) under voltage clamp (b) recorded consecutively, in the same Purkinje fiber, in response to 1-min exposures to K-free solution, indicated by the *upper lines*. Dashed line in (a) indicates the steady resting potential in 4 mM K, low Cl solution, which was at the lower level, −33 mV; this was chosen as the holding potential for (b) so that, in the steady state, the holding current was zero (*dashed line*). Note the different voltage calibrations in (a) and (b). Current flowing outwards across the cell membrane, i.e., hyperpolarizing current, is defined as positive. [From Gadsby and Cranefield (20), with permission.]

the upper line), used to temporarily inhibit the Na/K pump to cause a rise in [Na]$_i$ and, hence, a temporary stimulation of the pump on returning to 4 mM K solution. This fiber had already been impaled with two microelectrodes, but during this run the voltage-clamp amplifier was switched off and the membrane potential changes were recorded in the absence of any applied current (see lower trace, Fig. 2a). In 4 mM K, low Cl Tyrode's solution, the N-shaped, steady-state current–voltage relationship of these small, quiescent fibers often intersects the voltage axis at 3 points, 2 of which occur in regions of positive-slope conductance and are, therefore, possible resting potentials. The higher level of resting potential is near −90 mV, and the lower level is usually near −40 mV (19). The steady resting potential of the fiber in Fig. 2, at 4 mM [K]$_o$, was −33 mV (dashed line in Fig. 2a), and there was a rapid, maintained depolarization on switching to K-free solution. On returning to 4 mM [K]$_o$ a transient hyperpolarization was recorded, which reached a peak in a few seconds and then slowly decayed over the next minute or two. The voltage clamp was then switched on to hold the membrane potential at its steady resting, i.e., zero current, level and the brief exposure to K-free solution was repeated (Fig. 2b). With the membrane potential constant, a net inward current was recorded during the exposure to zero [K]$_o$ and, after switching back to 4 mM [K]$_o$, a transient outward current was recorded, which peaked in a few seconds and then declined with a single exponential time course. In terms of our working hypothesis, the inward shift of holding current in K-free solution is at least partly due to abolition of steady-state outward pump current, and the subsequent outward current transient in 4 mM [K]$_o$ reflects the temporarily increased pump rate during extrusion of the small Na load accumulated in the K-free fluid.

To verify that the transient outward current is, indeed, caused by enhanced pump activity, the experiment was repeated after blocking the Na/K pump by application of a maximal concentration of a cardiac steroid, in the present case, acetylstrophanthidin (Fig. 3). The time course of pump block by 5 μM acetylstrophanthidin in 4 mM K, low Cl solution is illustrated by the rapid, inward shift of holding current in the lower trace of Fig. 3a, reflecting abolition of the steady-state component of Na/K pump current. Pump block appears to be virtually complete by the end of the 2-min exposure to acetylstrophanthidin and is readily reversible on washing out the drug. The effects of 2 μM acetylstrophanthidin on the transient outward current following brief exposures to K-free solution are shown in Fig. 3b. The top line indicates the timing of the 2-min periods of K-free superfusion. The upper current record shows the usual inward shift of holding current on switching into zero [K]$_o$, and the usual transient outward current after returning to 4 mM [K]$_o$. The middle current record begins 3 min after the start of application of acetylstrophanthidin, the inward shift of the holding current (with respect to the control records, above and below) presumably reflecting abolition of steady-state pump current.

The switch to K-free solution, still in the presence of acetylstrophanthidin, caused a further, smaller inward shift in the holding current, but only to about the same level obtained in the previous and subsequent runs in the absence of

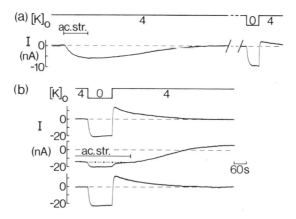

FIG. 3. Effects of micromolar concentrations of acetylstrophanthidin (applied during the periods indicated by the bars labeled ac. str.) on net membrane current in Purkinje fibers voltage-clamped at the lower resting potential in 4 mM K, low Cl solution. **(a)** Acetylstrophanthidin (5 μM) was applied for 2 min to a fiber held at −32 mV. After the net current had returned to zero (the break in the current trace indicates omission of a 5-min section of record), the fiber was exposed to zero [K]$_o$ for 1 min, as indicated by the *top line*. **(b)** Net current changes recorded in a fiber held at −40 mV during 3 consecutive 2-min exposures to K-free fluid. The timing of the step changes in [K]$_o$ (in mM) are indicated by the *upper line*. Acetylstrophanthidin (2 μM) was added 3 min before the start of the second run and was washed out at the end of the bar. The 60-sec time calibration applies to both (a) and (b). *Broken lines* in (a) and (b) indicate the levels of zero net current, and the *dot-dash line* in (b) indicates the steady current level in 4 mM [K]$_o$ in the presence of acetylstrophanthidin. [From Gadsby and Cranefield (20), with permission.]

acetylstrophanthidin (see below). The important point is that, in the presence of the cardiac steroid, on switching back to 4 mM $[K]_o$, the holding current returned monotonically to the level recorded just before the 2-min exposure to zero $[K]_o$, in spite of the increase in $[Na]_i$ that must be expected to have occurred during that exposure. The acetylstrophanthidin was then washed out and the holding current slowly moved in the outward direction, even becoming net outward for some minutes, presumably reflecting stimulation of unblocked pump sites by the raised $[Na]_i$ resulting from the prolonged pump inhibition in the presence of the drug.

Abolition by acetylstrophanthidin of the transient overshoot of the holding current argues strongly that the overshoot is caused by enhanced pump activity and is not caused, for example, by a temporary reduction in inward Na current, secondary to the diminished Na gradient resulting from the raised $[Na]_i$. Neither can the transient outward current reflect a temporary increase in a Ca-activated K conductance (30,37), or a Ca-activated nonspecific conductance (7,32), due to a rise in cytoplasmic Ca ion concentration, mediated by enhanced Na/Ca exchange (26) consequent to the raised $[Na]_i$, because none of these mechanisms is expected to be affected by acetylstrophanthidin. Having established that the transient outward current is caused by enhanced Na/K pump activity, can we rule out that it is an indirect effect, secondary to K depletion just outside the cell membrane due to rapid, pumped uptake of K into the cells? The records in Figs. 2 and 3 show clearly that, at holding potentials close to the lower level of resting potential, experimental reduction of $[K]_o$ results in a monotonic inward shift of the holding current with a time course (several seconds) that presumably reflects diffusion equilibration of K in the extracellular spaces. If the holding current at all $[K]_o$ levels < 4 mM is more inward than that at 4 mM $[K]_o$, then, obviously, the transient outward current cannot be attributed to temporary extracellular K depletion. We cannot argue that there is no pump-induced K depletion (see Discussion) but, simply, that some other mechanism must underlie the transient outward current; the simplest explanation is that it reflects the transient increment in pump current generated during electrogenic extrusion of the increment in $[Na]_i$.

The absence of any current overshoot on switching back to 4 mM $[K]_o$ from zero $[K]_o$ in the presence of 2 μM acetylstrophanthidin indicates that the Na/K pump was completely inhibited by that concentration of the cardiac steroid. Therefore, insofar as indirect effects of pump inhibition can be ignored, the amplitudes of the inward shifts in holding current on applying acetylstrophanthidin in Figs. 3a and b indicate the sizes of the steady-state pump current in those two preparations. Note that in both cases, a larger inward current shift was recorded on switching to K-free solution than on adding the cardiac steroid (Figs. 3a and b). Since the pump appears to be completely blocked by micromolar concentrations of acetylstrophanthidin, the additional inward current shift on withdrawal of $[K]_o$ presumably reflects reduction of outward K current flowing through inwardly rectifying K channels (1,33). Moreover, the records in Fig.

3b reveal that the absolute level of the holding current in K-free fluid was similar whether or not acetylstrophanthidin was present, suggesting that the pump was already completely inhibited by the absence of external K.

Although the inward shift of holding current caused by micromolar concentrations of cardiac steroids is consistent with the expected abolition of steady pump current, it should be mentioned that nanomolar concentrations sometimes cause a small outward current shift, as illustrated in Fig. 4. One possible explanation for this effect is that it reflects an increase in the steady pump current, so that this preliminary result might be taken to support other indirect evidence that low concentrations (in the nanomolar range) of certain cardiac steroids can stimulate, rather than inhibit, the Na/K exchange pump (2,5,11,23,34). Very recently, more direct evidence for a stimulatory effect on the Na/K pump of low concentrations of certain cardiac steroids has been obtained in experiments on partially purified Na/K ATPase from dog kidney (28). Moreover, Hamlyn's results suggest strongly that the mechanism for this effect is a disinhibition of the Na/K ATPase, rather than a direct stimulation of the native enzyme (cf. 34).

Measurements with Na-sensitive microelectrodes in sheep Purkinje fibers have shown that, on returning to K-containing solution following brief exposures to zero $[K]_o$, the increment in intracellular Na activity declines with a single exponential time course (10), and that the rate constants for the exponential

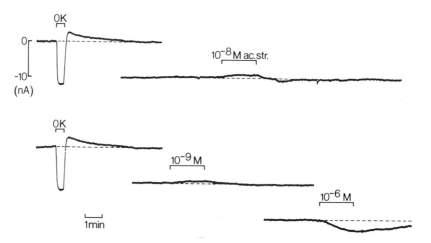

FIG. 4. Outward shift of net current caused by application of nanomolar concentrations of acetylstrophanthidin (during bars labeled ac. str.) in a Purkinje fiber held at its lower resting potential of −33 mV in 4 mM K, low Cl solution. As usual, the two 30-sec exposures to K-free fluid (bars labeled 0K) were followed by transient increments in pump current, while the application of 1 μM acetylstrophanthidin **(bottom right)** caused a net inward current shift, presumably due to reduction of steady, background pump current. One of several possible explanations for the small outward current shifts seen in the *upper records* at the **right** is that nanomolar concentrations of acetylstrophanthidin might stimulate, rather than inhibit, the Na/K exchange pump. [From Gadsby and Cranefield (22), with permission.]

decay of the increment in Na activity and of the increment in pump current are the same (14,25). These results suggest that, at least over a limited range of $[Na]_i$, (a) the increment in the Na/K pump rate is proportional to the increment in $[Na]_i$, while the pump rate constant is independent of $[Na]_i$, and (b) the coupling ratio of Na/K exchange remains constant, i.e., independent of $[Na]_i$ (cf. 39). In contrast to this independence of the rate constant for Na extrusion from the level of $[Na]_i$, the rate constant is markedly dependent on the level of $[K]_o$ below about 4 mM, as shown in Fig. 5. The graph shows the $[K]_o$ dependence of the average rate constants for the exponential decay of transient increments in pump current elicited at the K concentrations given on the abscissa. After equilibrating a fiber at a given K concentration, increments in pump current were elicited by several brief exposures to K-free fluid of various durations followed, each time, by return to the K-containing solution. The measurements were repeated over a range of K concentrations. The mean rate constant for pump-current decay at each $[K]_o$ level was normalized to the mean rate constant obtained in that fiber at 4 mM $[K]_o$; the points on the graph show those normalized rate constants averaged over 16 fibers. The Na/K pump seems to be half-maximally activated at a $[K]_o$ of approximately 1 mM, although it should be kept in mind that this number will represent an overestimate if any extracellular K depletion is caused by the enhanced pump activity that underlies the increments in pump current.

Armed with this knowledge of some of the basic characteristics of the current generated by the Na/K pump in canine Purkinje fibers, we are now in a position

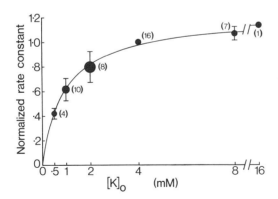

FIG. 5. Dependence on extracellular K concentration of the exponential rate constant for decay of increments in pump current elicited following brief exposures to K-free fluid. Rate constants were estimated from semilogarithmic plots of pump current increments recorded in the presence of different $[K]_o$ levels. Those rate constants were normalized with respect to the rate constant obtained at 4 mM $[K]_o$ in the same fiber, and the mean values are shown in the graph. *Circle diameters* = 2 × SEM, *vertical bars* indicate ± SD, and the numbers in parentheses indicate the number of fibers contributing to each mean. The smooth curve is a hyperbolic function, saturating at a normalized rate constant of 1.20 and half-maximal at $[K]_o = 0.94$ mM. [From Gadsby (18), with permission.]

to investigate the possible influences of changes in pump current on their electrical activity. Isenberg and Trautwein (31) had already shown that in sheep Purkinje fibers sudden reduction of the steady pump current, on inhibition of the pump by application of another fast-acting cardiac steroid, dihydroouabain, caused marked prolongation of the action potential, indicating that pump current normally plays a significant role in repolarization of the action potential. The results illustrated in Fig. 6 show that temporary stimulation of the Na/K pump, following a 6-min exposure to K-free solution, causes a marked, transient reduction in action potential duration associated with a transient increase in the resting, or maximum diastolic, potential. This fiber was stimulated at a rate of 1.2 Hz while in the 4 mM $[K]_o$, Cl-containing Tyrode's solution, but the stimulator was switched off while the fiber was depolarized in K-free fluid. The representative action potentials at the bottom were photographed at various times before and after exposure to zero $[K]_o$, and their durations, normalized

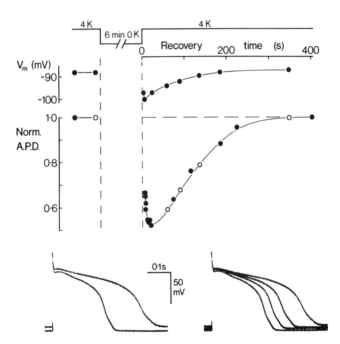

FIG. 6. Changes in action potential duration (APD) recorded in 4 mM K, Cl-containing Tyrode's solution, caused by a preceding 6-min exposure to K-free fluid, as indicated by the *top line*. The fiber was stimulated at 1.2 Hz except while depolarized in the K-free solution. The **top graph** shows diastolic potentials measured in 4 mM $[K]_o$, and the **bottom graph** shows corresponding durations of action potentials, normalized to the steady control duration which is indicated by the *broken horizontal line*. *Open circles* give the durations of the representative action potentials illustrated below the graphs; the time calibration for these action potentials marks the zero potential level. The pump-induced shortening of the action potential occurs simultaneously with the increase in membrane potential. [From Gadsby and Cranefield (21), with permission.]

to the control duration obtained before K-free superfusion, are indicated by the open circles in the graph. The two action potentials superimposed at the left are a control action potential recorded before Na loading (longer duration and more positive diastolic potential) and one recorded 60 sec after the return to 4 mM $[K]_o$. At the right, the shortened action potential recorded after 60 sec is superimposed on those recorded after 90, 140, and 350 sec of recovery at 4 mM $[K]_o$. The upper set of points in the graph shows the level of membrane potential recorded between action potentials, and it is clear that the resting potential and the duration of the action potential both recover their control values over a similar period of time, presumably the time required for elimination of the increment in $[Na]_i$. Although the transient hyperpolarization could be explained by a temporary depletion of extracellular K, due to pump stimulation, the concomitant shortening of the action potential could not be, because it is well known that lowering $[K]_o$ lengthens the action potential of Purkinje fibers (33,41,43). The simplest explanation, therefore, is that both the hyperpolarization and action-potential shortening are caused by a transient increase in pump current of the kind already described.

Although briefly exposing small preparations to K-free solution provides a convenient and reliable means of experimentally raising $[Na]_i$ in order to study the effects of the resulting, temporary stimulation of the Na/K pump, such events are of course extremely unlikely to occur in nature. An alternative and more physiological way to raise $[Na]_i$ is to increase Na influx (instead of decreasing Na efflux). This can be accomplished simply by increasing the frequency with which action potentials are elicited; i.e., by stimulating rapidly. The same approach was used by Vassalle (42) in his study of the effects of rapid stimulation (overdrive) on spontaneously active Purkinje fibers (see below). Figure 7 shows some results from our experiments, in which brief periods of rapid stimulation (labeled overdrive for convenience although, strictly, the term is reserved for stimulation of an automatically active preparation at a rate higher than the prevailing spontaneous rate) were applied to small Purkinje fibers that were otherwise continuously driven at a relatively low rate. The chart recording of membrane potential in panel A indicates how these experiments were conducted: stimulation at the basal rate of 0.67 Hz was interrupted by brief periods of stimulation at 2 Hz, lasting 1 min for the experiment shown in Fig. 7A and 2 min for that of B, and sample action potentials elicited at the lower rate were photographed from the storage oscilloscope at various times. After the period of overdrive, the action potentials at the lower drive rate were temporarily shortened with respect to the control duration, and, at the same time, the resting potential was slightly increased. These effects are illustrated in panel B, where action potential durations, normalized to the control duration before the overdrive, are plotted against time after termination of that 2-min period of rapid drive. Immediately after the overdrive, the action potential was shortened to about 70% of its control duration, but it recovered fully along an approximately exponential time course with a half-time of about 70 sec. The shorter of the

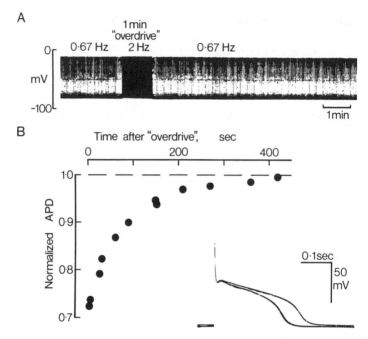

FIG. 7. Effects on action potentials of brief periods of rapid drive (labeled overdrive). **A:** Slow chart recording of action potentials elicited in 6 mM K, Cl-containing solution before, during, and after a 1-min period of rapid drive. **B:** Time course of recovery of action potential duration (APD), normalized to the steady, control duration, determined in another Purkinje fiber after 2 min of rapid drive at 2 Hz; the basal drive rate was 0.67 Hz. *Inset* shows superimposed action potentials from this experiment. The longer one is a control action potential photographed 9 min after stopping the rapid drive, while the shorter one, arising from the slightly more negative resting potential, was recorded just 25 sec after stopping the overdrive. The time calibration for these action potentials marks the zero potential level. [From Gadsby and Cranefield (22), with permission.]

two superimposed action potentials shown in the inset was recorded 25 sec after termination of the 2-min overdrive, and the longer one, arising from the more positive resting potential, was recorded about 9 min later, after the control duration had been regained. Presumably, both the transient increase in resting potential and the transient abbreviation of the action potential are attributable to a temporary increase in pump current, that increment in pump current decaying as $[Na]_i$ relaxes back to the steady-state level appropriate to the lower rate of the drive.

This conclusion, that a similar, temporary increase in pump current underlies the transient hyperpolarization and shortening of the action potential observed both after brief exposures to zero $[K]_o$ and after brief periods of rapid drive, receives further support from the results shown in Fig. 8. In this experiment, the voltage-clamp amplifier was switched on immediately after termination of a 10-min period of rapid drive to record the slowly decaying outward current

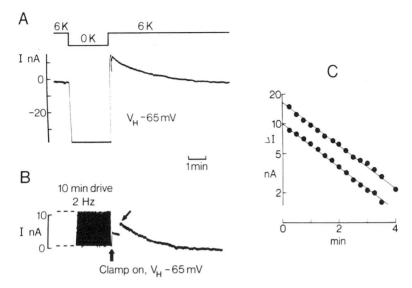

FIG. 8. The slowly decaying current transients that follow a period of rapid drive or a brief exposure to K-free fluid decline with the same exponential time course. The current record in **A** begins 2 min after the end of that in **B** and shows the increment in pump current following a 135-sec exposure to zero $[K]_o$ (the inward current in K-free fluid was off scale at this holding potential). **B:** At the *vertical arrow* the voltage clamp amplifier was switched on after a 10-min period of drive at 2 Hz; *diagonal arrow* marks a doubling of the sensitivity of the chart recorder to correspond to the current scale at the *left*. The *parallel straight lines* in **C** were fitted by eye to the semilogarithmic plots of the declining transient currents illustrated in **A** and **B**. Cl-containing solutions were used in this experiment; the time calibration is common to both **A** and **B**. [From Gadsby and Cranefield (22), with permission.]

(Fig. 8B) that, presumably, is responsible for transient shortening of the action potential of the kind just described. That outward current decayed with an exponential time course, as revealed by the lower set of points in the semilogarithmic plot at the right (Fig. 8C). Shortly after recording the current changes illustrated in panel B, with the clamp amplifier still switched on, the fiber was briefly exposed to K-free solution (Fig. 8A) to elicit a transient increment in pump current on returning to the K-containing fluid (cf. Figs. 2 and 3). That slowly decaying increment in pump current (Fig. 8A) is plotted as the upper set of points in the semilogarithmic graph of panel C, and it is clear that not only do the outward current transients elicited by the two different methods decay exponentially, but they do so with the same half-time, in this instance 82 sec. This result strongly suggests that both decaying outward currents reflect the same underlying process, namely, temporarily enhanced electrogenic Na extrusion during elimination of the increments in $[Na]_i$ induced by the two different experimental techniques.

The time courses of the increments in pump current thus seem generally appropriate to account for the observed transient reduction in action potential duration and increase in resting potential. However, to demonstrate that the magnitude of the increment in pump current is sufficient to account for the observed degree of action potential shortening requires measurement, in the same fiber, of both the reduction in action-potential duration and the increment in pump current that follow a given exposure to K-free solution. Figures 9 and 10 illustrate some results from an experiment of this kind, carried out on a small Purkinje fiber, already impaled with two microelectrodes and superfused with 6 mM K, Cl-containing Tyrode's solution while being regularly stimulated at 1 Hz via external platinum electrodes. In the first part of the experiment, we determined the extent of action-potential shortening caused by the enhanced pump activity following a 2-min exposure to K-free fluid, as indicated by the upper line in Fig. 9. The chart recording of membrane potential, beneath it, shows that the fiber depolarized and became inexcitable in K-free solution but rapidly repolarized on returning to 6 mM [K]$_o$. The two higher-speed chart records (one obtained shortly after switching back to 6 mM K solution, and

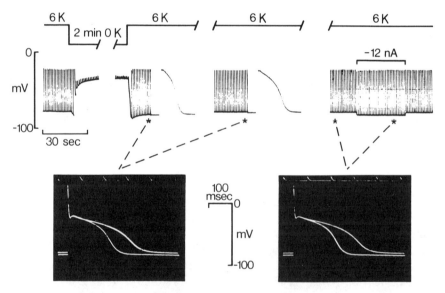

FIG. 9. Comparison of action potential shortening and membrane hyperpolarization caused by enhanced pump activity with that caused by current injection via a second microelectrode, in a Purkinje fiber exposed to 6 mM K, Cl-containing Tyrode's solution, and driven at 1 Hz. The *top line* shows the timing of the 2-min exposure to K-free fluid, and indicates the omission of 3 sections of record of duration: 90 sec, 10 min, and 3 min, respectively. The poor frequency response of the pen attenuates the action potential upstrokes in the chart recording. Nevertheless, the two high-speed chart records (for which the time calibration represents 500 msec) clearly reveal the temporary reduction in action potential duration that follows the exposure to zero [K]$_o$. *Asterisks* just beneath the chart records show when the action potentials, superimposed below, were displayed on the storage oscilloscope. Note that the pump-induced hyperpolarization and shortening of the action potential are both mimicked by injection of a steady (i.e., time- and voltage-independent) 12-nA hyperpolarizing current, as indicated by the *bar* above the chart record. [From Gadsby and Cranefield (22), with permission.]

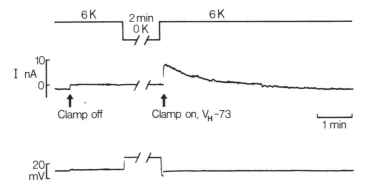

FIG. 10. Estimation of the size of the pump current increment elicited in 6 mM K solution following a 2-min exposure to K-free fluid. These records were obtained from the same fiber, but 150 min later than those shown in Fig. 9. **Top line** indicates the [K]$_o$ changes, **middle trace** shows applied current, and **bottom trace** shows membrane potential. At the *first arrow* the clamp was switched off so that the fiber could depolarize in K-free fluid as it did in Fig. 9 (the depolarization here was largely off scale; bottom trace). The clamp was switched on again at the *second arrow*, shortly after returning to 6 mM [K]$_o$, to record the transient increase in pump current, which measured about 10 nA in total amplitude. [From Gadsby and Cranefield (22), with permission.]

the other approximately 10 min later) show that, as usual, the action potential was transiently shortened after the exposure to zero [K]$_o$. The photograph at the lower left of Fig. 9 was taken from the storage oscilloscope and shows a shortened action potential, arising from a more negative resting potential, recorded about 15 sec after the return to 6 mM [K]$_o$, superimposed on the longer, control action potential recorded much later, after the resting potential had returned to its original, steady-state level. In the second part of the experiment, as illustrated by the records at the right-hand side of Fig. 9, we determined the approximate size of hyperpolarizing current required to cause the magnitude of hyperpolarization and shortening of the action potential seen during the enhanced pump activity. The photograph at the lower right of Fig. 9 was obtained a few minutes after that on the left and shows that injection of a steady (i.e., time- and voltage-independent), 12-nA hyperpolarizing current via the second intracellular microelectrode caused effects closely similar to those seen after the 2-min exposure to K-free fluid. The final part of the experiment involved direct measurement of the increment in pump current generated in the same fiber following 2 min of superfusion with K-free solution, as shown in Fig. 10. The upper line indicates the timing of the changes in [K]$_o$, the middle trace shows clamp current, and the lower trace shows the membrane potential. The clamp was switched off (first arrow) to allow the fiber to depolarize during the exposure to K-free solution just as it had in the first part of the experiment (see Fig. 9), but it was switched on again a few seconds after the return to 6 mM [K]$_o$ (second arrow) and a typical, exponentially decaying increment in pump current was recorded. The peak amplitude of that transient increase in pump current was about 10 nA, i.e., of appropriate magnitude to account for the shortening of the action potential and increase in resting potential illustrated in Fig. 9.

Results presented thus far document that a moderate increase in pump current can cause a significant increase in the resting potential of quiescent or driven Purkinje fibers, and a substantial reduction of the action potential duration in regularly stimulated fibers. As previously mentioned, Vassalle (42) showed that the transient increase in the rate of electrogenic Na extrusion resulting from a period of rapid overdrive could temporarily slow, or even abolish, spontaneous electrical activity in Purkinje fibers. A similar result is obtained if a brief period of exposure to K-free solution, rather than a period of overdrive, is used to raise $[Na]_i$. Thus, the chart recording in Fig. 11a shows that exposing a spontaneously active Purkinje fiber to K-free fluid for 2 min resulted, on returning to 4 mM $[K]_o$, in a period of temporary quiescence after which spontaneous action potentials reappeared, albeit at a lower rate. The spontaneous rate gradually increased and reached the initial, steady rate after about 4 min of recovery in the 4 mM K solution. The middle and lower records show, respectively, that

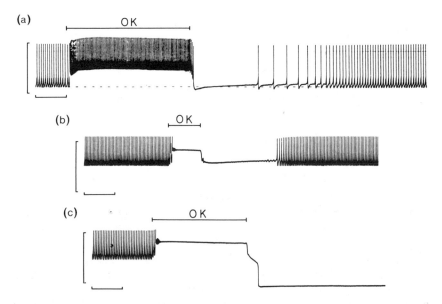

FIG. 11. After-effects of brief exposures to K-free fluid on spontaneous action potentials arising from either the higher (a) or the lower (b and c) level of membrane potential. The periods of exposure to K-free solution are indicated by the *bars* above the chart recordings. (a) Effects on normal automaticity in 4 mM K, Cl-containing Tyrode's solution. *Dashed line* is drawn at the control level of maximum diastolic potential, −87 mV. (b) and (c) show effects on slow response action potentials in a Purkinje fiber exposed to low Cl solutions; $[K]_o$ was 2 mM in (b) and 2.5 mM in (c). *Vertical calibration bars* all represent 100 mV, and their upper ends indicate the zero potential level; the horizontal calibrations are 30 sec. [From Gadsby and Cranefield (21), with permission.]

abnormal, slow-response action potentials (for review see ref. 9) arising in partially depolarized fibers (i.e., arising near the lower level of resting potential; see 19) can be temporarily (Fig. 11b), or even permanently (Fig. 11c), abolished by the transient increase in pump current that follows brief periods of exposure to K-free fluid. The long-term abolition of slow-response action potentials in Fig. 11c presumably occurred because the transient increment in pump current was able to cause a sufficiently large, upward shift of the N-shaped steady-state current–voltage relationship to switch the membrane potential to the more negative resting level.

These effects of an increase in pump current, which cause hyperpolarization and temporary or long-term abolition of spontaneously arising action potentials, are potentially antiarrhythmic because, in the heart *in situ*, they would tend to cause a general increase in diastolic membrane potentials (moving them further from threshold) and a specific reduction in the excitability of cells in any focus of abnormal automaticity. Vassalle (42) has presented the arguments for electrogenic-pump-mediated suppression of all subsidiary (latent) pacemakers in the heart, by virtue of the overdrive exerted by the fastest, and therefore the dominant, pacemaker in the sinoatrial node. Suppression of a particular, abnormal kind of rhythmic activity, called triggered activity, by the increase in pump current associated with enhanced electrogenic Na extrusion is strongly suggested by the results illustrated in Figs. 12 to 14.

These experiments were carried out with small strips of atrial muscle, about 2 to 4 mm long and up to 1 mm wide, dissected from the wall of the canine coronary sinus. The coronary sinus is a muscle-lined tube that drains the blood from the coronary circulation back into the right atrium. An interesting property of atrial cells in the coronary sinus is that when they are electrically stimulated

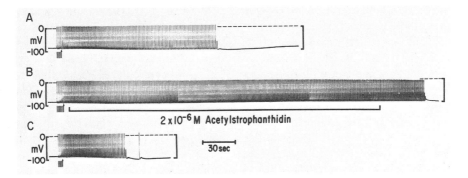

FIG. 12. Chart recordings of bursts of triggered action potentials recorded from a small strip of coronary sinus muscle exposed to 4 mM K, low Cl solution containing 0.3 μM norepinephrine, showing the effects of acetylstrophanthidin. **A:** A control burst, elicited by the 5 stimuli indicated by the *vertical lines* below the record (the last stimulus is indicated by the *arrow*). **B:** Effects on the triggered burst of applying 2 μM acetylstrophanthidin during the period indicated below the chart recording. The record in **C** shows another control burst of triggered activity initiated 18 min after washing out the acetylstrophanthidin. [From Wit et al. (45), with permission.]

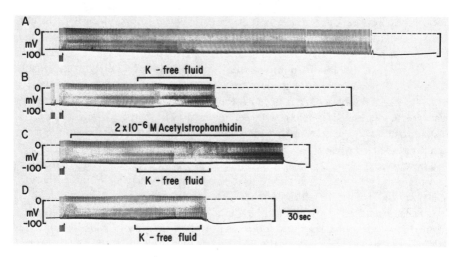

FIG. 13. After-effects of brief exposures to K-free fluid on bursts of triggered activity. *Vertical lines* below the beginning of each record mark the application of stimulus pulses, and the *arrows* indicate the last pulse in each series. 0.3 μM norepinephrine was present throughout the experiment. **A:** A control burst of triggered activity in 4 mM [K]$_o$, low Cl fluid. **B:** A 1-min exposure to K-free fluid (indicated by the *bar* over the record) was followed, on reexposure to 4 mM [K]$_o$, by a marked hyperpolarization that led to termination of the triggered activity. The record in **C** shows that brief exposure to K-free fluid in the presence of 2 μM acetylstrophanthidin (indicated by the *horizontal bars* below and above the record, respectively) did not cause the sudden hyperpolarization and termination previously seen in the absence of acetylstrophanthidin. The record in **D** was obtained after washing out the acetylstrophanthidin for 27 min and shows, once again, rapid termination of triggered activity following brief exposure to K-free fluid. [From Wit et al. (45), with permission.]

in the presence of moderately low concentrations of norepinephrine, each action potential is followed by a delayed after-depolarization (44). The amplitude of the after-depolarization increases as the frequency of stimulation is increased, and with sufficiently high rates of stimulation (or with premature impulses at sufficiently short intervals) the after-depolarization can exceed threshold and thus trigger a premature action potential. That action potential is, itself, followed by a suprathreshold after-depolarization, which triggers another action potential and so on, thereby giving rise to a train of nondriven, triggered action potentials. Such a burst of triggered activity, elicited in the presence of 0.3 μM norepinephrine, is shown in Fig. 12A. The 5 stimulated action potentials that initiated the burst are indicated by the vertical lines at the beginning of the record; the arrow marks the last driven beat. After an initial speeding up, there is a gradual slowing of the triggered rate, accompanied by a hyperpolarization, that eventually leads to termination of the triggered burst and to quiescence. To investigate the possibility that this gradual hyperpolarization and slowing are caused by enhanced electrogenic Na extrusion secondary to the rise in [Na]$_i$ resulting from the extra Na influx associated with the burst of triggered action potentials, we first tested the effects of inhibiting the Na/K pump with acetylstrophanthidin.

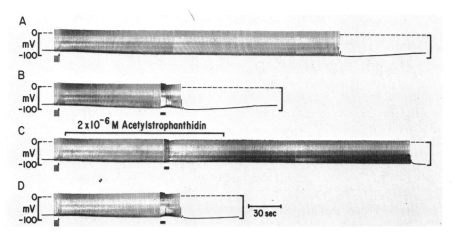

FIG. 14. After-effects of brief periods of rapid overdrive on triggered activity. The preparation was exposed to 4 mM K, low Cl solution containing 0.3 μM norepinephrine throughout the experiment. The 5-sec periods of overdrive (cycle length, 350 msec) are indicated by the *thick bars* under records **B, C,** and **D. A:** A control burst of activity, initiated at the *vertical arrow.* The record in **B** shows that a 5-sec period of overdrive was followed by sudden hyperpolarization and premature termination of the burst. **C** shows that, in the presence of 2 μM acetylstrophanthidin, the rapid overdrive was followed neither by rapid hyperpolarization nor by sudden quiescence. The record in **D** was obtained 17 min after washing out the acetylstrophanthidin and shows that the overdrive was once again followed by a rapid hyperpolarization that terminated the triggered activity. [From Wit et al. (45), with permission.]

The record in Fig. 12B shows that application of 2 μM acetylstrophanthidin, immediately after the start of the triggered burst, largely abolished the gradual hyperpolarization and slowing that led to termination of the burst under control conditions, and substantially prolonged the triggered activity. The lower record (Fig. 12C) was obtained almost 20 min after washing out the acetylstrophanthidin and shows once again the gradual hyperpolarization and slowing that lead to normal termination of triggered activity. This result is consistent with our working hypothesis that electrogenic Na extrusion can play a role in the normal termination of triggered bursts. If that is so, then experimental interventions that accelerate the Na/K pump still further might cause premature termination of triggered bursts. The results of Fig. 13 confirm this suggestion. The record in Fig. 13A shows a control burst of triggered activity that displays the usual gradual hyperpolarization and slowing leading to quiescence. In the second record (Fig. 13B), the preparation was exposed to K-free solution for 1 min, to raise $[Na]_i$, so that on returning to 4 mM $[K]_o$, the Na/K pump would be temporarily stimulated. That transient increase in pump rate caused a sudden marked hyperpolarization that, within a few seconds, led to premature termination of the triggered activity. The third record (Fig. 13C) confirms that the precipitous hyperpolarization underlying the premature termination was indeed due to enhanced pump activity, because it did not occur when the 1-min exposure to K-free solution was repeated in the presence of 2 μM acetylstrophanthidin. The bottom record (Fig. 13D) shows that after washing out the acetylstrophanthidin for 27 min, the enhanced pump activity following the brief exposure to zero $[K]_o$ once again caused sudden hyperpolarization and quiescence. Figure 14 shows that essentially similar results were obtained if a brief period of rapid overdrive, instead of brief superfusion with K-free fluid, was used to cause a temporary stimulation of the Na/K pump. Thus, a 5-sec period of overdrive at a cycle length of 350 msec was followed by a marked hyperpolarization that led to premature termination of the triggered burst (Fig. 14B and D) in comparison to the duration of the control burst (Fig. 14A). Abolition of this after-effect of the overdrive by exposure to 2 μM acetylstrophanthidin (Fig. 14C) confirms that it was due to enhanced pump activity.

The results of Figs. 12 to 14 indicate that the gradual hyperpolarization and slowing of the triggered rate that normally lead to termination of bursts of triggered activity are caused by increased activity of the Na/K pump. It is unlikely that this hyperpolarization observed during the triggered burst reflects a depletion of extracellular K ions secondary to pump stimulation, because K depletion is expected to occur only when net K flux is inwardly directed (when the cellular K content must be increasing), i.e., only after termination of the burst (45). Therefore, the hyperpolarization leading to quiescence most probably results from an increase in pump current.

If such bursts of triggered action potentials in atrial cells of the canine coronary sinus are controlled by mechanisms similar to those underlying some forms of paroxysmal atrial tachycardia, then our results suggest that the duration of these tachycardias might be limited by activity of the electrogenic Na/K pump. In other words, the effects of enhanced electrogenic Na extrusion are likely to be antiarrhythmic.

DISCUSSION

The major conclusion from this work is that changes in the rate of electrogenic Na extrusion, i.e., changes in the size of the pump current, can have pronounced effects on the electrical activity of cardiac cells. Comparison of the sizes of the current changes, measured under voltage clamp, caused by inhibiting the pump with micromolar concentrations of cardiac steroids or by stimulating the pump by temporarily increasing $[Na]_i$, suggests that the pump current can be roughly doubled in magnitude after about 2 min of Na-loading in K-free solution (18,21). Such changes in pump current have marked effects on the duration of the action potential in Purkinje fibers. Complete abolition of the pump current can even prevent full membrane repolarization, the membrane potential remaining near the lower resting level (31), whereas after a 2-min exposure to K-free solution, the resulting increase in pump current reduces action potential duration by about one-quarter or one-third (e.g., Fig. 9; ref. 21). Although such drastic changes can easily be dismissed as unphysiological, it should be kept in mind that, as outlined in the introduction, the pump current seems to be proportional to the pumped rate of Na efflux and, for $[Na]_i$ to be steady, Na efflux must equal Na influx. Pump current can therefore be expected to scale with Na influx which, in turn, depends on action potential frequency. If doubling the impulse rate were to double Na influx then, in the steady state (after about 5 min), the pump current would also be doubled in size. It is well known that an increase in drive rate is associated with a maintained reduction in action potential duration, at least part of which develops rather slowly, and a reduction in drive rate causes prolongation of the action potential (for review, see ref. 3); presumably, changes in the size of the pump current contribute to these alterations in action potential shape (21,22).

Throughout the period of temporarily enhanced Na/K pump activity that follows a brief exposure to K-free solution or a brief episode of rapid drive, $[Na]_i$ declines and $[K]_i$ rises back toward the steady-state level. Net K movement must be inward in order for $[K]_i$ to rise so that, depending on the magnitude of net K influx and the rate of diffusion equilibration of K ions in the extracellular space, K ions are expected to be depleted, to some extent, just outside the cell membrane. However, neither the transient shortening of the action potential in driven Purkinje fibers (Figs. 6, 7, and 9), nor the temporary quiescence of spontaneously active preparations (Fig. 11), caused by enhanced pump activity, can be accounted for by extracellular K depletion because experimental reduction of $[K]_o$ is known to cause prolongation of the action potential, and an increase in the frequency of spontaneous action potentials (41,43).

The pump-induced transient hyperpolarization, or transient outward current under voltage clamp, recorded near the lower level of resting potential (Figs. 2 to 4) also cannot be attributed to K depletion because, in the same records, lowering $[K]_o$ is seen to cause effects in the opposite direction. It is clear therefore that, at least qualitatively, these effects of enhanced pump activity in small bundles of canine Purkinje cells must be attributed to increased pump current and not to extracellular K depletion; note that this need not be the case for larger bundles of Purkinje fibers, even those from dog hearts (see comment by Fozzard, in ref. 22).

The small size of the Purkinje fiber preparations used in these experiments is indicated by their physical dimensions, by the low amplitudes of the voltage-clamp currents, and by the relatively rapid time course of apparent K equilibration in the extracellular space in response to step changes in $[K]_o$ (e.g., Figs. 2 and 3). The morphological findings on small canine Purkinje fibers that a large fraction of the surface membrane of the cells faces spaces wider than 1 μm (13; T. D. Pham and D. C. Gadsby, *unpublished observations*), are consistent with rapid diffusion equilibration of extracellular K, as are the findings of small K-depletion currents during hyperpolarizing voltage steps in these fibers (6; D. C. Gadsby, *unpublished observations*). The rapid extracellular K equilibration in these small Purkinje fibers used in combination with the fast-flow system, in turn permitted reliable investigation of even small increases in pump activity, e.g., following 1- or 2-min exposures to K-free solution, so that the associated net K influx would be correspondingly small. Thus, small fluxes together with relatively rapid extracellular equilibration should limit the extent of extracellular K depletion during enhanced pump activity in the experiments presented here. Indeed, the graph in Fig. 5 suggests that the pump rate constant is half-maximal at a superfusate K concentration of 1 mM and, since the K concentration just outside the cell membrane must be greater than zero under those conditions, 1 mM must represent an upper limit for the extent of K depletion accompanying the enhanced pump activity induced at 1 mM $[K]_o$. In fact, the K depletion is likely to be much smaller than that for the following reasons. The fact that the holding current in K-free fluid is little altered by the presence of a maximal concentration of acetylstrophanthidin (Fig. 3b) argues that the pump is practically completely inhibited in the K-free solution. Even if pump activity were reduced to only 10 or 20% on switching from 4 mM to zero $[K]_o$, then the results plotted in Fig. 5 indicate that the K concentration just outside the cell membrane in K-free fluid (i.e., the extent of K accumulation due to the net K efflux) would, under those conditions, be only about 0.1 mM (or even less, if the graph of Fig. 5 were corrected for possible effects of extracellular K depletion by shifting it slightly to the left, toward lower $[K]_o$ values). Now, at a fixed holding potential, passive K efflux into 1 mM $[K]_o$ is expected, on account of inward rectification, to be somewhat greater than that into K-free solution (roughly 20% greater, on the basis of the results in Fig. 3 and simplifying assumptions) and in the steady state, at 1 mM $[K]_o$, it must exactly be balanced by pumped K influx; a transient doubling of the pump rate at 1 mM $[K]_o$, therefore, would give rise, temporarily, to a net K influx equivalent in magnitude to the passive K efflux into 1 mM $[K]_o$, i.e., somewhat larger than the passive, net K efflux into K-free solution. That net K influx during enhanced pump activity (doubling the steady-state pump rate) at 1 mM $[K]_o$ would be expected to cause extracellular K depletion roughly 20% greater in extent than the ~0.1 mM K accumulation estimated to occur in K-free solution. In other words, immediately following a 2.5- to 3-min exposure to K-free fluid (18), the K concentration just outside the cells of a fiber bathed in 1 mM $[K]_o$ solution should be at least 0.8 mM, and should rise to reach 1 mM in the steady state, since there can be no depletion (or accumulation) when the K fluxes are in balance.

ACKNOWLEDGMENTS

Much of the work reported here was done in collaboration with Dr. Paul F. Cranefield to whom I am indebted for constant advice and encouragement. The experiments of Figs. 12 to 14 were done in collaboration with Dr. Andrew L. Wit. I thank Drs. Frank Brink and C. M. Connelly for helpful discussions.

The preparation of this article and the research reported in it were supported by USPHS grant No. HL-14899 and by an Established Fellowship of the New York Heart Association.

REFERENCES

1. Almers, W. (1971): The potassium permeability of frog muscle membrane. PhD Thesis, University of Rochester, Rochester, New York. University Microfilms, Ann Arbor, Michigan #72-14,689.
2. Blood, B. E. (1978): Glycoside induced stimulation of membrane Na-K ATP-ase—Fact or artifact? In: *Biophysical Aspects of Cardiac Muscle,* edited by M. Morad, pp. 379–389. Academic Press, New York.
3. Carmeliet, E. E. (1977): Repolarization and frequency in cardiac cells. *J. Physiol. (Paris),* 73:903–923.
4. Cohen, C. J., Fozzard, H. A., and Sheu, S.-S. (1982): Increase in intracellular sodium ion activity during stimulation in mammalian cardiac muscle. *Circ. Res.,* 50:651–662.
5. Cohen, I., Daut, J., and Noble, D. (1976): An analysis of the actions of low concentrations of ouabain on membrane currents in Purkinje fibers. *J. Physiol. (Lond.),* 260:75–103.
6. Cohen, I., Falk, R., and Kline, R. P. (1983): Voltage-clamp studies on the canine Purkinje strand. *Proc. R. Soc. Lond. Ser. B,* 217:215–236.
7. Colquhoun, D., Neher, E., Reuter, H., and Stevens, C. F. (1981): Inward current channels activated by intracellular Ca in cultured cardiac cells. *Nature,* 294:752–754.
8. Connelly, C. M. (1959): Recovery processes and metabolism of nerve. *Rev. Mod. Phys.,* 31:475–484.
9. Cranefield, P. F. (1975): *The Conduction of the Cardiac Impulse: the Slow Response and Cardiac Arrhythmias.* Futura Publishing Company, Mt. Kisco, New York.
10. Deitmer, J. W., and Ellis, D. (1978): The intracellular sodium activity of cardiac Purkinje fibres during inhibition and re-activation of the Na-K pump. *J. Physiol. (Lond.),* 284:241–259.
11. DeWeer, P. (1970): Effects of intracellular adenosine-5'-diphosphate and orthophosphate on the sensitivity of sodium efflux from squid axon to external sodium and potassium. *J. Gen. Physiol.,* 56:583–620.
12. DeWeer, P. (1975): Aspects of the recovery processes in nerve. In: *Neurophysiology (MTP Int. Rev. Sci., Physiol. Ser. 1, Vol. 3),* edited by C. C. Hunt, pp. 231–278. Butterworths, London.
13. Eisenberg, B. R., and Cohen, I. S. (1983): The ultrastructure of the cardiac Purkinje strand in the dog: a morphometric analysis. *Proc. R. Soc. Lond. Ser. B.,* 217:191-213.
14. Eisner, D. A., Lederer, W. J., and Vaughan-Jones, R. D. (1981): The dependence of sodium pumping and tension on intracellular sodium activity in voltage-clamped sheep Purkinje fibres. *J. Physiol. (Lond.),* 317:163–187.
15. Eisner, D. A., Lederer, W. J., and Vaughan-Jones, R. D. (1981): The effects of rubidium ions and membrane potential on the intracellular sodium activity of sheep Purkinje fibres. *J. Physiol. (Lond.),* 317:188–205.
16. Ellis, D. (1977): The effects of external cations and ouabain on the intracellular sodium activity of sheep heart Purkinje fibers. *J. Physiol. (Lond.),* 273:211–240.
17. Furchgott, R. F. (1955): The pharmacology of vascular smooth muscle. *Pharmacol. Rev.,* 7:183–265.
18. Gadsby, D. C. (1980): Activation of electrogenic Na^+/K^+ exchange by extracellular K^+ in canine cardiac Purkinje fibers. *Proc. Natl. Acad. Sci. USA,* 77:4035–4039.
19. Gadsby, D. C., and Cranefield, P. F. (1977): Two levels of resting potential in cardiac Purkinje fibers. *J. Gen Physiol.,* 70:725–746.
20. Gadsby, D. C., and Cranefield, P. F. (1979): Direct measurement of changes in sodium pump current in canine cardiac Purkinje fibers. *Proc. Natl. Acad. Sci. USA,* 76:1783–1787.
21. Gadsby, D. C., and Cranefield, P. F. (1979). Electrogenic sodium extrusion in cardiac Purkinje fibers. *J. Gen. Physiol.,* 73:819–837.
22. Gadsby, D. C., and Cranefield, P. F. (1982): Effects of electrogenic sodium extrusion on the membrane potential of cardiac Purkinje fibers. In: *Normal and Abnormal Conduction in the Heart,* edited by A. Paes de Carvalho, B. F. Hoffman, and M. Lieberman, pp. 225–247. Futura Publishing Company, Mt. Kisco, New York.
23. Ghysel-Burton, J., and Godfraind, T. (1975): Stimulation and inhibition by ouabain of the sodium pump in guinea-pig atria. *Br. J. Pharmacol.,* 55:249P.
24. Glitsch, H. G. (1982): Electrogenic Na pumping in the heart. *Annu. Rev. Physiol.,* 44:389–400.

25. Glitsch, H. G., Kampmann, W., and Pusch, H. (1981): Activation of active Na transport in sheep Purkinje fibres by external K or Rb ions. *Pfluegers Arch.*, 391:28–34.
26. Glitsch, H. G., Reuter, H., and Scholz, H. (1970): The effect of the internal sodium concentration on calcium fluxes in isolated guinea-pig auricles. *J. Physiol. (Lond.)*, 209:25–43.
27. Glynn, I. M., and Karlish, S. J. D. (1975): The sodium pump. *Annu. Rev. Physiol.*, 37:13–56.
28. Hamlyn, J. M., Cohen, N., and Blaustein, M. P. (1983): Stimulation of dog kidney Na^+ K-ATPase by low doses of cardiotonic steroids is due to disinhibition of the enzyme. *Circulation*, (Part II), 68:III 63.
29. Hodgkin, A. L., and Horowicz, P. (1959): The influence of potassium and chloride ions on the membrane potential of single muscle fibres. *J. Physiol. (Lond.)*, 148:127–160.
30. Isenberg, G. (1977): Cardiac Purkinje fibres. $[Ca^{2+}]_i$ controls steady state potassium conductance. *Pfluegers Arch.*, 371:71–76.
31. Isenberg, G., and Trautwein, W. (1974): The effect of dihydro-ouabain and lithium-ions on the outward current in cardiac Purkinje fibers; evidence for electrogenicity of active transport. *Pfluegers Arch.*, 350:41–54.
32. Kass, R. S., Lederer, W. J., Tsien, R. W., and Weingart, R. (1978): Role of calcium ions in transient inward currents and aftercontractions induced by strophanthidin in cardiac Purkinje fibres. *J. Physiol. (Lond.)*, 281:187–208.
33. Noble, D. (1965): Electrical properties of cardiac muscle attributable to inward-going (anomalous) rectification. *J. Cell. Comp. Physiol.*, 66 (suppl. 2): 127–135.
34. Noble, D. (1980): Mechanism of action of therapeutic levels of cardiac glycosides. *Cardiovasc. Res.*, 14:495–514.
35. Noma, A., and Irisawa, H. (1975): Contribution of an electrogenic sodium pump to the membrane potential in rabbit sinoatrial node cells. *Pfluegers Arch.*, 358:289–301.
36. Rang, H. P., and Ritchie, J. M. (1968): On the electrogenic sodium pump in mammalian non-myelinated nerve fibres and its activation by various external cations. *J. Physiol. (Lond.)* 196:183–221.
37. Siegelbaum, S. A., and Tsien, R. W. (1980): Calcium-activated transient outward current in calf cardiac Purkinje fibres. *J. Physiol. (Lond.)*, 299:485–506.
38. Sommer, J. R., and Johnson, E. A. (1968): Cardiac muscle: A comparative study of Purkinje fibers and ventricular fibers. *J. Cell Biol.*, 36:497–526.
39. Thomas, R. C. (1969): Membrane current and intracellular sodium changes in a snail neurone during extrusion of injected sodium. *J. Physiol. (Lond.)* 201:495–514.
40. Thomas, R. C. (1972): Electrogenic sodium pump in nerve and muscle cells. *Physiol. Rev.*, 52:563–594.
41. Vassalle, M. (1965): Cardiac pacemaker potentials at different extra- and intracellular K concentrations. *Am J. Physiol.*, 208:770–775.
42. Vassalle, M. (1970): Electrogenic suppression of automaticity in sheep and dog Purkinje fibers. *Circ. Res.*, 27:361–377.
43. Weidmann, S. (1956): *Elektrophysiologie der Herzmuskelfaser.* Huber, Bern.
44. Wit, A. L., and Cranefield, P. F. (1977): Triggered and automatic activity in the canine coronary sinus. *Circ. Res.*, 41:435–445.
45. Wit, A. L., Cranefield, P. F., and Gadsby, D. C. (1981): Electrogenic sodium extrusion can stop triggered activity in the canine coronary sinus. *Circ. Res.*, 49:1029–1042.

Electrogenic Transport: Fundamental Principles and Physiological Implications, edited by Mordecai P. Blaustein and Melvyn Lieberman. Raven Press, New York © 1984.

Effects of Membrane Potential on Sodium-Dependent Calcium Transport in Cardiac Sarcolemma Vesicles

Robin T. Hungerford, George E. Lindenmayer, William P. Schilling, and Eldwin Van Alstyne

Departments of Pharmacology and Medicine, Medical University of South Carolina, Charleston, South Carolina 29425

The contraction–relaxation sequence of the myocardial cell requires the cyclic influx and efflux of calcium. One means by which calcium is moved across the plasma membrane (sarcolemma) of these cells is believed to involve sodium/calcium exchange (1,10,16,17,29,30). This reaction couples the movement of calcium across the sarcolemma in one direction to the movement of sodium in the opposite direction. The stoichiometry of the process has been suggested to be greater than 2 sodiums moved per calcium such that the exchange is electrogenic (12,19–21).

Sodium/calcium exchange appears to be able to move calcium either into or out of the cell. The direction depends, in part, on the intracellular and extracellular concentrations (strictly speaking, chemical activities) of sodium and calcium. If the exchange is electrogenic, membrane potential also serves as a key determinant of the direction of calcium flux (21). When the exchange reaction is at equilibrium, internal free calcium is linked to the other determinants by the following equation (7):

$$[Ca]_i = \frac{[Ca]_o [Na]_i^r}{[Na]_o^r e^{-(r-2) E_m F / RT}} \quad (1)$$

where the subscripts i and o refer to intracellular and extracellular cation concentrations, respectively; r = the number of sodiums exchanged per calcium; E_m is the membrane potential and F, R, and T have their usual meanings.

A method for the direct evaluation of pathways for calcium movement across the sarcolemma has been made possible by the recent availability of highly enriched sarcolemma preparations from cardiac tissue (5,14,33). In 1979, Reeves and Sutko (26) and Pitts (25) provided evidence for the existence of a sodium/calcium exchange system in membrane preparations from heart, which were enriched to varying extents with sarcolemma markers. In addition, Pitts (25) showed that sodium/calcium exchange was promoted by vesicles in the preparation that manifest Na, K-ATPase activity (another putative sarcolemma marker) and concluded that the stoichiometry of the exchange was three sodiums/calcium. Van Alstyne et al. (33) confirmed that sodium-stimulated calcium movements are promoted by a highly enriched sarcolemma preparation, and Bartschat and Lindenmayer (2) described a calcium/calcium exchange reaction promoted by vesicles in the same preparation. The stoichiometry of the latter reaction was one-to-one and a comparison of certain properties of calcium movements stimulated by sodium and calcium suggested that calcium/calcium and sodium/calcium exchange may be manifestations of a single sarcolemma system. Philipson and Nishimoto (24) also found that sarcolemma-enriched preparations from heart promote calcium/calcium exchange. Other studies showed that sodium/calcium exchange, as promoted by sarcolemma-enriched preparations, could contribute to membrane potential and was affected by membrane potential (4, 8,15,18,23,27), which is consistent with the reaction's electrogenicity. Furthermore, Bartschat et al. (3) reported that depolarization of vesicles in a sarcolemma-enriched preparation caused the rapid uptake of very large amounts of calcium through a reaction that required sodium. These studies support the feasibility of using sarcolemma-enriched preparations from the heart for the study of the effects of membrane potential on sodium/calcium exchange.

The procedure for isolation of sarcolemma-enriched preparations currently used by this laboratory (33) yields membrane preparations with enrichments of 27- to 40+-fold of putative sarcolemma markers over the starting homogenate. Succinic dehydrogenase (marker for inner mitochondrial membrane) and Ca-adenosine triphosphatase (ATPase) (assay conditions for the sarcoplasmic reticulum system) are slightly reduced in the preparation over what is present in the starting homogenate. Total centrifugation time is 130 min (at the desired force) and total process time for 100 g of tissue is about 5.5 hr. Total protein recovered in the sarcolemma-enriched preparation is 10 to 11 mg per 100 g starting tissue. With regard to the study of cation fluxes across the sarcolemma, it is important to note that the preparation is composed of membrane vesicles that manifest osmotic reactivity. This is consistent with the presence of sealed vesicles in the preparation. These and other considerations generated the conclusions that about 67 to 75% of the preparation consists of sealed right-side-out vesicles (33), and the remaining 33 to 25% consists of vesicles (possibly inside-out) with an apparent permeability of about 10 times that of the other species (31,32).

During a 10-sec assay at 37°C, we found that sodium-dependent calcium uptake by a sarcolemma-enriched fraction was characterized by two components. One component was apparently complete after 2 sec and the other linear over 2 to 10 sec (ref. 13; see also Fig. 1A). The time course of the fast component has subsequently been measured down to 70 msec with rapid-quench techniques (13). The rapid-quench data were optimally fit by two exponential profiles, and the total profile is similar to that obtained with the metallochromic indicator,

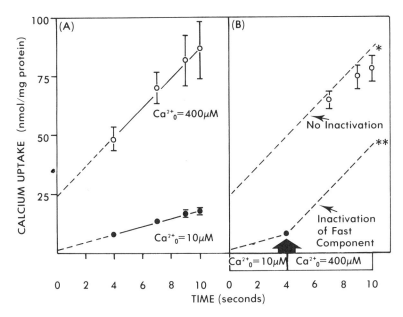

FIG. 1. Sodium-dependent calcium uptake by sarcolemma-enriched preparation. Sodium-dependent calcium uptake is defined as the difference in uptake in the presence and absence of an outwardly directed sodium gradient. Aliquots of the preparation were loaded 15 to 18 hr at 5°C with 10 mM Tris-Cl, pH 7.4, and 160 mM NaCl. After prewarming to 37°C, the reaction was started by a 30-fold dilution of the loaded preparation into a reaction medium to yield final extravesicular concentrations of 10 mM Tris-Cl, pH 7.4, 5.3 mM NaCl, 154.7 mM LiCl, and either 10 or 400 μM CaCl$_2$ with ^{45}Ca. In one set of experiments **(A)**, the calcium concentration was constant throughout the course of the assays. The reactions were terminated at the times indicated by the addition of an ice-cold stopping solution containing 160 mM KCl, 1 mM CaCl$_2$, 0.2 mM LaCl$_3$ and 10 mM Tris-Cl, pH 7.4 at 4°C. The suspension was rapidly filtered and washed three times with 5-ml aliquots of the stopping solution. The filters were assayed for radioactivity by liquid scintillation techniques. Sodium-independent calcium uptake was determined in parallel assays in the absence of a sodium gradient, and the data presented were corrected to yield sodium-dependent calcium uptake. In the other set of experiments **(B)**, a jump in calcium concentration from 10 to 400 μM was made at 4 sec, and calcium uptake was measured at 7, 9, and 10 sec. Since the data after the calcium jump were essentially identical to those for 400 μM calcium in **A**, it was concluded that detection of the slow component (i.e., described by the time course of uptake over 4 to 10 sec) was not due to inactivation of the fast component (i.e., extrapolated value at time zero). Lines drawn in **A** were derived from regression analysis of the data to yield slope (i.e., rate of slow component) and intercept (extent of uptake via fast component). *Regression line from **A** for Ca^{2+} = 400 μM. **Regression lines from **A** for Ca$_o^{2+}$ = 10 μM (0 to 4 sec) and Ca$_o^{2+}$ = 400 μM (4 to 10 sec).

arsenazo III, by Caroni et al. (8,9). The two components may be independent and reflect sodium-dependent processes for calcium movements in different vesicles in the preparation. The two components appeared to be differentiated to some extent by multiple freeze–thaw cycles, which gradually render the vesicles permeable (Fig. 2). Thus, it is possible that the components are promoted by species of vesicles that differ in their sensitivity to the freeze–thaw treatment. It should be noted, however, that both components co-purify with the Na, K-

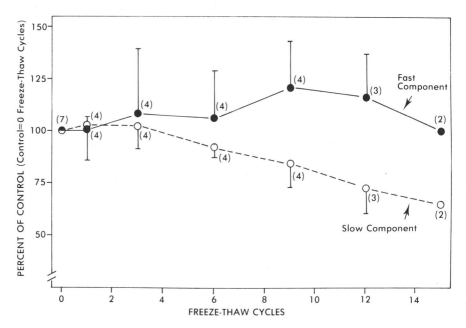

FIG. 2. Effect of freeze–thaw cycles on the fast and slow components of sodium-dependent calcium uptake. Aliquots of the sarcolemma-enriched preparation were loaded 15 to 18 hr at 5°C with 10 mM Tris-Cl, pH 7.4, and 160 mM NaCl. After prewarming to 37°C, the suspensions were added to a medium containing final concentrations of 154.7 mM LiCl, 5.3 mM NaCl, 10 mM Tris-Cl, pH 7.4, and 0.04 mM $CaCl_2$ with ^{45}Ca. The experiment was conducted on aliquots of the preparation subjected to repetitive freeze–thaw cycles as indicated on the abscissa. Freeze–thaw was effected by immersing the tube containing the loaded preparation into a Dry Ice-acetone mixture and then in water at 37°C.

ATPase (i.e., specific ouabain binding sites) when the sarcolemma-enriched preparation is obtained with different sucrose densities (Figs. 3 and 4). This suggests that vesicles manifesting the two components have similar levels of Na, K-ATPase. An alternate interpretation is that the fast and slow components are present in the same vesicle but are independent. This would require that the fast component exhibit some form of inactivation so that the slow process can be expressed at later times. Such a mechanism has been proposed for fast and slow potassium-stimulated calcium movements in synaptosome preparations (22). This possibility for sodium-dependent calcium movement in the sarcolemma preparation appears, however, to have been eliminated by results of the experiments shown in Figs. 1 A and B, since a time-dependent inactivation was not detected. A third explanation is that the two components might be interdependent in that $[Ca]_i$ delivered by the fast component is more slowly bound to sites inside the vesicle. This would require that the fast component continue to deliver calcium at a rate controlled by the binding process. It is interesting in this regard that temperature differentially affects the two components. Reduction of the assay temperature inhibits the fast component to a greater extent than

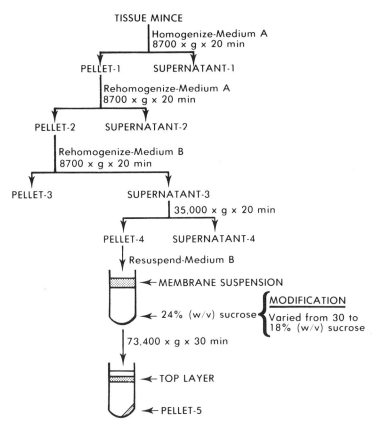

FIG. 3. Modification of the isolation scheme of Van Alstyne et al. (33) to yield fractions of different enrichments of ouabain binding sites and sodium-dependent calcium uptake. Medium A: 15 mM $NaHCO_3$, pH 7.0. Medium B: 10 mM Tris-Cl, pH 7.4.

the slow, with the result that below 15°C, only the slow component is detectable (Fig. 5; see also ref. 28). It is conceivable that below 15°C, delivery of calcium into the vesicle is slower than the binding and, therefore, is rate limiting for the overall process of calcium uptake.

Other explanations for the two-component profile are inhibition of a single component by the buildup of intravesicular negative charge by electrogenic sodium/calcium exchange, or inhibition caused by the buildup of intravesicular calcium (15). Neither appear to explain adequately the two components of sodium-dependent calcium *influx* that we observe. Vesicles in the sarcolemma-enriched preparation can be polarized by the use of a potassium gradient in the presence of valinomycin (Fig. 6; refs. 3,31). Any net charge moved by the electrogenic sodium/calcium exchange should, under these conditions, be dissipated by valinomycin-facilitated potassium movement. However, when sodium-dependent calcium uptake was examined in the presence of potassium Nernst

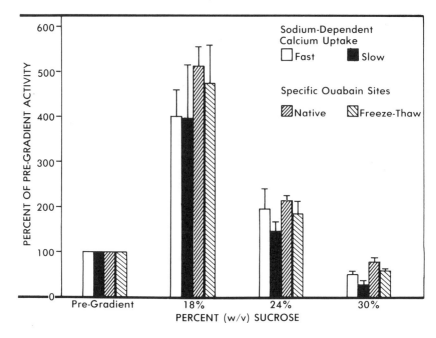

FIG. 4. Coenrichment of two components of sodium-dependent calcium uptake and ouabain binding sites. Pellet 4 (Fig. 3), on resuspension, was layered over 18, 24, or 30% (wt/vol) sucrose. After centrifugation, the top layers were removed, washed, and assayed for the fast and slow components of sodium-dependent calcium uptake as described in the legend of Fig. 2, except that the preparation was not subjected to freeze–thaw cycles. Assays for specific ouabain binding sites of the preparation before and after rendering the preparation permeable by freeze–thaw cycles were conducted as described by Van Alstyne et al. (33). [Note that the freeze–thaw cycles in Fig. 2 were carried out on preparations previously loaded with physiological concentrations of salt; it takes more freeze–thaw cycles to render these preparations permeable than for preparations loaded with low-ionic-strength buffer (i.e., 10 mM Tris-Cl, pH 7.4), which is what was used for the ouabain binding studies (33).]

potentials of either 0 or −67.2 mV over 10 sec (i.e., prior to significant loss of the potential on addition of valinomycin; Fig. 6), both the fast and slow components were still detectable (Fig. 7). This suggests that the two-component profile of sodium-dependent calcium uptake is not the result of a charge accumulation. In other experiments where sodium-dependent calcium efflux was measured, two components of efflux were detected when the extravesicular medium contained 1.0 mM EGTA (2). Thus, the two-component profile was observed under conditions that precluded calcium accumulation in the extravesicular space and inhibition of sodium-dependent calcium efflux.

The extent of sodium-dependent calcium uptake via the fast component and the rate of uptake via the slow component differed in their apparent sensitivity to membrane potential (Fig. 8). Increasingly negative inside potentials depressed uptake via the fast component to a greater extent than the slow. These data suggest that sodium-dependent calcium movement promoted by these prepara-

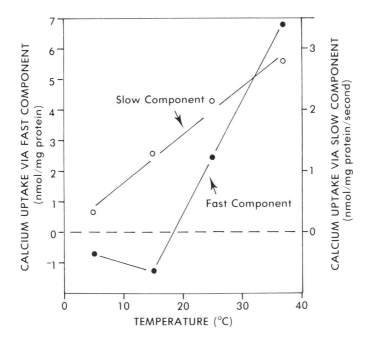

FIG. 5. Effect of temperature on the two components of sodium-dependent calcium uptake. The assays were performed as described in Fig. 2, except that the preparations were not subjected to freeze–thaw cycles, and the assay temperature was varied between 5 and 37°C.

tions is, in fact, electrogenic. This result is consistent with those of others (4,8, 15,18,23,27), but the result (Fig. 8) also generates the question as to how membrane potential affects the reaction.

As presented above, there are 5 determinants of the equilibrium state for electrogenic sodium/calcium exchange (assuming that the stoichiometry remains constant): $[Na]_o$, $[Ca]_o$, $[Na]_i$, $[Ca]_i$ and E_m. Equation (1) therefore requires $[Ca]_i$ to change with membrane potential when the other three determinants of the reaction, $[Na]_o$, $[Ca]_o$, and $[Na]_i$, are kept constant. Membrane potential is thought to affect some membrane channels through which ions move by modulating gating particles that control the movement of ions into and out of the channel (11). But calcium and sodium movements through sodium/calcium exchange are generally envisioned as occurring through a carrier mechanism (1,6,30). In other words, some membrane constituent reacts with sodium and calcium possibly in a sequential manner. The constituent may shuttle back and forth across the membrane leading to the coupled, net movement of the cations in response to their gradients existing across the membrane. Given that the sodium/calcium exchange is electrogenic and, therefore, sensitive to membrane potential, changes in membrane potential must presumably affect one or more properties of the carrier.

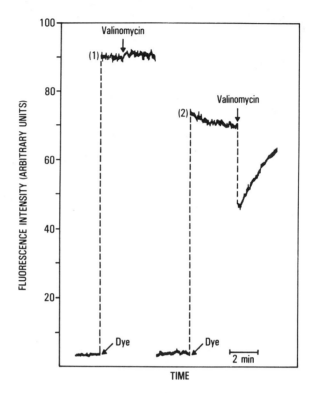

FIG. 6. Change in fluorescence intensity of the voltage-sensitive dye, 3′3′-dipropyl-2,2′-thiadicarbocyanine. Aliquots of the sarcolemma-enriched preparation were incubated at 4°C with a medium containing 150 mM KCl, 10 mM NaCl, and 10 mM Tris-Cl, pH 7.4, for 15 to 18 hr. Aliquots of the loaded suspension were then diluted into a reaction medium containing, for trace 1, 150 mM KCl, 10 mM Tris-Cl, pH 7.4 (i.e., same as loading conditions) and, for trace 2, 1.5 mM KCl, 158.5 mM NaCl, and 10 mM Tris-Cl, pH 7.4, at 37°C. After a 1- to 2-min incubation (*baseline trace*), the dye (1 μM final concentration) was added to the preparation, which produced a rapid increase in fluorescence intensity (*dashed line at dye arrow*). After approximately 2 to 4 min, valinomycin (100 nM final concentration) was added as indicated. Instrumentation was as described previously (3,31).

At equilibrium, calcium movement promoted by electrogenic sodium/calcium exchange can be viewed simplistically as:

$$[Ca]_o \underset{k_{-1}}{\overset{k_1}{\rightleftharpoons}} [Ca]_i$$

(Rate constants, k_1 and k_{-1}, would encompass all rate constants involved in the movement of sodium coupled to calcium influx and efflux, respectively.) By definition, equilibrium requires that the rate of calcium influx equal the rate of calcium efflux:

$$k_1[Ca]_o = k_{-1}[Ca]_i \tag{2}$$

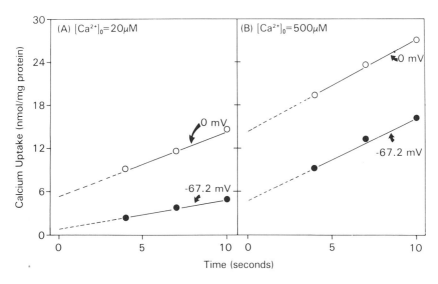

FIG. 7. The effect of membrane potential on sodium-dependent calcium uptake by the sarcolemma-enriched preparation at two calcium concentrations. Aliquots of the preparation were incubated 15 to 18 hr at 5°C with 10 mM Tris-Cl, pH 7.4, 100 mM KCl, 0.04 mM CaCl$_2$ with ^{45}Ca, and 50 mM NaCl. After prewarming to 37°C, these suspensions were added to a medium containing final concentrations of 1.6 mM NaCl, 20 μM **(A)** or 500 μM **(B)** CaCl$_2$ with ^{45}Ca, 10 mM Tris-Cl, pH 7.4, 0.3 μM valinomycin, and either 8.06 or 100 mM KCl to yield Nernst potentials of −67.2 or 0 mV, respectively. The concentration of LiCl was adjusted to keep the ionic strength of the extravesicular medium equal to the intravesicular medium. The data presented have been corrected for calcium movements in the absence of sodium (parallel assays).

which, on rearrangement, yields:

$$\frac{[\text{Ca}]_i}{[\text{Ca}]_o} = \frac{k_1}{k_{-1}} = K_{eq} \quad (3)$$

where K_{eq} is the equilibrium constant for the reaction. Thus, membrane potential must affect the rate constants, k_1 and k_{-1}, in a disproportionate manner to alter K_{eq} and, thereby, $[\text{Ca}]_i$, at equilibrium. There are at least 4 mechanisms by which this may occur (i.e., assuming the $[\text{Na}]_o$, $[\text{Ca}]_o$ and $[\text{Na}]_i$ remain constant). First, membrane potential modulates the affinity of the sodium/calcium exchange process for intracellular and/or extracellular sodium and/or calcium. Second, membrane potential modulates the rate of transport for sodium and calcium across the membrane once they have been bound to the sodium/calcium exchange process. Third, membrane potential changes the stoichiometry of the reaction. Fourth, membrane potential modulates combinations of the above.

Our initial attempt to explore these hypotheses examined sodium-dependent calcium uptake over a range of extravesicular calcium concentrations at membrane potentials of −67.2 and 0 mV (potassium Nernst potentials). The extent of the fast component and the rate of the slow component saturated at higher

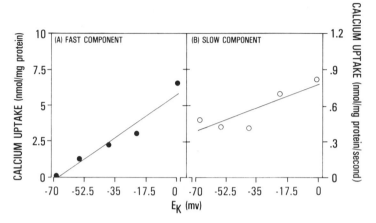

FIG. 8. Effect of membrane potentials on two components of sodium-dependent calcium uptake by the sarcolemma-enriched preparation. Aliquots of the membrane preparation were incubated 15 to 18 hr at 5°C with 10 mM Tris-Cl, pH 7.4, 100 mM KCl, 0.04 mM CaCl$_2$ with ^{45}Ca, and 50 mM NaCl. After prewarming to 37°C, the suspensions were added to media containing final concentrations of 1.6 mM NaCl, 0.04 mM CaCl$_2$ with ^{45}Ca, 10 mM Tris-Cl, pH 7.4, 0.3 μM valinomycin, and various concentrations of KCl to yield potassium Nernst potentials as indicated. Lithium chloride was adjusted to keep the ionic strength of the extravesicular medium equal to that of the intravesicular medium. Data presented have been corrected for sodium-independent calcium movement (parallel assays carried out in absence of sodium). Sodium-dependent uptake of calcium by the fast component **(A)** and the rate of uptake by the slow component **(B)** are the means of 3 experiments.

calcium concentrations (Fig. 9). Values for $K_{0.5}$ (extravesicular calcium yielding 50% of response at saturation), maximal extent (fast component) and maximal rate (slow component) were determined by fitting the data to the Michaelis-Menten equation. Membrane potential affected the $K_{0.5}$ for calcium activation of both components; the more negative potential was associated with a higher $K_{0.5}$ (Table 1). No effect of membrane potential was noted on the maximal rate of the slow component. Membrane potential did affect the maximal extent of the fast component (Fig. 9; Table 1). This would be expected for electrogenic sodium/calcium exchange if our assessment of the fast component reflects equilibrium of the reaction (Eq. 1). Conversely, it should be noted, the fast component

TABLE 1. *Effect of membrane potential on $K_{0.5}$ and maximal extent/rate of calcium uptake*

Membrane potential (mV)	Fast component		Slow component	
	$K_{0.5}$ (μM)	Maximal extent (nmol/mg)	$K_{0.5}$ (μM)	Maximal rate (nmol/mg/sec)
−67.2	147.5	7.0	43.1	1.17
0	37.3	15.4	8.8	1.28

Values obtained from analysis of data presented in Fig. 9.

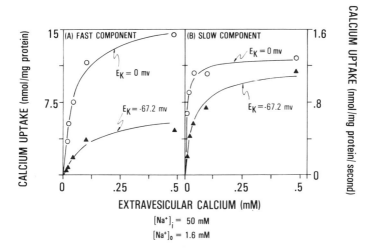

FIG. 9. Sodium-dependent calcium uptake versus [Ca]$_o$ at two membrane potentials. The assays were performed essentially as described in the legend for Fig. 7. The data were fit as a function of [Ca]$_o$ to the Michaelis-Menten equation, which yielded the values presented in Table 1 for $K_{0.5}$, maximal extent of uptake (fast component) and maximal rate of uptake (slow component).

saturates with calcium, which is inconsistent with the assumption that equilibrium values for this component were measured.

These data suggest that hyperpolarization of the vesicles (inside negative) decreases the sensitivity of the reaction to extravesicular calcium, whereas depolarization increases the sensitivity. Extrapolation to the cardiac cell suggests, therefore, that during the plateau of the action potential, calcium influx would be enhanced because the exchange system is more sensitive to extracellular calcium. Conversely, at the resting membrane potential calcium efflux would be favored, since the exchange system is less sensitive to extracellular calcium.

Finally, a more complete understanding of the effects of membrane potential on sodium/calcium exchange requires measurements of the following: (a) initial velocities of the fast component; (b) sodium fluxes, since it remains possible that some of the sodium-dependent calcium movement is not coupled to calcium movement and/or that the stoichiometry of sodium/calcium exchange is altered by membrane potential; (c) the reaction with varying [Ca]$_i$, [Na]$_o$, and [Na]$_i$; (d) the reaction in other types of sarcolemma-enriched preparations, because it is possible that some regions of the surface membrane with characteristics different from those observed are not present in the preparation used; and (e) the contributions of right-side-out and inside-out vesicles to the profiles obtained.

ACKNOWLEDGMENTS

This work was supported in part by grants HL 23802 and GM 20387 from the USPHS. The authors thank Amey Maybank and Linda Vinson for isolation of the sarcolemma-enriched preparations.

REFERENCES

1. Baker, P. F., Blaustein, M. P., Hodgkin, A. L., and Steinhardt, R. A. (1969): The influence of calcium on sodium efflux in squid axons. *J. Physiol. (Lond.)*, 200:431–458.
2. Bartschat, D. K., and Lindenmayer, G. E. (1980): Calcium movements catalyzed by vesicles in highly enriched sarcolemma preparations from canine ventricle: Calcium-calcium and sodium-calcium countertransport. *J. Biol. Chem.*, 255:9626–9634.
3. Bartschat, D. K., Cyr, D. L., and Lindenmayer, G. E. (1980): Depolarization-induced calcium uptake by vesicles in a highly enriched sarcolemma preparation from canine ventricle. *J. Biol. Chem.*, 255:10044–10047.
4. Bers, D. M., Philipson, K. D., and Nishimoto, A. Y. (1980): Sodium-calcium exchange and sidedness of isolated cardiac sarcolemma vesicles. *Biochim. Biophys. Acta*, 601:358–371.
5. Besch, H. R., Jones, L. R., and Watanabe, A. M. (1976): Intact vesicles of canine cardiac sarcolemma. Evidence from vectorial properties of Na^+,K^+-ATPase. *Cir. Res.* 39:586–595.
6. Blaustein, M. P. (1974): The interrelationship between sodium and calcium fluxes across cell membranes. *Rev. Physiol. Biochem. Pharmacol.*, 70:33–82.
7. Blaustein, M. P., and Hodgkin, A. L. (1969): Effects of cyanide on the efflux of calcium from squid axons. *J. Physiol. (Lond.)*, 200:497–527.
8. Caroni, P., Reinlib, L., and Carafoli, E. (1980): Charge movements during the Na^+-Ca^{2+} exchange in heart sarcolemma vesicles. *Proc. Natl. Acad. Sci. USA*, 77:6354–6358.
9. Caroni, P., Fabrizio, V., and Carafoli, E. (1981): The cardiotoxic antibiotic doxorubicin inhibits the Na^+/Ca^{2+} exchange of dog heart sarcolemma vesicles. *FEBS Lett.*, 130:184–186.
10. Glitsch, H. G., Reuter, H., and Scholz, H. (1970): The effect of the internal sodium concentration on calcium fluxes in isolated guinea-pig auricles. *J. Physiol. (Lond.)*, 209:25–43.
11. Hodgkin, A. L., and Huxley, A. F. (1952): A quantitative description of membrane current and its application to conduction and excitation in nerve. *J. Physiol. (Lond.)*, 117:500–544.
12. Horackova, M., and Vassort, G. (1979): Sodium-calcium exchange in regulation of cardiac contractility. *J. Gen. Physiol.*, 73:403–424.
13. Hungerford, R. T., and Lindenmayer, G. E. (1983): Manuscript in preparation.
14. Jones, L. R., Besch, H. R., Jr., Fleming, J. W., McConnaughey, M. M., and Watanabe, A. M. (1979): Separation of vesicles of cardiac sarcolemma from vesicles of cardiac sarcoplasmic reticulum. *J. Biol. Chem.*, 254:530–539.
15. Kadoma, M., Froehlich, J., Reeves, J., and Sutko, J. (1982): Kinetics of sodium ion induced calcium ion release in calcium ion loaded cardiac sarcolemma vesicles: Determination of initial velocities by stopped-flow spectrophotometry. *Biochemistry*, 21:1914–1918.
16. Langer, G. A. (1972): Effects of digitalis on myocardial ionic exchange. *Circulation*, 46:180–187.
17. Langer, G. A., and Serena, S. D. (1970): Effects of strophanthidin upon contraction and ionic exchange in rabbit ventricular myocardium. Relation to control of active state. *J. Mol. Cell. Cardiol.*, 1:65–90.
18. Miyamoto, H., and Racker, E. (1980): Solubilization and partial purification of the Ca^{2+}/Na^+ antiporter from the plasma membrane of bovine heart. *J. Biol. Chem.*, 255:2656–2658.
19. Mullins, L. J. (1977): Mechanism for Na/Ca transport. *J. Gen. Physiol.*, 70:681–695.
20. Mullins, L. J. (1979): The generation of electric currents in cardiac fibers by Na/Ca exchange. *Am. J. Physiol.*, 236:C103–C110.
21. Mullins, L. J. (1981): *Ion Transport in the Heart*. Raven Press, New York.
22. Nachshen, D. A., and Blaustein, M. P. (1980): Some properties of potassium-stimulated calcium influx in presynaptic nerve endings. *J. Gen. Physiol.*, 76:709–728.
23. Philipson, K. D., and Nishimoto, A. Y. (1980): Na^+-Ca^{2+} exchange is affected by membrane potential in cardiac sarcolemma vesicles. *J. Biol. Chem.*, 255:6880–6882.
24. Philipson, K. D., and Nishimoto, A. Y. (1981): Efflux of Ca^{2+} from cardiac sarcolemma vesicles. *J. Biol. Chem.*, 256:3698–3702.
25. Pitts, B. J. R. (1979): Stoichiometry of sodium-calcium exchange in cardiac sarcolemma vesicles. Coupling to the sodium pump. *J. Biol. Chem.*, 254:6232–6235.
26. Reeves, J. P., and Sutko, J. L. (1979): Sodium-calcium ion exchange in cardiac sarcolemma vesicles. *Proc. Natl. Acad. Sci. USA*, 76:590–594.
27. Reeves, J. P., and Sutko, J. L. (1980): Sodium-calcium exchange activity generates a current in cardiac membrane vesicles. *Science*, 208:1461–1463.

28. Reeves, J., Trumble, W., Sutko, J., Kadoma, M., and Froehlich, J. (1981): Calcium transport mechanisms in cardiac sarcolemma vesicles. In: *Calcium and Phosphate Transport Across Biomembranes,* edited by F. Bronner and M. Peterlik, pp. 15–18. Academic Press, New York.
29. Reuter, H. (1974): Exchange of calcium in the mammalian myocardium. *Circ. Res.,* 34:599–605.
30. Reuter, H., and Seitz, N. (1968): The dependence of calcium efflux from cardiac muscle on temperature and external ion composition. *J. Physiol. (Lond.),* 195:451–470.
31. Schilling, W. P. (1982): Development of new approaches for the study of mechanisms responsible for the generation and maintenance of resting membrane potentials in the heart. PhD Thesis, Medical University of South Carolina, Charleston, South Carolina.
32. Schuil, D. W. (1982): Characterization of solute movements across the isolated mammalian myocardial sarcolemma. PhD Thesis, Medical University of South Carolina, Charleston, South Carolina.
33. Van Alstyne, E., Burch, R. M., Knickelbein, R. G., Webb, J. G., Hungerford, R. T., Gower, E. J., Poe, S. L., and Lindenmayer, G. E. (1980): Isolation of sealed vesicles highly enriched with sarcolemma markers from canine ventricle. *Biochim. Biophys. Acta,* 602:131–143.

Electrogenic Transport: Fundamental Principles and Physiological Implications,
edited by Mordecai P. Blaustein and Melvyn
Lieberman. Raven Press, New York © 1984.

Functional Significance of Electrogenic Pumps in Neurons

David O. Carpenter and Robert A. Gregg

Center for Laboratories and Research, New York State Department of Health, Albany, New York 12201

In neurons, as in other types of cells, the principal function of the Na^+ pump is to maintain ion-concentration gradients. However, large transient changes in membrane potential may occur across neuronal membranes as a result of alterations in Na^+-pump activity (34). These changes are due to currents generated by the Na^+ pump on the basis of sensitivity to cardiac glycosides and temperature and dependence on internal Na^+ and external K^+.

In neurons electrogenic Na^+-pump-dependent potentials have resulted from posttetanic activity in small-diameter axons (28,29) and crayfish stretch-receptor neurons (32). They have been induced by injecting Na^+ directly into neuronal cell bodies to stimulate transport (19,33), by experiments designed to study the temperature dependence of membrane potential changes (5,9,11,23), and by comparing the labeled Na^+ influx from squid giant axons to the digitalis-sensitive membrane potential (1).

Although much data indicate that all cells probably have an electrogenic Na^+ pump, questions remain regarding the physiologic function, if any, of the electrogenic portion of pump activity. Another major question concerns the steady state, rather than transient, contributions of electrogenic pumps to membrane potential.

Our studies have focused on the temperature dependence of membrane potential and neuronal excitability. Results to date suggest that an electrogenic Na^+ pump is the mechanism that generates potentials in specific cold thermoreceptors, but it also imparts a temperature dependence to membrane potential in many other cell types, which may not utilize such information.

MATERIALS AND METHODS

Most procedures have been previously described. The methods for the *Aplysia* experiments were described by Carpenter and Alving (9), Carpenter (5), and Willis et al. (36). Methods for rat scrotal nerve studies have been described by Pierau et al. (25,26). Cat sensorimotor cortex methods were described by

Barker and Carpenter (3). Rat brain slice studies were done as described by Hori et al. (17), except that a Neslab RTE4 refrigerated circulating bath was used to alter the perfusate temperature by circulating warm or cool water through the jacket of the recording chamber. This procedure resulted in a perfusate temperature change of about 1°C/min in either direction.

RESULTS

The effects of a relatively abrupt temperature change on the membrane potential and discharge of 4 identified neurons of the marine mollusk *Aplysia* are illustrated in Fig. 1. At the beginning of the trace only cell R_{15} shows a discharge at about 25°C. This discharge is due to an endogenous pacemaker mechanism, characteristic of this and several other neurons. When the temperature is lowered to about 5°C within about 30 sec, cells R_6 and R_{14} depolarize and begin to discharge; the pleural giant cell (PGC) depolarizes but remains silent; and R_{15}

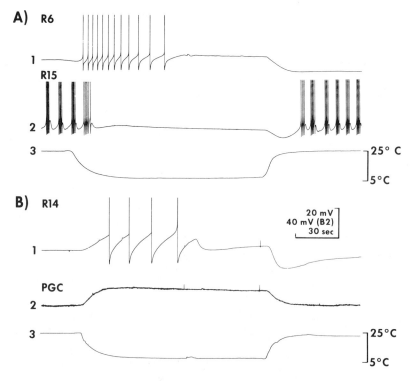

FIG. 1. Transient responses of 4 *Aplysia* neurons to sudden temperature change. *Traces 1 and 2* are intracellular DC recordings of responses from different identified neurons, cells R_6, R_{15}, R_{14}, and PGC (pleural giant cell). *Trace 3* shows the temperature. Temperature was changed by switching between two seawater reservoirs maintained at 5 and 25°C, respectively. [From Wiederhold and Carpenter (35), with permission.]

ceases to discharge without much obvious potential change. On rewarming, all 4 neurons undergo a clear hyperpolarization, which is as great as 20 to 30 mV. After this transient hyperpolarization R_{15} resumes bursting discharge.

These results, all obtained while perfusing the ganglion with seawater containing a four-times-normal concentration of Mg^{2+} to depress synaptic activity, represent a variety of responses to temperature change. Further analysis showed that the various patterns can all be explained by two separate temperature-dependent processes that have opposite polarities, different rapidity of changes after the temperature alteration, and different relative contributions in various cells. The first process is an electrogenic Na^+ pump, which tends to cause hyperpolarization of the cell as the temperature increases. The second process relates to the temperature dependence of passive permeabilities.

Evidence for the pump contribution is shown in Figs. 2 to 5. Figure 2 shows two neurons, one of which discharges at 5°C but not at 25°C, whereas the other has the reverse pattern. Addition of ouabain (10^{-4} M) produces little change in the activity of either cell at 5°C, when the pump would be expected to have little activity, but a dramatic increase in the discharge rate at 25°C, when pump should be maximally active within its physiologic range. Assuming that ouabain alters only the Na^+ pump, these results indicate that the pump functions as a maintained outward-current generator at 25°C. The discharge patterns of three pacemaker neurons are shown in Fig. 3. Under control conditions these neurons are models of warm (A), biphasic (B), and cold (C) thermoreceptors. In all 3 cells, after ouabain treatment, the discharge increases with warming. Thus the mechanism that makes cell C a cold receptor is present in all of these neurons, but in cells A and B its effects are masked by other temperature-dependent mechanisms.

Removal of K^+ is a very useful tool for study of electrogenic pumps. If membrane potential is described by the Goldman-Hodgkin-Katz equation and if the cell is more permeable to K^+ than to other ions, lowering extracellular K^+ should hyperpolarize the cell. However, Na^+-pump activity requires external K^+. Removal of K^+ should block the electrogenic pump potentials and tend to depolarize the cell. The results of perfusion with K^+-free seawater on a silent cell (R_2) are shown in Fig. 4. Figure 4A shows the effects at 6 and 22°C. At 6°C, when the pump is not very active, K^+-free seawater causes the expected hyperpolarization. At 22°C the initial membrane potential is greater by 10 mV, and lack of exogenous K^+ results in depolarization and initiation of spontaneous discharge in this normally silent cell. These results are consistent with an inference that the membrane potential follows that predicted from the passive permeabilities at 6°C but not at 22°C. Figure 4B shows the temperature dependence of membrane potential in this cell, reflecting the electrogenic pump activity.

The effects on a pacemaker neuron of perfusion with K^+-free seawater are shown in Fig. 5. This neuron has a highly patterned discharge at higher temperatures and a slow discharge at 5°C. In K^+-free seawater the pattern is lost at

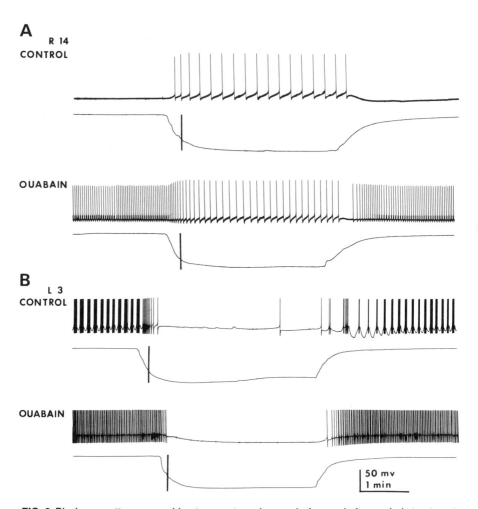

FIG. 2. Discharge patterns on sudden temperature changes before and after ouabain treatment in *Aplysia* neurons R_{14} and L_3. In each case the *upper trace* is the intracellular recording, and the *lower trace* is temperature, where the vertical bar rises from 5 to 20°C. In R_{14} exposure to ouabain (10^{-4} M) for 10 min changes the response from that of cold receptor to that of warm receptor. Cell L_3 in control conditions illustrates its normal bursting pattern, which ouabain alters to a more regular discharge. [From Willis et al. (36), with permission.]

high temperatures, and the cell fires furiously. The loss of pattern is not due to any impairment of the cell's ability to burst, since the pattern can be restored by hyperpolarizing current (Fig. 5C). The effect is totally reversible. These results support the conclusion that the pump contributes a maintained outward current at warmer temperatures, without which an uncontrolled discharge results.

Each of these figures shows a dramatic temperature dependence of membrane potential and discharge in the absence of pump activity. This dependence is characteristic of all *Aplysia* neurons, although it is most dramatic in pacemaker neurons. The depolarization on warming is not due to an active process; but, rather, it reflects the differences in temperature dependence of the passive Na^+ and K^+ conductances, g_{Na^+} and g_{K^+} (5,12,23). Differences in Q_{10} between g_{Na^+} and g_{K^+} were noted by Hodgkin and Keynes (15,16) in the squid giant axon.

Overall, these results in *Aplysia* neurons were surprising at the time the observations were made. These neuronal cell bodies have a much greater electrogenic potential and temperature dependence of membrane potential than does squid axon. However, consideration of the passive membrane properties of the two preparations showed that the difference was due to differences in the specific membrane resistance. The specific membrane resistance of squid giant axon is about 1,000 Ωcm^2 (10), whereas that of *Aplysia* neurons is at least 100,000 Ωcm^2 (5). Since an electrogenic pump generates potential as a result of the current crossing a resistance, the greater electrogenic potentials in *Aplysia* neurons can be explained by Ohm's law without proposing any difference in pump activity or coupling ratio.

The comparison of *Aplysia* neurons to squid axon led us to question whether there are, in general, consistent and significant differences in specific membrane resistance between cell bodies and axons. Since cell bodies and their associated dendrites are regions of synaptic input, a potential must be effectively generated there by relatively small currents. Unlike axons, which are conducting cables, the cell bodies are integrative and responsive components that need the high resistance. To test the generality of this view we studied the small cell bodies that give rise to the squid giant axon. These cells had a higher specific resistance (at least 27,000 Ωcm^2) and a temperature dependence of membrane potential similar to that of *Aplysia* neurons (6).

These observations in invertebrates led us to attempt an extrapolation to mammalian systems. The invertebrate results suggested not only possible mechanisms of generator potentials in cold and warm thermal afferents, but also that these mechanisms may be common to all neurons, whether or not the thermal information is used. Studies on neurons in the anesthetized cat cortex (Fig. 6) demonstrated that local brain temperature change alters the rate of neuron discharge in the sensorimotor cortex, a region not known to process thermal information. The temperature change was produced by perfusing water at various temperatures through metal coils placed near the carotid rete; this altered the temperature of the blood perfusing the brain relatively rapidly. Whereas in the intact mammalian brain some activity is synaptic, many neurons are pacemakers (37). These results are consistent with—but certainly do not prove—the hypothesis that mammalian neuronal cell bodies have a high specific membrane resistance and temperature-dependent processes like those of the *Aplysia* nervous system.

We have recently extended this approach to investigate the effects of temperature changes, cardiac glycoside application, and perfusion with K^+-free solution on an isolated, submerged, and perfused slice of rat prepyriform cortex. In this preparation one can evoke a population-response field potential reflecting

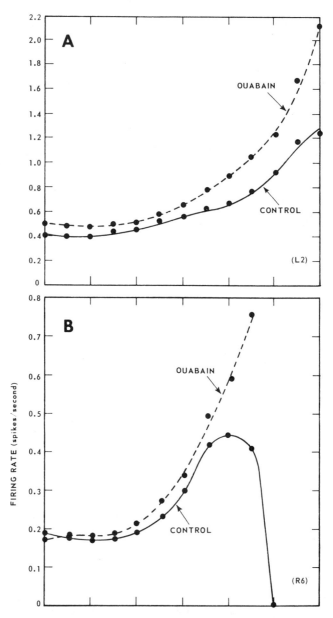

FIG. 3. Effects of temperature change on firing rate in 3 *Aplysia* neurons before and after exposure to ouabain. **A:** A warm receptor response, is from cell L_2; **B:** a peaked response; and **C:** a cold receptor response, are separate experiments on different cells R_6. Each point represents the average frequency over 2 min, taken at least 10 min after temperature change. Ouabain-modified responses were obtained after 10-min incubation with 10^{-4} M ouabain, which was washed out before the responses were recorded. In all 3 cases, ouabain treatment results in an increase in spike frequency with increasing temperature over entire range studied. [From Willis et al. (36), with permission.]

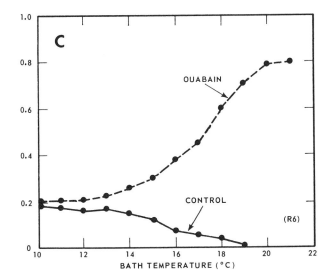

FIG. 3C. See legend opposite.

a monosynaptic excitatory input onto the pyramidal neurons. If a pump contributes to membrane potential in these neurons, as in *Aplysia* neurons, manipulations that block pump activity will depolarize the cell. Assuming that other critical factors, such as presynaptic release of neurotransmitter, are not as temperature-dependent, the depolarization will result in a decrease in the amplitude of the population excitatory postsynaptic potential, since the driving force is reduced.

The effect of temperature on the peak amplitude of the population response is illustrated in Fig. 7. Cooling to 20°C reduces the peak field potential by about 50%. The effect of K^+-free Ringer's perfusion on peak amplitude as a function of time is shown in Fig. 8. Initially the population response increases, as expected on the basis of the increased K^+ concentration gradient. It then falls rapidly to approximately 50% of control. In other experiments perfusion of 50% normal K^+ Ringer's solution resulted in a maintained increase in amplitude. Thus, the transient increase probably results from an increased K^+ concentration gradient at a time when there is still too much extracellular K^+ to block pump activity.

The results of both temperature changes and K^+-free solutions are consistent with an important and maintained contribution of an electrogenic pump to membrane potential on these neurons. Further support for this conclusion is shown in Fig. 9, where a 20-min perfusion of 1 μM ouabain in Ringer's solution effectively—and essentially irreversibly—eliminated the field potential normally induced by stimulation of the lateral olfactory tract.

Although the above-mentioned studies are all on neurons not specifically involved in thermoreception, the activity of the pump might be a determinant

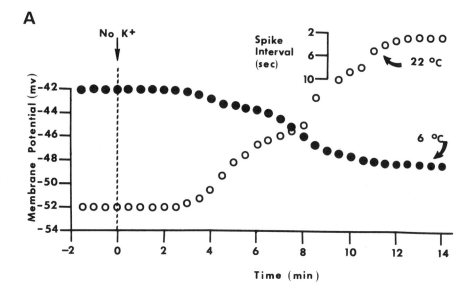

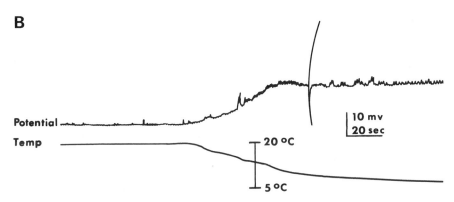

FIG. 4. Resting membrane potential in giant neuron of abdominal ganglion of *Aplysia* as a function of time after exposure to K+-free seawater **(A)** at 6 and 22°C and **(B)** during cooling from 20 to 5°C. In **A** the preparation was perfused with normal artificial seawater containing 10 mM K+ until the *arrow;* thereafter the solution lacked K+. At 22°C the neuron depolarized so much after 9 min of K+-free perfusion that it began to discharge. This made the resting membrane potential values unclear, and thus the frequency of discharge is indicated. [From Carpenter (8), with permission.]

of temperature-dependent discharge in cold receptors. Sudden temperature changes create transient changes in the response of afferent fibers, both those responding only to temperature and those responding to both temperature and mechanical stimulation. Figure 10 illustrates these transient changes in a mechano- and thermosensitive afferent fiber isolated from the pudendal nerve

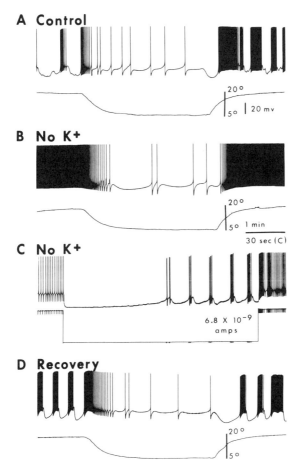

FIG. 5. Effect of K$^+$-free seawater on discharge and Na$^+$ pump activity in *Aplysia* cell R$_{15}$. In each panel the **top** trace illustrates the voltage recorded from one of two independently inserted microelectrodes. The **bottom** trace records either the temperature (**A, B,** and **D**) or the hyperpolarizing current (**C**) applied through the second electrode. In **C** the temperature was 20°C, and the trace was taken at twice the sweep speed used for **A, B,** and **D.** [From Carpenter (6), with permission.]

of a rat scrotum. These responses are very similar to those of *Aplysia* neurons (Fig. 7 in ref. 4). They are a clear transient acceleration of discharge on cooling (see Fig. 1, R$_6$ and R$_{14}$) and, when a neuron has spontaneous activity at both high and low temperatures, a transient pause on warming, even though the maintained discharge at higher temperatures may be greater.

In order to test more directly whether the transient responses in specific afferent fibers reflect electrogenic Na$^+$-pump activity, Pierau et al. (22) infiltrated ouabain into the receptor field area of a cold-sensitive mechanoreceptor. Al-

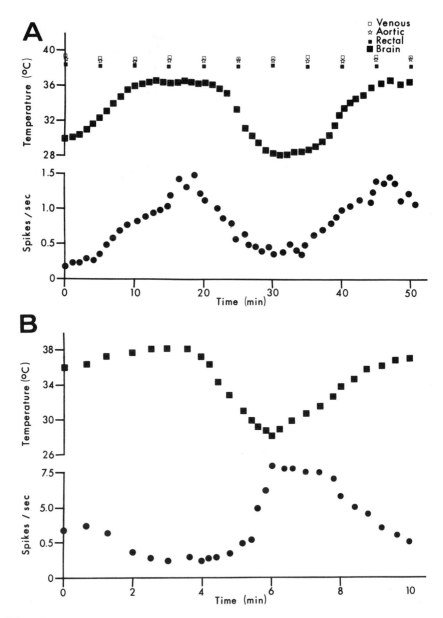

FIG. 6. Change in cat brain temperature and frequency of response of 2 neurons in the sensorimotor cortex. The patterns show that **(A),** the discharge rate (1-min average), is directly related to temperature; and **(B),** unit activity (25-sec average), is inversely related to temperature. Aortic, venous, and rectal temperatures **(A** only) do not change despite the change in brain temperature. [From Barker and Carpenter (3), with permission.]

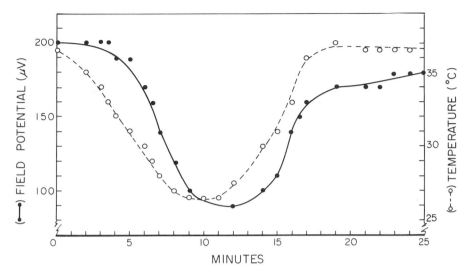

FIG. 7. Effect of temperature on field potential, which reflects monosynaptic excitation of pyramidal neurons of rat prepyriform cortex by stimulation of the lateral olfactory tract (LOT). The field potential was generated by bipolar stimulation of the LOT with two pulses, each 30 μsec in duration. Each point represents the maximal amplitude of the second of two stimuli delivered at a 100-msec interval.

though the receptor discharge as a function of temperature is unchanged by saline, with ouabain treatment the cell becomes a warm receptor (Fig. 11). This effect of ouabain on rat scrotal skin is essentially identical to the response in *Aplysia* (Fig. 2A).

To better control the medium, Pierau et al. (24) developed an isolated pudendal nerve and rat scrotal skin preparation. Figure 12 shows the effect of perfusion with K^+-free solution on activity recorded from a cold receptor in this preparation. As for *Aplysia* neurons, this cell has an accelerated spontaneous discharge in K^+-free medium despite the increased K^+-concentration gradient, indicating that an electrogenic pump contributes to normal maintenance of membrane potential.

DISCUSSION

Our results with isolated invertebrate neurons, mammalian brain slices, and intact mammalian peripheral afferent fibers and cortical neurons suggest that an electrogenic Na^+ pump is a common mechanism contributing to potential in these cells and tissues. Manipulations that block Na^+-pump activity (cooling, K^+-free solutions, application of cardiac glycosides) result in an immediate loss of potential at all but low temperatures, where the pump is not active, and a consequent depolarization. In *Aplysia* neurons the depolarization is not due to a change in the K^+-concentration gradient (9), and it is probably too rapid

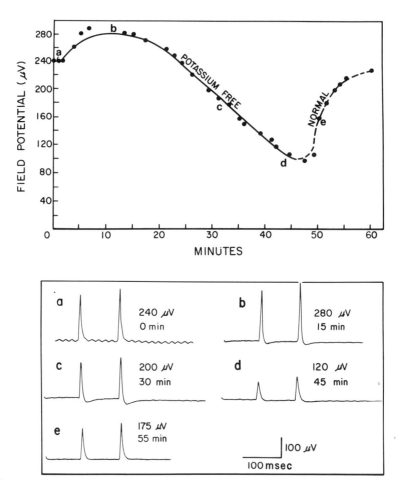

FIG. 8. Effects of perfusion of quarter normal K^+ Krebs-Ringer's solution on the LOT-evoked field potential in a rat prepyriform brain slice. **Top:** Maximal amplitudes of the second of paired responses. **Bottom:** Raw data for five time points on the field-potential curve. The second response is larger than the first because of frequency potentiation.

to be explained by a loss of concentration gradients in the other preparations as well.

Perhaps it should not be surprising that the electrogenic pump has similar effects in a variety of neurons. What is surprising is the magnitude of these effects in brain slices and afferent fibers. In this regard the various mammalian preparations we have studied are similar to *Aplysia* neurons rather than to squid axon. Although the actual operation of the pump in squid axon is apparently not different (1), the maximal potential that it can generate there is < 2 mV, given the low membrane resistance (16). Although membrane resistance, especially of complex central neurons, is difficult to determine accurately, that

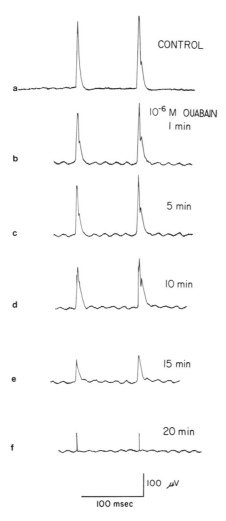

FIG. 9. Effect of ouabain on LOT-evoked field potentials in a rat prepyriform brain slice. The responses are totally and irreversibly abolished within 20 min at 1 μM ouabain.

of squid axon was shown by Cole and Hodgkin (10) to be about 1,000 Ωcm², whereas Carpenter (5) estimated that of *Aplysia* neurons to be not less than 100,000 Ωcm². Mammalian neurons may have specific resistances that approach or equal that of *Aplysia* neurons, as is suggested by their prominent pump potentials. As a general rule, axons may have a consistently lower resistance.

Although transient pump-dependent potentials are dramatic, it is clear that the electrogenic pump contributes directly to the maintenance of membrane potential in neuronal cell bodies. In squid axon the membrane potential is maintained for hours after pump blockade, even when action potentials are generated,

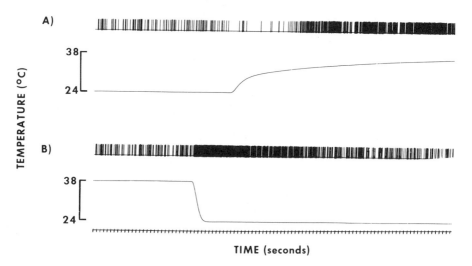

FIG. 10. Transient responses to sudden temperature changes in a mechanosensitive afferent fiber dissected from the pudendal nerve of a rat scrotum. **A** and **B** show responses to sudden warming and cooling, respectively. In each panel the *upper trace* is the output of a window discriminator, which gives a pulse for each action potential from the afferent fiber. The *lower trace* records temperature. [From Wiederhold and Carpenter (35), with permission.]

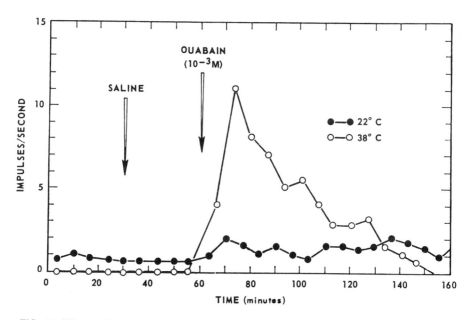

FIG. 11. Effects of temperature change and ouabain infiltration onto the receptive field area on static discharge of a single cold-sensitive mechanoreceptor of the scrotal skin of an anesthetized rat. The temperature was alternated between 22 and 38°C every 3 min. Each point represents the average impulse frequency over a 30-sec period beginning 2.5 min after the onset of the temperature change. [From Pierau et al. (24), with permission.]

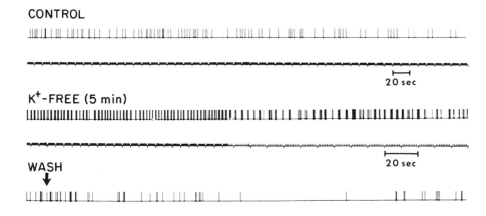

FIG. 12. Effect of K⁺-free Ringer's perfusion on discharge of a temperature-sensitive afferent fiber, recorded from an isolated, submerged, and perfused rat scrotal skin-nerve preparation. In each panel the *upper trace* is the output of a window discriminator and reflects single action potentials. The *lower trace* indicates time. Note the change in the time base in the middle of the K⁺-free trace, which is continuous with the wash trace. [From Pierau et al. (26), with permission.]

bringing Na^+ into the cell (20). In contrast, *Aplysia* and rat brain slice neurons cannot maintain a normal potential in the absence of an active pump, even when concentration gradients are maintained. Thus the electrogenic current is a critical component of resting potential. The regulatory mechanisms that determine membrane potential are not known exactly, but a major factor is probably the regulation of (Na^+-K^+)ATPase activity by the internal concentration of Na^+. In any case the electrogenic pump current in these cell bodies contributes as much to resting potential as do the ion-concentration gradients.

Given that the pump has a dramatic direct effect on excitability of neurons, some agents could regulate neuronal activity by altering the pump rate. Regulation of pump activity by neurotransmitter substances has been proposed in several studies (21,27,30), although the significance and mechanism of this regulation remain unclear. In at least some cases the transmitters appear only to increase membrane resistance (31). Glucocorticoid steroid hormones may have direct effects on (Na^+-K^+)ATPase activity. The cardiotonic steroid, chlormadione acetate, which inhibits (Na^+-K^+)ATPase activity (2), competes with ouabain in binding to dissociated dog heart muscle (22). Hall (13) demonstrated that methylprednisolone alters the excitability of cat spinal motor neurons by inducing hyperpolarization of the soma. Hall speculated that this change reflects electrogenic pump alteration.

Since Na^+ transport is very temperature-dependent, the electrogenic potential is an excellent candidate for a mechanism to create a cold receptor. To make a cold receptor one would want a neuron that has spontaneous discharge at cold temperatures but whose discharge is reduced and stopped as the cell is warmed. Many afferent fibers show a pacemaker discharge (35), and endogenous pacemaker discharge requires a high membrane resistance in conjunction with a relatively high and maintained g_{Na^+} (7). These are precisely the conditions that lead to pump activity and efficient production of electrogenic potentials. Our observations on specific and mechanosensitive, cold-sensitive afferent fibers from mammals are consistent with the hypothesis that afferent thermosensitivity is due to the activity of an electrogenic Na^+ pump, balanced against the effect of the temperature dependence of passive conductances to Na^+ and K^+.

Considerable controversy exists among sensory physiologists over the relative importance of transient versus maintained discharges in information processing (14). Although information can be transmitted in both ways, the transient responses are often more dramatic. Our studies have shown important differences in the time constants of change of the two principal temperature-dependent processes which determine neuronal thermosensitivity. The pump currents change rapidly with temperature, whereas the passive conductances have a short lag. The result is anomalous, transient responses—an excitation on cooling and a pause on warming—which may not reflect the discharge patterns at maintained temperatures. It seems likely that this information is used, since human temperature sensitivity is best for gradients (18).

The mechanisms that impart neuronal thermosensitivity are present in all of the neurons studied, whether or not the neurons are involved in mediating temperature information. Use of thermal information must therefore be a function of connective, not inherent, thermosensitivity.

In summary, electrogenic Na^+ pumps exist in neurons and probably all cells, but the physiologic effect of the pump varies with the pump rate (determined by Na^+ entry) and membrane resistance. In cold thermoreceptors the electrogenic pump has an important role in the generation of potential. In many cells the pump alters excitability after Na^+ entry and can possibly be modulated by transmitters and hormones. The electrogenic pump current is essential for maintenance of resting membrane potential in some neuronal cell bodies.

REFERENCES

1. Abercrombie, R. F., and DeWeer, P. (1978): Electric current generated by squid giant axon sodium pump: External K and internal ADP effects. *Am. J. Physiol.*, 235:C63–C68.
2. Akera, T. (1977): Membrane adenosinetriphosphatase: Digitalis receptor? *Science*, 198:569–574.
3. Barker, J., and Carpenter, D. O. (1970): Thermosensitivity of neurons in the sensorimotor cortex of the cat. *Science*, 169:597–598.
4. Carpenter, D. O. (1967): Temperature effects on pacemaker generation, membrane potential, and critical firing threshold in *Aplysia* neurons. *J. Gen. Physiol.*, 50:1469–1484.
5. Carpenter, D. O. (1970): Membrane potential produced directly by the Na^+ pump in *Aplysia* neurons. *Comp. Biochem. Physiol.*, 35:371—385.
6. Carpenter, D. O. (1973): Electrogenic sodium pump and high specific resistance in nerve cell bodies of the squid. *Science*, 179:1336–1338.

7. Carpenter, D. O. (1973): Ionic mechanisms and models of endogenous discharge of *Aplysia* neurones. In: *Neurobiology of Invertebrates: Mechanisms of Rhythm Regulation,* edited by J. Salánki, pp. 35–58. Akadémiai Kiodó, Budapest.
8. Carpenter, D. O. (1977): Membrane excitability. In: *Mammalian Cell Membranes, vol. 4, Membranes and Cellular Functions,* edited by G. A. Jamieson and D. M. Robinson, pp. 184–206. Butterworths, London.
9. Carpenter, D. O., and Alving, B. O. (1968): A contribution of an electrogenic Na^+ pump to membrane potential in *Aplysia* neurons. *J. Gen. Physiol.,* 52:1–21.
10. Cole, K. S., and Hodgkin, A. L. (1939): Membrane and protoplasm resistance in the squid giant axon. *J. Gen. Physiol.,* 22:671–687.
11. Gorman, A. L. F., and Marmor, M. F. (1970): Contributions of the sodium pump and ionic gradients to the membrane potential of a molluscan neurone. *J. Physiol. (Lond.),* 210:897–917.
12. Gorman, A. L. F., and Marmor, M. F. (1970): Temperature dependence of the sodium-potassium permeability ratio of a molluscan neurone. *J. Physiol. (Lond.),* 210:919–931.
13. Hall, E. D. (1982): Acute effects of intravenous glucocorticoid on cat spinal motor neuron electrical properties. *Brain Res.,* 240:186–190.
14. Hensel, H. (1973): Neural processes in thermoregulation. *Physiol. Rev.,* 53:948–1017.
15. Hodgkin, A. L., and Keynes, R. D. (1955): Active transport of cations in giant axons form *Sepia* and *Loligo. J. Physiol. (Lond.),* 128:28–60.
16. Hodgkin, A. L., and Keynes, R. D. (1955): Experiments on the injection of substances into squid giant axons by means of a micro syringe. *J. Physiol. (Lond.),* 131:592–616.
17. Hori, N., Auker, C. R., Braitman, D. J., and Carpenter, D. O. (1981): Lateral olfactory tract transmitter: Glutamate, aspartate, or neither? *Cell. Mol. Neurobiol.,* 1:115–120.
18. Johnson, K. O., Darien-Smith, I., and LaMotte, C. (1973): Peripheral neural detrminants of temperature discrimination in man: A correlative study of responses to cooling skin. *J. Neurophysiol.,* 36:347–370.
19. Kerkut, G. A., and Thomas, R. C. (1965): An electrogenic sodium pump in snail nerve cells. *Comp. Biochem. Physiol.,* 14:167–183.
20. Keynes, R. D., and Lewis, P. R. (1951): The sodium and potassium content of cephalopod nerve fibres. *J. Physiol. (Lond).,* 114:151–182.
21. Koketsu, K., and Nakamura, M. (1976): The electrogenesis of adrenaline-hyperpolarization of sympathetic ganglion cells in bullfrogs. *Jpn. J. Physiol.,* 26:63–77.
22. LaBella, F. S., Bihler, I., and Kim, R. S. (1979): Progesterone derivative binds to cardiac ouabain receptor and shows dissociation between sodium pump inhibition and increased contractile force. *Nature,* 278:571–573.
23. Marmor, M. F. (1971): The effects of temperature and ions on the current-voltage relation and electrical characteristics of a molluscan neurone. *J. Physiol. (Lond.),* 218:573–598.
24. Pierau, F.-K., Torrey, P., and Carpenter, D. O. (1974): Mammalian cold receptor afferents: Role of an electrogenic sodium pump in sensory transduction. *Brain Res.,* 73:156–160.
25. Pierau, F.-K., Torrey, P., and Carpenter, D. O. (1975): Afferent nerve fiber activity responding to temperature changes of scrotal skin of the rat. *J. Neurophysiol.,* 38:601–612.
26. Pierau, F.-K., Torrey, P., and Carpenter, D. (1975): Effect of ouabain and potassium-free solution on mammalian thermosensitive afferents *in vitro. Pfluegers Arch.,* 359:349–356.
27. Pinsker, H., and Kandel, E. R. (1969): Synaptic activation of an electrogenic sodium pump. *Science,* 163:931–935.
28. Rang, H. P., and Ritchie, J. M. (1968): On the electrogenic sodium pump in mammalian non-myelinated nerve fibres and its activation by various external cations. *J. Physiol. (Lond.),* 196:183–221.
29. Ritchie, J. M., and Straub, R. W. (1957): The hyperpolarization which follows activity in mammalian non-medullated fibres. *J. Physiol. (Lond.),* 136:80–97.
30. Sawada, M., Enomoto, K., Maeno, T., and Blankenship, J. E. (1980): Ionic mechanism of inhibition of long duration in *Aplysia* synapse. *J. Neurosci. Res.,* 5:537–553.
31. Shirasawa, Y., and Koketsu, K. (1978): An analysis of 5-HT hyperpolarization of sympathetic ganglion cells. *Jpn. J. Pharmacol.,* 28:57–60.
32. Sokolove, P. G., and Cooke, I. M. (1971): Inhibition of impulse activity in a sensory neuron by an electrogenic pump. *J. Gen. Physiol.,* 57:125–162.
33. Thomas, R. C. (1969): Membrane current and intracellular sodium changes in a snail neuron during extrusion of injected sodium. *J. Physiol. (Lond.),* 201:495–514.

34. Thomas, R. C. (1972): Electrogenic sodium pump in nerve and muscle cells. *Physiol. Rev.*, 52:563–594.
35. Wiederhold, M. L., and Carpenter, D. O. (1982): Possible role of pacemaker mechanisms in sensory systems. In: *Cellular Pacemakers, vol. 2, Function in Normal and Disease States*, edited by D. O. Carpenter, pp. 28–58. John Wiley, New York.
36. Willis, J. A., Gaubatz, G. L., and Carpenter, D. O. (1974): The role of the electrogenic sodium pump in modulation of pacemaker discharge of *Aplysia* neurons. *J. Cell. Physiol.*, 84:463–472.
37. Wong, R. K. S., and Schwartzkroin, P. A. (1982): Pacemaker neurons in the mammalian brain: Mechanisms and function. In: *Cellular Pacemakers, vol. 1, Mechanisms of Pacemaker Generation*, edited by D. O. Carpenter, pp. 237–254. John Wiley, New York.

Electrogenic Transport: Fundamental Principles and Physiological Implications, edited by Mordecai P. Blaustein and Melvyn Lieberman. Raven Press, New York © 1984.

Physiological Role of Electrogenic Pumps in Smooth Muscle

John A. Connor

University of Illinois, Department of Physiology and Biophysics, Urbana, Illinois 61801; and Bell Laboratories, Murray Hill, New Jersey 07974

The muscle layers of the small intestine generate a complex pattern of electrical activity as shown in Fig. 1A. This activity is composed of slow, periodic depolarizations, 10 to 30 mV in amplitude, which last from 2 to 5 sec. Often calcium-dependent action potentials are superimposed on the depolarization as well as on subthreshold oscillations, but neither of these latter forms of activity is obligatory with the slow depolarization. The spontaneous frequency depends on the region of intestine from which a given preparation is taken, with highest frequencies occurring in the duodenum. For *in vitro* preparations from cat, the periods range from approximately 3 to 10 sec. The slow depolarizations, variously called slow waves or control potentials, are propagated in the longitudinal direction of the intact gut for distances of several centimeters, the slow waves being nearly synchronous around the gut circumference.

Contractions of sufficient magnitude to be detected by implanted transducers are observed only when the slow waves trigger bursts of spikes in the vicinity, and the strength of contraction varies with the number of spikes per burst (2). The spikes themselves do not propagate but appear to be initiated locally by the slow-wave depolarization. Whether or not a spike is actually generated at a given site depends on slow-wave amplitude and other, poorly understood, factors that govern the excitability of the muscle cells. Consequently, the forms of mechanical activity seen in the small intestine can vary from relative quiescence through states in which rings of contraction occur periodically at relatively fixed points in space (segmentation) to peristaltic activity in which rings of strong contractions travel down the intestine at the velocity of the slow wave (cf. 9,19).

In cat intestine the two layers of the muscle coat, longitudinal and circular, are relatively easy to separate and the layers show markedly different electrical properties. Separation of the layers is not readily done in other preparations such as rabbit, a factor that has lead at least partially to differing interpretations of the roles of each layer in generating electrical activity (10a,16a). In the cat, circular muscle preparations are generally quiescent, although spontaneous spike activity is sometimes noted, whereas the longitudinal layer continues to generate

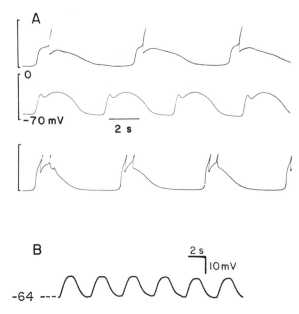

FIG. 1. Microelectrode recordings showing intestinal slow waves in preparations with longitudinal and circular layers intact **(A)** and in a preparation with longitudinal muscle in isolation **(B)**. The top two records of **A** were recorded from cells within the longitudinal muscle layer, whereas the third was taken from a cell within the circular muscle.

a somewhat modified form of slow-wave activity. Microelectrode recordings have shown that waveforms and spontaneous frequency observed in isolated longitudinal muscle are similar to that of intact musculature (see Fig. 1B) but that the slow waves are smaller (mean amplitude 13 ± 7 mV, compared to 27 ± 7.3 mV), and found only in localized regions a few millimeters square, separated by quiescent areas in which resting potentials are normal (12,22).

The potential changes generated in the isolated longitudinal muscle are believed to be the primary pacemaking activity in the small intestine. The propagating waveforms observed in cells of either layer where the musculature is intact (see Fig. 1A) are a complex summation of the pacemaking event and activity triggered by it in both muscle layers. Interactions between the muscle layers and the electrical properties of circular muscle have been examined in detail but are not a primary focus here (cf. 12,13,24,29,30).

This chapter describes a hypothesis in which the relatively small pacemaker potentials are generated by an electrogenic transport system that is apparently gated between high and low operating rates by internal Ca^{2+} levels.

RESULTS

It has long been known that treatments that interfere with cell metabolism, anoxia or metabolic inhibitors and uncouplers, quickly disrupt slow-wave activity in intestinal muscle without seriously affecting the ability to generate action potentials under applied stimulation (4,15,16,21). Slow-wave generation is also blocked, considerably more rapidly, by exposure to ouabain (15,23) and potassium-free bathing media (10), both of which are thought to act directly on the Na-K pump. When carried out with standard microelectrode recording techniques, these experiments demonstrated that maintenance of pump activity was necessary for slow-wave generation but could not show whether blockage occurred directly from interruption of the pump or through an intermediary such as a change in resting potential, membrane resistance, or ion concentration. We examined the effects of ouabain and K-free bathing solutions on small bundles of longitudinal muscle using the double-sucrose gap (14). This technique enables rapid changes in bathing solution to be made as well as the monitoring of membrane voltage and resistance. Slow-wave amplitudes measured in the double-sucrose gap were variable between 5 and 15 mV, a range very similar to microelectrode determinations on isolated longitudinal muscle. Also, the percentage of otherwise healthy preparations which showed slow-wave activity was small, only about 20%, a value roughly comparable to the amount of active tissue in sheets of isolated longitudinal muscle. Typical slow-wave and spike records are shown in Fig. 2. Figure 3A shows the effects of ouabain (10^{-5} M) and Fig. 3B the effects of exposure to K-free saline. In both cases the slow waves were disrupted very rapidly and membrane voltage stabilized in all preparations at a value slightly more positive than the slow-wave crest. The K-free block was readily reversible for short exposures, whereas the ouabain blockage generally did not reverse within the time span of the sucrose-gap experiments. Ouabain at 10^{-6} M blocked slow waves with a much greater latency but did not bring

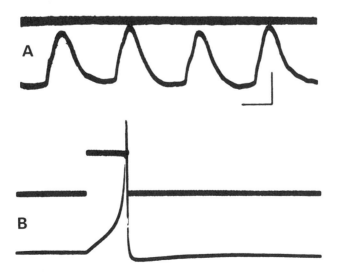

FIG. 2. A: Slow-wave activity recorded from a small strip of longitudinal muscle in the double-sucrose gap, node size approximately 100 × 120 μm. **B:** Longitudinal muscle spike recorded in the double-sucrose gap. *Calibration:* 2 sec, 5 mV in **A**; 10 mV, 10 nA, 200 msec in **B**.

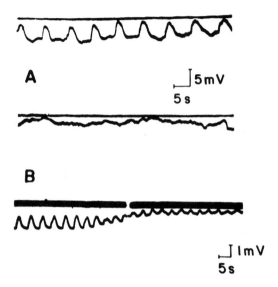

FIG. 3. A: Action of ouabain (10^{-5} M) on slow-wave activity and resting potential. *Top:* Control saline. *Bottom:* Recording begun 45 sec after ouabain application. **B:** Effect of K-free saline. Preparation was exposed to K-free saline at beginning of trace. [From Connor et al. (14), with permission.]

about as large a depolarization as 10^{-5} M. Applying hyperpolarizing current to restore membrane potential to its initial value did not lead to resumption of slow-wave activity following either treatment. Neither treatment caused measurable changes in membrane resistance. We concluded therefore that changes in membrane voltage and resistance were not intermediaries through which the slow-wave block occurred. Also, because of the rapid onset of the block, it did not appear that changes in internal concentrations of Na and K were involved either.

Exposure to K-free media for extended periods results in increased intracellular sodium in smooth muscle (3,5). Following restoration of normal K, active transport of Na and K ions resumes at a higher rate than normal, producing a marked hyperpolarization. Figure 4 illustrates an experiment in which a longitudinal muscle preparation (a quiescent one) had been exposed to K-free saline for 30 min to bring about Na loading. It was then exposed to normal saline, and the hyperpolarization shown in Fig. 4A resulted. When K-free saline was again applied, the membrane again depolarized (Fig. 4B), thus illustrating the voltage swing that cycling the Na-K transport system between two extreme values is capable of producing in this tissue. In normally spontaneous preparations loaded with Na in this fashion, slow-wave activity remained suppressed after the restoration of normal K for many minutes, until the initial conditions were restored. Applying steady current to nullify the hyperpolarization did not bring about the recovery of activity during this period of high transport rate.

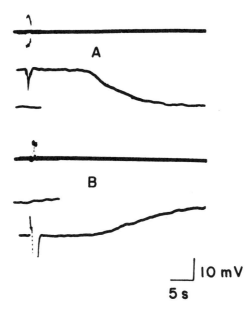

FIG. 4. Voltage swing produced by cycling longitudinal muscle between K-free saline and control saline. **A:** Preparation had been pretreated by bathing it in K-free saline for 30 min and then for a brief period (2 min) in normal saline. At the start of the trace, the preparation was again exposed to K-free saline, which resulted in the depolarization shown. **B:** Hyperpolarization that resulted from reexposure to normal saline. The vertical deflections in the traces of **A** and **B** mark the times at which solution changes were begun. Final voltage levels are indicated by the partial sweeps of **A** and **B**. [From Connor et al. (14), with permission.]

There was no significant difference between spontaneous and quiescent preparations with respect to the voltage swing produced by the procedures illustrated in Fig. 4.

Changes in the concentration of external Na^+ or Cl^- had very little effect on slow-wave generation over the short run. In our sucrose-gap chamber, washout of the extracellular space was completed within approximately 1 min; however, isosmotic replacement of 50% of the external NaCl by sucrose produced no significant change in amplitude on rate of rise of the slow wave over measurement periods as long as 20 min. More complete replacement of NaCl resulted in marked changes in the slow-wave amplitude and waveform but only after periods of 10 to 15 min, times that are more reflective of internal ionic changes than extracellular changes (cf. 6). Calcium concentration changes had little effect on the amplitude of the slow waves in isolated longitudinal muscle but did have effects on frequency, which will be described in a later section. In addition to the ion-substitution data, which give little indication that ion-specific conductance changes generate slow waves, we were unable to detect changes in membrane resistance during the time course of a slow wave.

In a portion of this study, a voltage-clamp circuit was used to hold membrane voltage constant in spontaneous preparations. Under these conditions, periodic inward currents were observed which had roughly the same frequency as the free-running slow waves (see Fig. 5A). It was shown through numerical manipulations that the slow-wave time course could be reconstructed from the voltage-clamp current time course and the nodal membrane resistance and capacitance, with the exception of such features as spikes or small voltage-dependent responses (see Fig. 5B). These findings have been viewed with considerable skepticism because of the possibility of inhomogeneities in membrane voltage within the experimental node of the sucrose gap.

It is quite possible, as Anderson (1) has shown with the mesotubarium preparation, that a certain population of nodal cells is, to a degree, electrically isolated

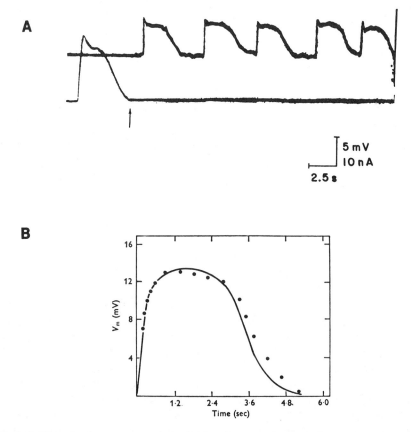

FIG. 5. A: Records of membrane voltage (*top*) and current (*bottom*). One spontaneous slow wave is shown before voltage-control loop was closed (*arrow*) with remaining record showing spontaneous current transients (inward-going current plotted upward). **B:** Comparison of slow-wave voltage computed from a current transient and values of R_m and C_m measured in the same preparation (*continuous curve*) with a measured slow wave (*dots*). [From Connor et al. (14), with permission.]

from the population that is monitored from the voltage pool of the double-sucrose gap. This situation would be realized if the two populations were separated by a relatively high internal resistance, the internal resistance partially decoupling the two populations by acting as one element of a voltage divider and attenuating the voltage changes produced in one region by activity in the other. This would mean that a spontaneous, semi-isolated population of cells would be affected very little, whether the voltage of the monitored population were clamped to various levels or free to change, since most of the voltage difference between the two regions is dropped across the internal resistance. However, it is an unavoidable corollary of this type of model that the amplitude of the semi-isolated activity is many times larger than the amplitude of what appears in the monitored population; the greater the isolation, the greater the difference. To generate the correspondence between clamp current and slow-wave voltage observed in our preparations would have required the existence of uncontrolled voltages many times larger than anything observed in the isolated longitudinal muscle either in the sucrose gap or with microelectrodes. In the case of Anderson's records (1), the spontaneous currents under voltage clamp would have been capable of generating voltage changes less than 10% of those actually observed under unclamped conditions. Spontaneous currents that are similar in many respects to those described here have also been observed by Ohba et al. (26–28) in smooth muscle from guinea pig stomach.

If the observations of spontaneous current transients are sound, they imply that the pacemaker mechanism is not primarily a voltage-sensitive function as the pacemaking mechanism in some neurons and in heart Purkinje cells appears to be (cf. 11,25). The frequency of slow waves, however, depends on membrane voltage, decreasing with hyperpolarization and increasing somewhat with steady depolarization; hyperpolarization >30 mV blocked spontaneous activity in most preparations (14). It is also possible to entrain slow waves to frequencies higher than normal with pulse stimulation (see Fig. 6). Pulsed current stimulus could not produce slow-wave activity where none existed spontaneously, even though spike activity could be driven readily in these preparations.

In an effort to determine whether a plausible pacemaking mechanism could be found, which had only a secondary dependence on membrane voltage, we were led to investigate the effects of calcium ions on slow-wave generation. Changes in this ion concentration in the bathing medium have very little effect on the amplitude of slow waves in the isolated longitudinal muscle but do have a pronounced effect on frequency. This is illustrated in Fig. 7, where microelectrode penetration was maintained during and after exposure to nominally Ca-free saline.

As in most cases studied, there was a slight slow-wave amplitude increase after exposure to Ca-free solution. The most significant change, however, was a prolongation of the period between slow-wave depolarizations. Exposure to Ca-free solution for periods longer than 10 min resulted in a loss of resting potential (10 to 15 mV), a decline in slow-wave amplitude as shown in Fig.

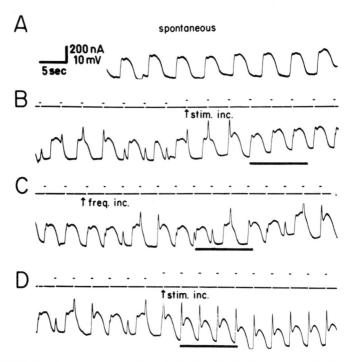

FIG. 6. Records showing spontaneous (**A**) and entrained (**B–D**) slow-wave activity in the double-sucrose gap. Stimulus current pulses are shown in the *upper traces* of **B–D**. In the initial portions of **B** and **D** the stimulus amplitude was not sufficient to entrain the preparation to the driving frequency and the electrotonic potentials drift in phase with respect to the spontaneous waves. Stimulus and slow waves are locked in phase in the latter parts of **B** and **D** and the initial part of **C**. *Horizontal bars* below the voltage traces indicate the duration of two spontaneous cycles from part **A**. [From Connor et al. (13), with permission.]

7C, and an ultimate block of activity. Where block occurred, the membrane voltage simply remained at the resting level and continued the very slow upward drift. The Ca-conductance blocker, verapamil (10 μM), produced a similar change in frequency as shown in Fig. 8, as did exposure to Mn, Co, or high Mg saline (24). These same treatments produced a slow-wave amplitude decline in intact preparations as well as the effect on frequency. The effect on amplitude can be explained by the loss of Ca-dependent activity in the circular muscle (see 12,24,30). Treatment with Ca-free saline and calcium channel blockers resulted in loss of baseline tension and either reduction or abolition of phasic contractions associated with the slow waves.

High Ca saline reduced the period between slow waves and extended the depolarization period somewhat, as shown in Fig. 9 for 3 times normal Ca. Very high levels of Ca blocked slow waves with membrane voltage appearing to lock at a value near the slow-wave crest rather than the trough as with

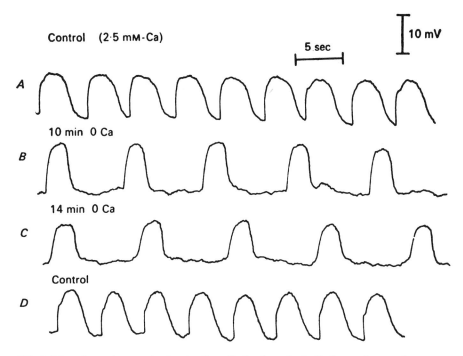

FIG. 7. Microelectrode records showing the effects of nominally Ca-free saline on slow waves in isolated longitudinal muscle. All traces are from the same cell. [From Connor et al. (12), with permission.]

low Ca. This sequence is illustrated in Fig. 10. Baseline tension was increased by exposure to high Ca.

The studies on the effects of Ca summarized above suggested that the intracellular level of this ion might, in some way, be controlling the pacemaker mechanism. We performed a number of experiments using chlorotetracycline (CTC)

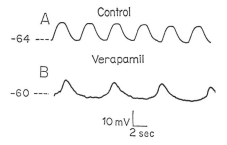

FIG. 8. Microelectrode recordings showing the effects of 10 μM verapamil in isolated longitudinal muscle. Trace in **B** was recorded after a 4.5-min incubation in verapamil. Both traces are from the same cell. [From Mangel et al. (24), with permission.]

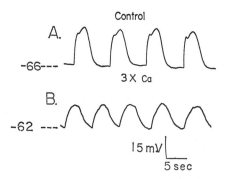

FIG. 9. Effects of 3 times normal external calcium on slow waves recorded from intact musculature. Trace **B** was recorded following a 3-min incubation in high calcium. Both traces are from the same cell. [From Mangel et al. (24), with permission.]

fluorescence as an indicator to see if intracellular Ca^{2+} was fluctuating during the slow-wave cycle. This membrane-permeable compound was chosen over the more standard calcium indicators, arsenazo III and aequorin, which must be injected into cells. CTC has been used in many cell preparations to monitor calcium changes (7,8,17,20). Data from one of twenty muscle segments are shown in Fig. 11: The top traces are of membrane voltage changes sensed by an extracellular electrode (depolarizing voltage gives a downward deflection), whereas the lower traces are CTC fluorescence. In all cases, averages of from 15 to 60 slow-wave cycles showed a periodic change in light intensity of the same frequency as slow-wave activity.

When the fluorescence signal was gathered from the longitudinal muscle layer (Fig. 11A), the waveform was always of a sawtooth shape. In contrast, fluores-

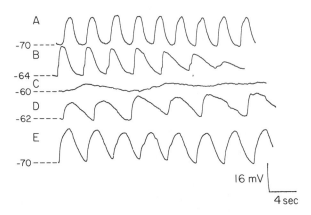

FIG. 10. Effects of high external calcium (35 mM) on slow waves. **A:** Control saline records; **B:** 2 min in high Ca; **C:** 5 min in high Ca; **D:** 17 min after return to normal saline; **E:** 23 min after return to normal saline. All microelectrode recordings are from the same cell. [From Mangel et al. (24), with permission.]

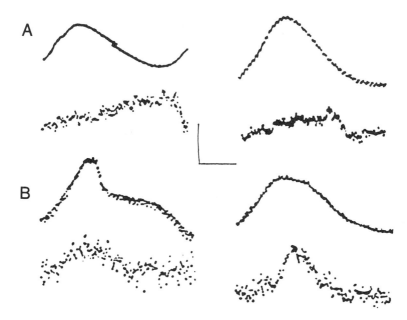

FIG. 11. Averaged waveforms of extracellular voltage (*upper traces*) and CTC fluorescence (*lower traces*). Upward deflections indicate positive-going voltage and increasing fluorescence. Sweeps were triggered just after initial slow-wave depolarization (most negative point on extracellular records) and sweep duration included initial portion of the following wave. **A:** Preparation mounted so that the light signal was collected from the longitudinal layer, and voltage recorded at surface of underlying circular muscle. Record on *left* (average of 60 sweeps) was taken under control conditions; record on *right* (average of 15 sweeps) is from same preparation after exposure to 10^{-4} M atropine. **B:** Light signal gathered from circular muscle layer and voltage recorded at surface of underlying longitudinal muscle. Record on *left* (average of 29 sweeps) is control and record on *right* (average of 25 sweeps) is in 10^{-4} M atropine. *Time scale:* **A,** 2 sec; **B,** 1 sec. Voltage and fluorescence are uncalibrated. [From Mangel et al. (24), with permission.]

cence signals from the circular muscle (Fig. 11B) were somewhat sinusoidal, with the maximum occurring at a different phase of the cycle. For the records shown, the average was triggered just after the initial depolarization of the slow wave (a downward deflection in the extracellular voltage record). The duration of each sweep was sufficient to include the following slow wave, the onset of which occurred between half and three-quarters of the distance to the end of the sweep. The apparent rising phase of the second wave was smeared out considerably in the averaged traces due to irregularities in slow-wave period. Characteristically, the CTC fluorescence from the longitudinal layer rose steadily until around the time of slow-wave onset, when there was an abrupt reset, and the cycle began again. The records on the right in Fig. 11 are repeats of the experiment, done in 10^{-4} M atropine, which is sufficient to eliminate measurable contractions in this muscle and therefore should provide a control for movement artifacts.

DISCUSSION

The results summarized here have led us to consider the following hypothesis on the generation of slow-wave activity in the small intestine. First, the immediate cause of the pacemaker potentials in longitudinal muscle is due to an electrogenic transport system that changes its rate of operation. Because of the sensitivity to ouabain and K-free saline, it seems likely that the primary current is generated by the Na-K pump and not by some other transport system. Slow-wave depolarization would then be produced by the pump shifting from a high to a low transport rate. Second, the primary cue for the change in operating rate is not membrane voltage, but voltage does have an effect on the period of oscillation. Third, internal calcium levels are either directly or indirectly involved in regulating the Na-K transport rate, and therefore involved in the primary control. The dependency would have to be one in which high internal calcium inhibits Na-K transport. Fourth, calcium conductance of the cell membranes is voltage dependent over membrane voltage ranges usually considered subthreshold (-40 to -70 mV), with conductance decreasing with hyperpolarization. Such dependence has been demonstrated in molluscan giant neurons, where calcium indicators such as arsenazo III can be introduced into the cytoplasm (11,18). At the present time, however, there is no direct evidence that this is true in the smooth muscle cells.

The general scheme outlined in Fig. 12 is suggested as one of several possibilities that would generate oscillatory activity of the type observed. It is presumed that net calcium influx normally exceeds the combined extrusion and uptake capacity of the cells, resulting in a buildup of internal Ca^{2+}.

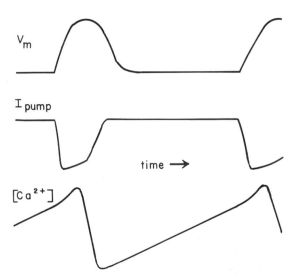

FIG. 12. Possible timing scheme for slow-wave generation with fluctuations in calcium concentration as the driving mechanism.

After some level of Ca^{2+} is reached, the increase both inhibits Na-K transport and accelerates Ca transport either across the plasma membrane or to internal storage sites. With regard to the net electrogenic effect of these two changes, it is probably reasonable to assume that the increase in outward Ca flux is much smaller than the decrease in net outward flux brought about by inhibiting the Na-K pump so that a depolarization would result. The cycle is reset as the Ca concentration is restored to a low level. An applied depolarization would accelerate the steady Ca influx giving a faster rate of accumulation and a decreased cycle time, whereas hyperpolarization would decrease the influx and allow the Na-K pump to remain on longer. Thus, the voltage sensitivity necessary to synchronize populations of cells would be conferred through a voltage-dependent Ca conductance. Low external Ca or Ca conductance blockers should also reduce Ca influx, giving a frequency reduction.

SUMMARY

Rhythmic electrical activity in the small intestine of cat appears to be initiated by small pacemaker potentials originating in the longitudinal muscle layer. These depolarizing potentials (slow waves) have durations of from 2 to 5 sec and are generally <15 mV in amplitude. Frequency depends on the region of intestine and varies between 3 and 10 waves per min. Electrical activity of the intact muscle coat is the summation of this pacemaker activity with triggered activity and is of larger amplitude and more complicated waveform. Exposure to ouabain (10^{-5} M) or K-free saline causes a rapid block of pacemaker slow-wave activity and depolarization to levels more positive than the slow-wave crest. Repolarization does not restore spontaneity. Amplitude and frequency are relatively insensitive to changes in extracellular Na and Cl concentrations, and conductance changes, if they occur during the slow-wave cycle, are too small to be detected in the double-sucrose gap. Changes in extracellular Ca have pronounced effects on frequency, being decreased in low Ca and increased in high Ca. Calcium channel blockers have the same effects as low Ca saline, implying that the changes in frequency are mediated through internal Ca levels. Experiments using chlorotetracycline fluorescence provided data that intracellular Ca^{2+} levels oscillate even under conditions where measurable contractile activity is absent. We have proposed that oscillations in the activity of the Na-K pump are responsible for the pacemaker slow waves and that internal Ca^{2+} is an important cycle-to-cycle controlling factor in the oscillation.

REFERENCES

1. Anderson, N. C. (1977): Limitations and possibilities in smooth muscle voltage-clamp. In: *Excitation-Contraction Coupling in Smooth Muscle,* edited by R. Casteels, pp. 81–89. Elsevier North-Holland, Amsterdam.
2. Bass, P., and Wiley, J. N. (1965): Electrical and extraluminal contractile-force activity of the duodenum of the dog. *Am. J. Digest. Dis.,* 10:183–200.

3. Bolton, T. B. (1973): Effects of electrogenic sodium pumping on the membrane potential of longitudinal smooth muscle from terminal ileum of guinea-pig. *J. Physiol. (Lond.)*, 228:693–712.
4. Bortoff, A. (1961): Slow potential variations of small intestine. *Am. J. Physiol.*, 201:203–208.
5. Casteels, R. G., Droogmans, G., and Hendricks, H. (1971): Electrogenic sodium pump in smooth muscle cells of the guinea-pig's taenia coli. *J. Physiol. (Lond.)*, 217:197–313.
6. Casteels, R., Droogmans, G., and Hendricks, H. (1973): Effect of sodium and sodium substitutes on the active ion transport and on the membrane potential of smooth muscle cells. *J. Physiol. (Lond.)*, 228:733–748.
7. Caswell, A. H., and Hutchinson, J. D. (1971): Visualization of membrane bound cations by a fluorescent technique. *Biochem. Biophys. Res. Commun.*, 42:43–49.
8. Chandler, D. E., and Williams, J. A. (1978): Intracellular divalent cation release in pancreatic acinar cells during stimulus-secretion coupling. I. *J. Cell Biol.*, 76:371–385.
9. Code, C. F., Szurszewski, J. H., Kelly, K. A., and Smith, I. B. (1968): A concept of control of gastrointestinal mobility. In: *Handbook of Physiology: Alimentary Canal, Sect. 6, vol. V*, edited by C. F. Code, pp. 2881–2896. American Physiological Society, Washington, D.C.
10. Connor, C., and Prosser, C. L. (1974): Comparison of ionic effects on longitudinal and circular muscle of cat jejunum. *Am. J. Physiol.*, 226:1212–1218.
10a. Connor, J. A. (1979): On exploring the basis for slow potential oscillations in the mammalian stomach and intestine. *J. Exp. Biol.*, 81:153–173.
11. Connor, J. A. (1982): Mechanisms of pacemaker discharge in invertebrate neurons. In: *Cellular Pacemakers*, Vol. 1, edited by D. O. Carpenter, pp. 187–217. John Wiley & Sons, New York.
12. Connor, J. A., Kreulen, D., Prosser, C. L., and Weigel, R. (1977): Interaction between longitudinal and circular muscle in intestine of cat. *J. Physiol. (Lond.)*, 273:665–689.
13. Connor, J. A., Mangel, A. W., and Nelson, B. (1979): Propagation and entrainment of slow waves in cat small intestine. *Am. J. Physiol.*, 237:C237–C246.
14. Connor, J. A., Prosser, C. L., and Weems, W. A. (1974): A study of pacemaker activity in intestinal smooth muscle. *J. Physiol. (Lond.)*, 240:671–701.
15. Daniel, E. E. (1965): Effects of intra-arterial perfusions on electrical activity and electrolyte contents of dog small intestine. *Can. J. Physiol. Pharmacol.*, 43:551–577.
16. Daniel, E. E., Honour, A. J., and Bogoch, A. (1960): Electrical activity of the longitudinal muscle of dog small intestine studied *in vivo* using microelectrodes. *Am. J. Physiol.*, 198:113–118.
16a. Daniel, E. E., and Sarna, S. (1978): The generation and conduction of activity in smooth muscle. *Ann. Rev. Pharmacol. Toxicol.*, 18:145–166.
17. Fabiato, A., and Fabiato, F. (1979): Use of chlorotetracycline fluorescent to demonstrate Ca^{2+}-induced release of Ca^{2+} from the sarcoplasmic reticulum of skinned cardiac cells. *Nature*, 281:146–148.
18. Gorman, A. L. F., and Thomas, M. V. (1978): Changes in the intracellular concentration of free calcium ions in a pacemaker neuron, measured with the metallochromic indicator dye arsenazo III. *J. Physiol. (Lond.)*, 275:357–376.
19. Grivel, M. L., and Ruckebusch, Y. (1972): The propagation of segmental contractions along the small intestine. *J. Physiol. (Lond.)*, 227:611–625.
20. Hallett, M., Scheider, A. S., and Carbone, E. (1972): Tetracycline fluorescence as calcium-probe for nerve membrane with some model studies using erythrocyte ghosts. *J. Membr. Biol.*, 10:32–44.
21. Job, D. D. (1969): Ionic basis of intestinal electrical activity. *Am. J. Physiol.*, 217:1534–1541.
22. Kobayashi, M., Nagai, T., and Prosser, C. L. (1966): Electrical interaction between muscle layers of cat intestine. *Am. J. Physiol.*, 21:1281–1291.
23. Liu, J., Prosser, C. L., and Job, D. D. (1969): Ionic dependence of slow waves and spikes in intestinal muscle. *Am. J. Physiol.*, 217:1542–1547.
24. Mangel, A. W., Connor, J. A., and Prosser, C. L. (1982): Effects of alterations in calcium levels on cat small intestinal slow waves. *Am. J. Physiol.*, 243:C7–C13.
25. McAllister, R. E., Noble, D., and Tsien, R. W. (1975): Reconstruction of the electrical activity of cardiac Purkinje fibres. *J. Physiol. (Paris)*, 51:1–59.
26. Ohba, M., Sakamoto, Y., and Tomita, T. (1975): The slow wave in the circular muscle of the guinea-pig stomach. *J. Physiol. (Lond.)*, 253:505–516.
27. Ohba, M., Sakamoto, Y., and Tomita, T. (1976): Spontaneous rhythmic activity of the smooth

muscle of the guinea-pig stomach and effects of ionic environment. In: *Smooth Muscle Pharmacology and Physiology,* edited by M. Worcel and G. Vassort, pp. 301–316. INSERM, Paris.
28. Ohba, M., Sakamoto, Y., and Tomita, T. (1977): Effects of sodium, potassium and calcium ions on the slow wave in the circular muscle of the guinea-pig stomach. *J. Physiol. (Lond.),* 267:167–180.
29. Taylor, A. B., Kreulen, D., and Prosser, C. L. (1977): Electron microscopy of the connective tissues between longitudinal and circular muscle of small intestine of cat. *Am. J. Anat.,* 150:427–441.
30. Weigel, R. J., Connor, J. A., and Prosser, C. L. (1979): Two roles of calcium during the spike in circular muscle of small intestine of cat. *Am. J. Physiol.,* 237:C247–C256.

Electrogenic Transport: Fundamental Principles and Physiological Implications, edited by Mordecai P. Blaustein and Melvyn Lieberman. Raven Press, New York © 1984.

Membrane Electrical Parameters of Normal Human Red Blood Cells

Joseph F. Hoffman and *Philip C. Laris

Department of Physiology, Yale University School of Medicine, New Haven, Connecticut 06510

The aim of this chapter is to describe some of the electrical properties of the human red cell plasma membrane and also to briefly discuss the methods used for their determination. This latter aspect is particularly important because in small cells, such as the human red blood cell, the use of conventional microelectrode techniques irreparably damages the membrane, thus precluding analysis of the electrical characteristics by these means (8). However, it is possible to measure membrane potentials (E_m) indirectly with a fluorescence technique utilizing the dye, diS-C_3(5), i.e., 3,3′-dipropylthiadicarbocyanine iodide, as previously described (6). This dye bears a net positive charge and since the membrane is permeable to the dye, the distribution of the dye between cells in suspension and their bulk phase depends on E_m and tracts changes in E_m (1,9). Since the dye technique is based on the fact that the fluorescence of the dye inside cells is quenched relative to its fluorescence in the bulk phase, changes in the relative fluorescence of a cell suspension consequentially indicate changes in E_m (11). There are now several methods available for converting changes in fluorescence intensity to mV (2,6,9).

By means of the fluorescence dye technique, we have previously shown that E_m in human red blood cells is approximately −9 mV (inside negative with respect to the outside) in accordance with expectations based on Donnan considerations; that is, that E_m is equivalent to the chloride equilibrium potential in these cells (6).

We have also demonstrated that the Na/K pump in human red cells is electrogenic, operating to transfer net charge across the membrane (4,5). These measurements of the influence of the Na/K pump on E_m, together with estimates of the current flow through the pump (taken as one-third the ouabain-sensitive Na efflux), have been used to calculate, by Ohm's Law, the membrane resistance (R_m) of these cells. Since the contribution of the Na/K pump to E_m is small

---

* Permanent Address: Department of Biological Sciences, University of California, Santa Barbara, California 93106.

in human red cells, the pump in these cells had been studied under conditions where the Na/K pump activity was high, i.e, when cellular Na was elevated. In this report these studies are extended to include measurements of the influence of the Na/K pump on E_m as well as estimations of R_m at *normal* cellular Na levels.

METHODS

Human red cells obtained by venipuncture were washed three times with a solution containing 150 mM NaCl and 20 mM Hepes at pH 7.4 (NaCl-Hepes). The cells were then incubated at 37°C as 5% suspensions in media containing either 150 mM NaCl, 150 mM KCl, or mixtures of 150 mM NaCl and 150 mM choline chloride together with 20 mM Hepes (pH 7.4), 11.1 mM D-glucose, 1.9 mM Na_2HPO_4, 3 mM adenosine and chloramphenicol. After 1 hr the cell suspensions were centrifuged and the media were replaced with fresh solutions (sometimes containing trace amounts of $^{24}NaCl$, when Na efflux was to be measured). The incubation was continued for an additional 3 hr, and the suspensions were then stored overnight at 4°C. The next day the incubation media were replaced again and the cells were incubated for 1 hr. The suspensions were then centrifuged and the cells were resuspended in NaCl-Hepes. In some cases 5% suspensions of the cells were treated with 125 μM 4,4'-diisothiocyanostilbene-2,2'-disulfonate (DIDS) in NaCl-Hepes at room temperature for 30 min. After this treatment the cells were washed four times at 4°C with NaCl-Hepes. All cell suspensions were finally adjusted to a 50% hematocrit and stored on ice until use.

Fluorescent measurements were made at 37°C on 0.17% suspensions of red cells in NaCl-Tris containing 1.8 μM diS-C_3(5) using an Aminco DW-2 spectrofluorimeter in the split-beam mode with 1.8 μM diS-C_3(5) in ethanol as the reference cell. Excitation was at 622 μM with a 10 μM band width and a 3-mm cut-on filter (RG 665) was placed between the cuvettes and the photomultiplier tube. Cell Na was measured by flame photometry (Instrument Laboratory Model 143) with a lithium internal standard. Measurements of the outward rate constant for Na from ^{24}Na loaded cells were carried out in the presence and absence of ouabain, as previously described (6). The outward rate constant multiplied by the cellular content of Na represents the efflux of Na in mM/liter cells × hr.

RESULTS

A typical experiment measuring the influence of the Na/K pump on E_m is shown in Fig. 1. Approximately 30 min after the addition of dye, K was added to a final concentration of 9 mM and a decrease in fluorescence intensity indicative of hyperpolarization was observed. This change in fluorescence intensity was reversed on the addition of ouabain (final concentration 33 μM). Some changes

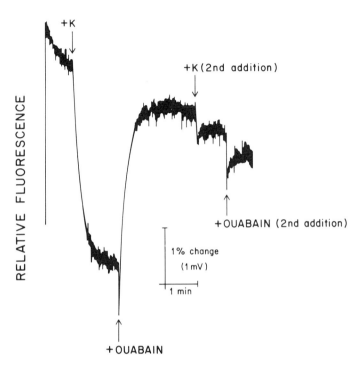

FIG. 1. Fluorescence intensity with time in a 0.17% suspension of red cells in NaCl-Hepes medium containing 1.8 μM diS-C_3(5). Each K addition increases external KCl by 9 mM. Each ouabain addition increases ouabain by 33 μM. One percent change in fluorescence intensity equals 1 mV. Intracellular Na = 27.7 mmol/liter cells.

in fluorescence were also seen when the same amounts of K and ouabain were added a second time. The changes recorded with the second addition of K were comparable to those seen when equal concentrations of NaCl were added initially or when the initial sequence was first ouabain followed by K. These changes, which represent dilution and mixing artifacts not attributable to the activity of the Na/K pump, were subtracted from the change in fluorescence, resulting from the first addition in all subsequent analysis.

The difference in fluorescence caused by activation of the Na/K pump is plotted in Fig. 2 as a function of intracellular Na (mmol/liter cells) with and without treatment with DIDS. As reported earlier (4), the influence of the pump on membrane potential is larger in the presence of DIDS, an inhibitor of Cl conductance (cf. 7). The percent change in fluorescence intensity resulting from Na/K pump activity showed approximately twofold variation in the prepa-

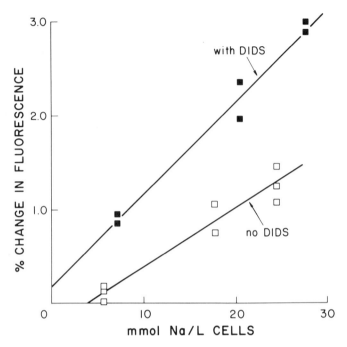

FIG. 2. The percent change in fluorescence intensity caused by activation of the Na/K pump by the addition of K to a final concentration of 9 mM as a function of intracellular Na (mmol/liter cells) after treatment of the cells with 125 μM DIDS or in the absence of such treatment. These values are corrected for dilution factors as described in the text.

rations studied. The data from various preparations (i.e., different blood samples) were normalized by expressing the changes from any one preparation as a percentage of the change observed with that preparation for 20 mmol/liter cells (found by interpolation). The data for DIDS-treated cells and untreated cells were handled separately. The data from 4 preparations of blood are presented in Fig. 3. The line of best fit for the data without DIDS is described by the equation, $y = -0.086 = 0.054x$; with DIDS the equation is $y = 0.171 + 0.04x$. The lines are not significantly different. The y intercept for each line is not significantly different from zero.

Of particular interest is the change in fluorescent intensity recorded in the absence of DIDS when cell Na is in the normal range (10.4 mmol/liter cells) as seen in Fig. 4. This change in fluorescent intensity is equivalent to a change of about 0.2 mV using the method described by Freedman and Hoffman (2) to calibrate the system. The ouabain-sensitive Na efflux with this concentration of intracellular Na is 3.5 mmol/liter cells × hr. Assuming the stoichiometry of the pump is 3 Na:2 K, the pump current is approximately 1.2 mmol/liter cells × hr or 2.1×10^{-9} A/cm². The estimated R_m is 1 to 2×10^5 ohm cm² in the absence of DIDS treatment and about 3×10^5 ohm cm² after such treatment.

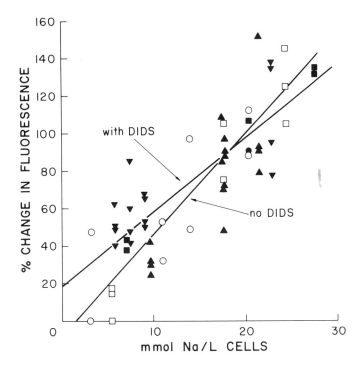

FIG. 3. Summary of the relationship between the change in fluorescent intensity caused by the activation of the Na/K pump by the addition of K to a final concentration of 9 mM as a function of intracellular Na (mmol/liter cells) after treatment with 125 μM DIDS or in the absence of such treatment. Four preparations are presented here. In each case the change at 20 mmol/liter intracellular Na, obtained by interpolation, is set at 100. DIDS-treated and untreated cells are handled separately. *Closed* symbols are DIDS treated; *open* are untreated.

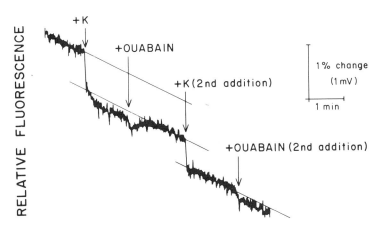

FIG. 4. Fluorescent intensity with time in a 0.17% suspension of red cells in NaCl-Hepes medium containing 1.8 μM diS-C_3(5), K, and ouabain additions as in Fig. 1. Intracellular Na = 10.4 mmol/liter cells (normal cellular Na in this sample of red blood cells).

DISCUSSION

In this report, the Na/K pump in human red blood cells is shown to be electrogenic at normal concentrations of cellular Na. An important conclusion that can be drawn from this result is that it is unlikely that the Na/K pump's contribution to E_m (0.1 to 0.2 mV) in normal human red blood cells is physiologically significant. On the other hand, the fact that it is electrogenic is presumably important in terms of the obligate exchange stoichiometry, the charge structure, and the reaction mechanism of the pump. Estimations of the R_m based on the changes in E_m with activation of the Na/K pump and the current flow through the pump (one-third of the ouabain-sensitive Na efflux) are in the same range as those estimated previously at high cell Na (4,5) and also agree with values 10^5 to 10^6 ohm cm^2 estimated from measurements of Cl conductance (7). Treatment of the cells with DIDS or substitution of Cl with SO_4 both result in higher values of R_m (4,5).

The relationship between the electrogenicity of the Na/K pump and the cellular content of Na appears to be the same in the presence and absence of DIDS (Fig. 3), suggesting that DIDS does not alter pump activity but only R_m. The relationship is described here as a linear function over the range 3 to 30 mmol Na/liter cells. Studies of Sachs (10) and Garay and Garrahan (3) indicate, however, that the ouabain-sensitive Na efflux in the presence of external Na is S-shaped in this range. The variation in the data presented here is too great to make the distinction between linearity and an S-shaped curve such as that described by these workers. To make this distinction, more studies would be required with varying intracellular Na between 0 and 10 mmol/liter cells. These measurements would be difficult to make using the methods employed here, because the signal to noise ratio at 9 mmol K/liter cells is about one to one in the absence of DIDS (exclusive of the dilution mixing artifacts).

SUMMARY

The studies reported in this chapter indicate that the Na/K pump's contribution to E_m of human red blood cells with normal cellular Na is no more than 0.1 to 0.2 mV. Estimates based on this magnitude of the electrogenic component and the pump current (taken as one-third the ouabain-sensitive Na efflux) indicate that R_m in these cells is about 1 to 2 $\times$ 10^5 ohm cm^2.

ACKNOWLEDGMENTS

The work reported in this chapter was supported by NIH grants HL-09906 and AM-17433.

REFERENCES

1. Cohen, L. B., and Hoffman, J. F. (1982): Optical measurements of membrane potential. In: *Techniques in the Life Sciences,* edited by P. F. Baker, vol. P118, pp. 1–13. Elsevier/North-Holland, Amsterdam.

2. Freedman, J. C., and Hoffman, J. F. (1979): The relation between dicarbocyanine dye fluorescence and the membrane potential of human red blood cells set at varying Donnan equilibria. *J. Gen. Physiol.,* 74:187–212.
3. Garay, R. P., and Garrahan, P. J. (1973): The interaction of sodium and potassium with the sodium pump in red cells. *J. Physiol. (Lond.),* 231:297–325.
4. Hoffman, J. F., Kaplan, J. H., and Callahan, T. J. (1979): The Na:K pump in red cells is electrogenic. *Fed. Proc.,* 38:2440–2441.
5. Hoffman, J. F., Kaplan, J. H., Callahan, T. J., and Freedman, J. C. (1980): Electrical resistance of the red cell membrane and the relation between net anion transport and the anion exchange mechanism. *Ann. N.Y. Acad. Sci.,* 341:357–360.
6. Hoffman, J. F., and Laris, P. C. (1974): Determination of membrane potentials in human and *amphibian* red blood cells by means of a fluorescent probe. *J. Physiol. (Lond.),* 239:519–552.
7. Knauf, P. A., Fuhrmann, G. F., Rothstein, S., and Rothstein, A. (1977): The relationship between anion exchange and net anion flow across the human red blood cell membrane. *J. Gen. Physiol.,* 69:363–386.
8. Lassen, U. V. (1977): Electrical potential and conductance of the red cell membrane. In: *Membrane Transport in Red Cells,* edited by J. C. Ellory and V. L. Lew, pp. 137–172. Academic Press, New York.
9. Rink, T. J., and Hladky, S. B. (1982): Measurement of red cell membrane potential with fluorescent dyes. In: *Red Cell Membranes—A Methodological Approach,* edited by J. C. Ellory and J. D. Young, pp. 321–334. Academic Press, New York.
10. Sachs, J. R. (1970): Sodium movements in the human red blood cell. *J. Gen. Physiol.,* 56:322–341.
11. Sims, P. J., Waggoner, A. S., Wang, C.-H., and Hoffman, J. F. (1974): Studies on the mechanism by which cyanine dyes measure membrane potential in red blood cells and phosphatidylcholine vesicles. *Biochemistry,* 13:3315–3330.

The Sodium Pump of Mouse Pancreatic β-Cells: Electrogenic Properties and Activation by Intracellular Sodium

H. P. Meissner and J. C. Henquin

I Physiologisches Institut und Medizinische Klinik, Universität des Saarlandes, 6650 Homburg/Saar, Germany

Experimental evidence for the presence of a sodium pump in pancreatic islet cells was first provided by Howell and Taylor (15), who showed that ouabain inhibits ^{42}K uptake by isolated islets. Since then, a Na-K-activated ouabain-inhibitable adenosine triphosphatase (ATPase) has been identified and characterized in islet-cell membranes (7,16–18). Electrophysiological studies further showed that conditions known to inhibit or to activate the sodium pump produced rapid changes in the membrane potential of insulin-secreting β-cells (2, 19,20), suggesting that the pump is electrogenic in this as in other tissues (29).

We recently published (13) a detailed study of the electrogenic characteristics of the sodium pump in pancreatic β-cells. Its contribution to the resting membrane potential and its role in the generation of the slow waves of membrane potential triggered by glucose were evaluated. The principal results of that study are summarized in this chapter. We further investigated the importance of Na ions for pump function and glucose-induced electrical activity.

METHODS

Details of the method used for recording the membrane potential of mouse pancreatic β-cells have been published previously (26). In brief, fed female NMRI mice were killed 2 hr after intraperitoneal injection of 20 mg/kg pilocarpine. Such treatment facilitates visualization and preparation of the islets without affecting the changes in membrane potential brought about by physiological stimuli of β-cells. A section of pancreas was fixed in a small chamber (1 ml) and continuously perifused (3.5 ml/min) at 37°C. After partial microdissection of an islet, single β-cells were impaled with microelectrodes filled with 2 M potassium citrate (tip resistance 200 to 400 MΩ). Membrane potential was continuously monitored on an oscilloscope and an ink recorder and stored on tape. The figures shown were obtained by playback of the tape to an ink recorder with high-frequency response (Brush Accuchart, Gould).

All experiments were started in the presence of 15 mM glucose. After impalement of a cell exhibiting electrical activity, the glucose concentration was decreased to 10 mM. The β-cells were identified by the typical electrical activity (20,26) that they display in the presence of these glucose concentrations. The impalement was considered successful, and the test solutions applied when the membrane potential exhibited slow waves of $\geq$ 10 mV in amplitude for at least 7 to 10 min.

The perifusion medium had the following ionic composition: 122 mM NaCl, 4.7 mM KCl, 2.5 mM $CaCl_2$, 1.1 mM $MgCl_2$, and 20 mM $NaHCO_3$. It was continuously gassed with a mixture of 95% O_2 and 5% CO_2 and the pH was 7.4. When KCl was omitted, the concentration of NaCl was adjusted to maintain isosmolarity. Nominally Na-free solutions were prepared by substituting choline salts for NaCl and $NaHCO_3$; these solutions were supplemented with 5 μM atropine.

RESULTS AND DISCUSSION

Effects of Ouabain and K Omission in the Presence of Nonstimulatory Glucose Concentrations

In agreement with earlier results (22), the average resting membrane potential of β-cells was not significantly different in a glucose-free medium (−67.8 mV) or in the presence of 3 mM glucose (−63.8 mV). Under these conditions, ouabain depolarized the membrane by an average of 7 mV (Table 1). The effect of ouabain was rapid (within 30 sec) and was complete after approximately 2 min. By contrast, substitution of the control medium by a K-free solution caused a marked hyperpolarization of the β-cell membrane (Table 1). This increase in membrane potential occurred with a delay corresponding only to the dead space of the system. Depolarization produced by ouabain and hyperpolarization produced by K omission were reversible on return to the control medium. In addition to these rapid changes in membrane potential, a slow and progressive depolarization was also observed, when ouabain or the K-free solution was applied for more than 5 to 6 min.

Increasing the concentration of glucose to 6 or 7 mM depolarized the β-cell membrane and induced electrical activity in a few cells. Addition of ouabain to the medium was again followed by membrane depolarization, whereas omission of K had no significant effect (Table 1). Whenever electrical activity was evoked by 6 to 7 mM glucose, ouabain or K omission increased it. Both experimental conditions could also induce activity in certain otherwise silent cells (13).

The rapid depolarization that ouabain produced when the perifusion medium contained 0 or 3 mM glucose strongly suggests that an electrogenic sodium pump directly contributes to the resting membrane potential of β-cells. Secondary changes expected to occur on inhibition of the pump are unlikely to play

TABLE 1. *Effects of ouabain or of K omission on the membrane potential and electrical activity of mouse β-cells perifused in the presence of low concentrations of glucose*

	Control medium	Ouabain (100 μM)	K-free medium
Glucose 0	−67.8 mV ± 1.5 (9)	Depolarization[b] −8.0 mV ± 0.9 (5)	Hyperpolarization[b] +17.3 mV ± 1.9 (6)
Glucose 3 mM	−63.8 mV ± 2.6 (9)	Depolarization[b] −6.2 mV ± 0.9 (6)	Hyperpolarization[b] +13.8 mV ± 1.4 (5)
Glucose 6 mM	−54.8 mV ± 1.5 (8)	Depolarization[a] −8.7 mV ± 1.5 (7)	No change + 2.0 mV ± 1.9 (7)
	No activity (7)	Activity (2/6)	Activity (1/6)
	Activity (1)	∫Activity (1/1)	∫Activity (1/1)
Glucose 7 mM	−52.7 mV ± 1.9 (9)	Depolarization[a] −7.0 mV ± 1.2 (6)	No change − 2.2 mV ± 1.4 (5)
	No activity (6)	Activity (3/4)	Activity (2/3)
	Activity (3)	∫Activity (2/2)	∫Activity (2/2)

The concentration of glucose remained the same throughout each experiment. The test solutions with ouabain or without K were applied for 4 min and their effect on the membrane potential was measured as the maximal change from the control membrane potential in the same cell. Electrical activity is defined as oscillations of the membrane potential with superimposed spikes. The two experimental conditions were not always tested in the same cells; this explains the different numbers of cells given in parentheses.
Values are means ± SEM.
Effects are significant at [a]$p < 0.005$ or [b]$p < 0.001$.

a major role in this fast depolarization. Omission of K did not mimic the effect of ouabain but produced a marked hyperpolarization. At low glucose, the K permeability of the β-cell membrane is high (3,21), and the hyperpolarization due to the large increase in K equilibrium potential probably masks any depolarization that could result from pump inhibition. This interpretation is supported by the absence of hyperpolarization, and even the occasional depolarization produced by K-free solutions in the presence of 6 to 7 mM glucose. Under these conditions, the potassium permeability of the β-cell membrane is markedly reduced (3,9).

Effects of Ouabain and K Omission During Glucose Stimulation

When the islets were perifused with a medium containing 10 mM glucose, the membrane potential of β-cells displayed repetitive slow waves with bursts of spikes superimposed on their plateau (20,26). As illustrated by Fig. 1, omission of K or addition of ouabain markedly modified this electrical activity triggered by glucose. In the K-free solution, the duration of the slow waves and of the intervals between them decreased markedly (Fig. 1A). During the first 2 min following addition of 100 μM ouabain (Fig. 1B), the slow waves lengthened and the amplitude of repolarization during the intervals decreased; later on, regular slow waves reappeared. With 500 μM ouabain, the repolarization phase of the slow waves was blocked; the membrane remained permanently depolarized at the plateau level and continuous spike activity appeared (Fig. 1C).

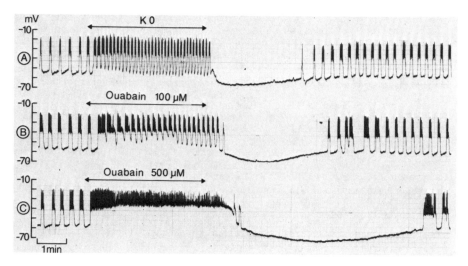

FIG. 1. Effect of K omission and of ouabain on the membrane potential of mouse β-cells perifused in the presence of 10 mM glucose. Potassium was omitted or ouabain added for 4 min as indicated by the *arrows*. Records **B** and **C** were obtained in the same cell.

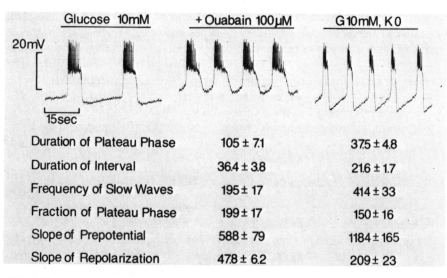

	+ Ouabain 100 μM	G 10mM, K 0
Duration of Plateau Phase	105 ± 7.1	37.5 ± 4.8
Duration of Interval	36.4 ± 3.8	21.6 ± 1.7
Frequency of Slow Waves	195 ± 17	414 ± 33
Fraction of Plateau Phase	199 ± 17	150 ± 16
Slope of Prepotential	588 ± 79	1184 ± 165
Slope of Repolarization	478 ± 6.2	209 ± 23

FIG. 2. Effects of ouabain addition or K omission on the slow waves of membrane potential triggered by 10 mM glucose in mouse β-cells. Experimental values were measured during the fourth minute after ouabain addition or K removal and are expressed as a percentage of the control values measured in the same cell 1 min before changing to the test condition. Results are given as means ± SEM for 10 different cells. The fraction of plateau phase is the fraction of time spent at a depolarized level, with spike activity. The prepotential is defined as the progressive depolarization (during the interval) that precedes the fast depolarization of the slow wave.

Figure 2 compares the major characteristics of control electrical activity and of activity recorded during the fourth minute after ouabain addition or K omission. Details on the dose dependency of ouabain effects and on the changes produced by various concentrations of K can be found in an earlier publication (13). One of the most prominent changes produced by ouabain was a sixfold increase in the slope of the prepotential (i.e., the progressive depolarization during the interval). The result of this slope increase was that the threshold at which the slow waves start was reached sooner than in control conditions (Fig. 2). Consequently, the duration of the intervals decreased, and the frequency of the slow waves increased. Since ouabain did not change the duration of the slow waves, the increment of the fraction of plateau phase was entirely due to shortening of the intervals.

Inhibition of the sodium pump by a K-free solution also increased the slope of the prepotential and the frequency of the slow waves, the duration of which was markedly shortened (Fig. 2). Since an increase in K permeability seems to be involved in the repolarization phase of the slow waves (1,14,20,24,28), their shortening in the absence of K can be explained by the marked increase in driving force for K outward current that results from K omission. This is consistent with the twofold increase in the rate of the repolarization phase of the slow waves under these conditions (Fig. 2). However, the decrease of that rate of repolarization by ouabain supports the earlier suggestion (20) that the activity of the sodium pump also contributes to the termination of the slow waves.

Effects of Ouabain Removal and K Reintroduction During Glucose Stimulation

In agreement with earlier findings (2,19,20), K reintroduction into a K-free solution or ouabain withdrawal was followed by a marked hyperpolarization of the β-cell membrane with suppression of electrical activity (Fig. 1). This hyperpolarization was transient, and normal slow waves resumed after a few minutes. As detailed in Table 2, the duration of hyperpolarization increased with the length of the period of pump inhibition and was consistently longer when this inhibition was due to the glycoside than to a K-free solution. As

TABLE 2. *Duration of the hyperpolarization of the β-cell membrane occurring after ouabain withdrawal or K reintroduction (sec)*

Duration of pump inhibition (min)	Ouabain (100 μM)	K-free medium
2	200 ± 11 (6)	123 ± 6 (8)
4	266 ± 17 (8)	178 ± 8 (9)
6	327 ± 21 (6)	217 ± 16 (5)
8	356 ± 28 (7)	248 ± 16 (7)

The experiments were performed in the presence of 10 mM glucose. The sodium pump was inhibited by ouabain or by omission of K for the indicated time.
Values are means ± SEM for the number of cells shown in parentheses.

illustrated in Fig. 1, it also increased with the concentration of ouabain used to inhibit the pump for a fixed period (13). The maximum level of hyperpolarization was always reached later after ouabain withdrawal than after K readmission, probably because of the progressive washing of the glycoside from the membrane.

Several lines of evidence indicate that the hyperpolarization that follows ouabain withdrawal and K reintroduction in the presence of 10 mM glucose can be ascribed to reactivation of the sodium pump. The hyperpolarization following K readmission was prevented by ouabain (13,19,20). Addition of ouabain at the time of maximum K-induced hyperpolarization immediately stopped it and caused a depolarization (13). The hyperpolarization that normally followed ouabain withdrawal did not occur if the medium was devoid of K (13). Furthermore, a decrease of the ^{86}Rb efflux from the islets was found after removal of ouabain from, or reintroduction of K to, the perifusion medium (13). This observation excludes the possibility that the hyperpolarization is due to an increase in K permeability of the β-cells. Finally, evidence could also be obtained that the hyperpolarization is not simply due to depletion of K ions just outside of the cell membrane (13).

In other tissues, the sodium pump exhibits some ion selectivity: Rb and K activate the pump with approximately the same effectiveness, but Cs is considerably less effective (4). Figure 3 compares the ability of these 3 ions to hyperpolarize the β-cell membrane when introduced into a K-free solution. Rubidium

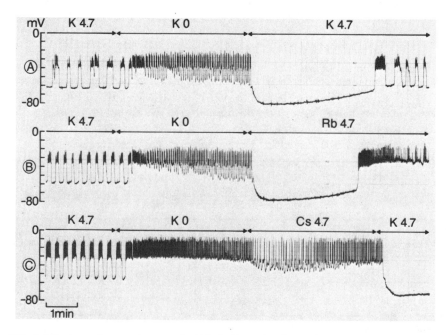

FIG. 3. Effects on the membrane potential of mouse β-cells produced by addition of K, Rb, or Cs to a K-free medium. The concentration of glucose was 10 mM in all solutions.

caused a marked hyperpolarization similar to that produced by K (Figs. 3A and B). Additional experiments will be necessary to determine whether the hyperpolarization induced by Rb differs in duration and amplitude from that induced by K. By contrast, Cs polarized the membrane only slightly for 2 to 3 min and did not block electrical activity. Subsequent replacement of Cs by K was rapidly followed by a marked hyperpolarization with complete suppression of activity (Fig. 3C). These results show that the sodium pump of β-cells exhibits an ion selectivity similar to that observed in other tissues.

Incidentally, in the experiment illustrated by Fig. 3C, K withdrawal caused a permanent depolarization with continuous spike activity and not the characteristic changes in slow waves described earlier and seen in Figs. 3A and B. That effect of K omission is usual at higher concentrations of glucose (13,20), but it can occasionally be observed in the presence of 10 mM glucose, when the cell is very active. Thus, in this cell, the fraction of plateau phase was 0.562, i.e., twice as much as the average for β-cells perifused with 10 mM glucose (13).

Influence of Sodium on the Membrane Potential of β-Cells Stimulated by Glucose

In the presence of 10 mM glucose, substitution of Na by choline hyperpolarized the β-cell membrane transiently. The duration of this hyperpolarization and its amplitude (exaggerated by a junction potential of 3 to 6 mV) varied from one cell to another (compare Figs. 4A and 5A). When Na was withdrawn

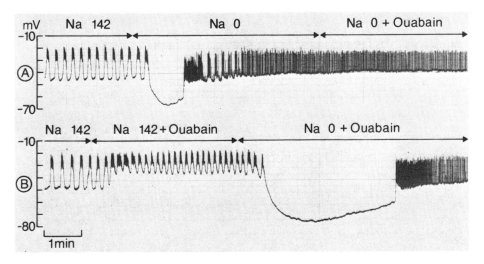

FIG. 4. Effects of Na omission on the membrane potential of mouse β-cells perifused in the presence of 10 mM glucose. Sodium was omitted, and ouabain (100 μM) was added as indicated by the *arrows*. Sodium-free solutions contained choline salts as substitutes and were supplemented with 5 μM atropine. The two records were obtained in the same cell.

from the extracellular medium after a period of inhibition of the sodium pump, (i.e., when the concentration of intracellular Na had been raised), the hyperpolarization was markedly lengthened (Fig. 4B). A first possible mechanism for this increase in membrane potential would be the suppression of a Na inward current, but such a current of sufficient magnitude does not appear to play an important role in the depolarizing effect of glucose (23,26). Furthermore, this would not explain why the phenomenon is only transient. It could also result from the activation of a Na/Ca exchange. Measurements of ^{45}Ca efflux from rat islets (11) are compatible with the electrogenicity (3 to 4 $Na^+/1$ Ca^{2+}) of such a process in β-cells as in other cells (5,27). Reversal of the sodium gradient by omission of external Na could thus generate a hyperpolarizing current, the amplitude of which may be increased if intracellular Na has been initially raised by pump blockade. In addition, the entering Ca may also activate the K permeability (1,10,22,28) and thereby contribute to the hyperpolarization. Elucidation of the exact mechanisms awaits further experiments.

After the initial hyperpolarization, the membrane depolarized slowly to an apparent threshold at which a fast depolarization occurred to a plateau onto which spike activity started (Fig. 4). As previously reported (6,23,25), the amplitude of these spikes was greater than in the presence of Na; this was not due to a change in their peak potential, but to their origin from a plateau several millivolts more negative than the plateau of the slow waves recorded under control conditions. In most cells, the burst pattern of electrical activity with

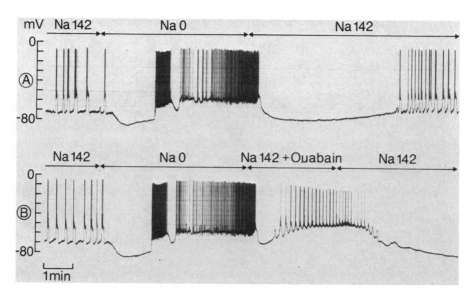

FIG. 5. Effects of Na omission and reintroduction on the membrane potential of mouse β-cells perifused in the presence of 10 mM glucose. Sodium was omitted and reintroduced and ouabain (200 μM) was added as indicated by the *arrows*. The two records were obtained in the same cell.

slow waves disappeared, and a continuous spike activity was observed in a Na-free medium (Figs. 4 and 5). This inhibition of the slow waves after removal of sodium from the external medium has been explained by an inactivation of the sodium pump due to a decrease of the intracellular Na concentration (23,25). In a few cells, however, the membrane potential slowly increased with time, and slow waves finally reappeared despite the absence of extracellular Na (Fig. 6A). The reason is unclear, but it could be the progressive inhibition of glucose metabolism in β-cells that is known to occur in the absence of extracellular Na (8). Such interference with glucose metabolism is expected to increase the K permeability of β-cells (9,12) and, therefore, to facilitate membrane repolarization. Earlier studies have shown that an increase in K permeability of β-cells (by high external Ca) is indeed able to counteract the depolarization due to inhibition of the pump blockade and to restore slow waves (13,28).

Substitution of the choline medium by a control medium containing Na produced three successive effects: a transient initial depolarization, followed by a long hyperpolarization with suppression of activity, and, finally, reappearance of normal slow waves (Fig. 5A). The initial depolarization is shown in greater detail in Figs. 7B and C. It usually increased in amplitude and duration when the initial period of perifusion without Na was prolonged. When Na was readmitted only at a low concentration (20 mM), this phase of depolarization was much less clear and sometimes apparently absent (Fig. 6B). It can thus reasonably be ascribed to a Na inward current triggered by the abrupt change in the Na

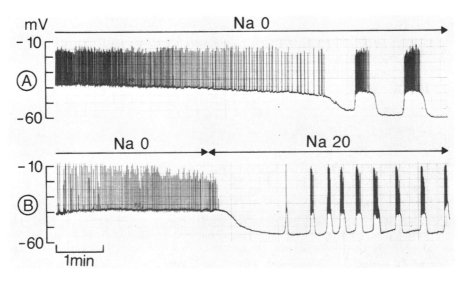

FIG. 6. Slow-wave activity in the absence of Na **(A)** and alteration of the activity pattern by addition of low Na concentrations (20 mM) to a Na-free medium **(B)**. In both experiments the β-cells were stimulated with glucose, 10 mM, throughout. Record **A** starts 4 min after and record **B** 2 min after removal of Na. The two records were obtained from different cells.

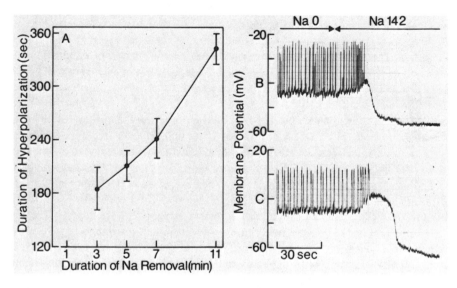

FIG. 7. Effects of Na (142 mM) on the membrane potential of mouse β-cells after perifusion with a Na-free medium for different periods of time. In all experiments the β-cells were stimulated with glucose, 10 mM, throughout. Graph **A** shows the relationship between the duration of Na removal (*abscissa*) and the length of the hyperpolarization of the membrane following reintroduction of Na (*ordinate*). Values are given as mean ± SEM; each point was obtained from 6 cells. Records **B** and **C** show the initial depolarization of the membrane after reintroduction of Na. In **B** the cell was perifused during 5 min and in **C** during 11 min with a Na-free solution. **B** and **C** were obtained from the same cell.

gradient. The following hyperpolarization that extended to −80 mV (Fig. 5A) may be explained by the reactivation of an electrogenic sodium pump by the increasing concentration of intracellular Na. Thus, when the choline-containing solution was replaced by a Na solution supplemented with ouabain, the hyperpolarization phase was largely inhibited (Fig. 5B). A few slow waves reappeared rapidly, followed by a depolarization to a level slightly less negative than the plateau level measured in the absence of Na. Even at this level, however, the spikes appeared in bursts. Removal of ouabain was finally followed by the expected hyperpolarization (Fig. 5B). If ouabain was added to the Na-free solution, no significant change in the membrane potential or the spike frequency was noted (Fig. 4A), but the subsequent reintroduction of Na (still in the presence of ouabain) produced only a depolarization of the β-cell membrane (not shown).

As shown in the graph of Fig. 7A the duration of the hyperpolarization following Na reintroduction was strongly dependent on the duration of the preceding period of Na removal. Thus, the hyperpolarization was lengthened when the period of Na withdrawal was augmented. Probably, the Na influx responsible for the immediate depolarization seen after the change from the Na-free to the Na-containing solution is larger (cf. Figs. 7B and C), and, therefore, the electrogenic pump is activated more after a longer period of Na removal.

CONCLUSIONS

Insulin-secreting pancreatic β-cells possess a truly electrogenic Na pump, activated by extracellular K or Rb and by intracellular Na. It contributes several millivolts to the resting membrane potential. Its blockade is followed by depolarization and appearance of electrical activity in the presence of glucose concentrations close to the threshold for stimulation of insulin release. The activity of the pump also modulates the slow waves of membrane potential triggered by higher concentrations of the sugar. In particular, it contributes to their repolarization phase and controls their frequency. Although no marked Na inward current is involved in the depolarizing effect of glucose, a sufficient influx of Na is necessary for appearance of the normal pattern of electrical activity. This requirement may be due to the necessity of maintaining an adequate concentration of intracellular Na to activate the pump.

ACKNOWLEDGMENTS

These studies were supported by the Deutsche Forschungsgemeinschaft SFB 38, Bonn-Bad Godesberg. We are grateful to Mr. W. Schmeer for skillful assistance and to Mrs. R. Stolz for editorial help. J. C. Henquin is "Chercheur Qualifié" of the FNRS, Brussels. His permanent address is Unité de Diabète et Croissance, University of Louvain School of Medicine, Brussels, Belgium.

REFERENCES

1. Atwater, I., Dawson, C., Ribalet, B., and Rojas, E. (1979): Potassium permeability activated by intracellular calcium ion concentration in the pancreatic β-cell. *J. Physiol. (Lond.)*, 228:575–588.
2. Atwater, I., and Meissner, H. P. (1975): Electrogenic sodium pump in β-cells of islets of Langerhans. *J. Physiol. (Lond.)*, 247:56P–58P.
3. Atwater, I., Ribalet, B., and Rojas, E. (1978): Cyclic changes in potential and resistance of the β-cell membrane induced by glucose in islets of Langerhans from mouse. *J. Physiol. (Lond.)*, 278:117–139.
4. Baker, P. F., and Connelly, C. M. (1966): Some properties of the external activation site of the sodium pump in crab nerve. *J. Physiol. (Lond.)*, 185:270–297.
5. Blaustein, M. P. (1974): The interrelationship between sodium and calcium fluxes across cell membranes. *Rev. Physiol. Biochem. Pharmacol.*, 70:33–82.
6. Dean, P. M., and Matthews, E. K. (1970): Electrical activity in pancreatic islet cells: effects of ions. *J. Physiol. (Lond.)*, 210:265–275.
7. Formby, B., Capito, K., and Hedeskov, C. J. (1976): (Na^+,K^+)-activated ATPase in microsomes from mouse pancreatic islets. *Acta Physiol. Scand.*, 96:143–144.
8. Hellman, B., Idahl, L. A., Lernmark, A., Sehlin, J., and Täljedal, I.-B. (1974): The pancreatic β-cell recognition of insulin secretagogues. Effects of calcium and sodium on glucose metabolism and insulin release. *Biochem. J.*, 138:33–45.
9. Henquin, J. C. (1978): D-glucose inhibits potassium fluxes from pancreatic islet cells. *Nature*, 271:271–273.
10. Henquin, J. C. (1979): Opposite effects of intracellular calcium and glucose on the potassium permeability of pancreatic islet cells. *Nature*, 280:66–68.
11. Henquin, J. C. (1979): The influence of calcium and sodium on calcium efflux from rat pancreatic islets. *J. Physiol. (Lond.)*, 296:103P.

12. Henquin, J. C. (1980): Metabolic control of the potassium permeability in pancreatic islet cells. *Biochem. J.*, 186:541–550.
13. Henquin, J. C., and Meissner, H. P. (1982): The electrogenic sodium-potassium pump of mouse pancreatic β-cells. *J. Physiol. (Lond.)*, 322:529–552.
14. Henquin, J. C., Meissner, H. P., and Preissler, M. (1979): 9-aminoacridine- and tetraethylammonium-induced reduction of the potassium permeability in pancreatic β-cells. Effects on insulin release and electrical properties. *Biochim. Biophys. Acta*, 587:579–592.
15. Howell, S. L., and Taylor, K. W. (1968): Potassium ions and the secretion of insulin by islets of Langerhans incubated in vitro. *Biochem. J.*, 108:17–24.
16. Kemmler, W., and Löffler, G. (1977): Na, K-ATPase in rat pancreatic islets. *Diabetologia*, 13:135–238
17. Lernmark, A., Nathans, A., and Steiner, D. F. (1976): Preparation and characterization of plasma membrane-enriched fractions from pancreatic islets. *J. Cell Biol.*, 71:606–623.
18. Levin, S. R., Kasson, B. G., and Driessen, J. F. (1978): Adenosine triphosphatases of rat pancreatic islets. Comparison with those of rat kidney. *J. Clin. Invest.*, 62:692–701.
19. Matthews, E. K., and Sakamoto, Y. (1975): Pancreatic islet cells: Electrogenic and electrodiffusional control of membrane potential. *J. Physiol. (Lond.)*, 246:439–457.
20. Meissner, H. P. (1976): Electrical characteristics of the beta-cells in pancreatic islets. *J. Physiol. (Paris)*, 72:757–767.
21. Meissner, H. P., Henquin, J. C., and Preissler, M. (1978): Potassium dependence of the membrane potential of pancreatic β-cells. *FEBS Lett.*, 94:87–89.
22. Meissner, H. P., and Preissler, M. (1979): Glucose-induced changes of the membrane potential of pancreatic β-cells: Their significance for the regulation of insulin release. In: *Treatment of Early Diabetes*, edited by R. A. Camerini-Davalos and B. Hanover, pp. 97–107. Plenum Press, New York.
23. Meissner, H. P., and Preissler, M. (1980): Ionic mechanisms of the glucose-induced membrane potential changes in β-cells. *Horm. Metab. Res. [Suppl.]*, 10:91–99.
24. Meissner, H. P., Preissler, M., and Henquin, J. C. (1980): Possible ionic mechanisms of the electrical activity induced by glucose and tolbutamide in pancreatic β-cells. In: *Diabetes 1979*, edited by W. K. Waldhausl, pp. 166–171. Excerpta Medica, Amsterdam.
25. Meissner, H. P., and Schmeer, W. (1981): The significance of calcium ions for the glucose-induced electrical activity of pancreatic β-cells. In: *The Mechanism of Gated Calcium Transport Across Biological Membranes*, edited by S. T. Ohnishi and M. Endo, pp. 157–165. Academic Press, New York.
26. Meissner, H. P., and Schmelz, H. (1974): Membrane potential of beta-cells in pancreatic islets. *Pfluegers Arch.*, 351:195–206.
27. Mullins, L. J. (1979): The generation of electric currents in cardiac fibers by Na/Ca exchange. *Am. J. Physiol.*, 236:C103–C110.
28. Ribalet, B., and Beigelman, P. M. (1979): Cyclic variation of K^+ conductance in pancreatic β-cells: Ca^{2+} and voltage dependence. *Am. J. Physiol.*, 237:C137–C146.
29. Thomas, R. C. (1972): Electrogenic sodium pump in nerve and muscle cells. *Physiol. Rev.*, 52:563–594.

Electrogenic Transport: Fundamental Principles and Physiological Implications, edited by Mordecai P. Blaustein and Melvyn Lieberman. Raven Press, New York © 1984.

Electrical Kinetics of Proton Pumping in *Neurospora*

Clifford L. Slayman and Dale Sanders

Department of Physiology, Yale University School of Medicine, New Haven, Connecticut 06510

Biological conversion of scalar electronic energy—whether that of covalent bonds or of free electrons excited, for example, by photons—into the vectorial energy of spatial electrochemical gradients is commonly referred to as active transport. Remarkably, no active transport mechanisms have yet been demonstrated for chemical species other than ions. Sugars, amino acids, and other neutral substrate molecules are accumulated in living cells either by coupling to the downhill movement of certain ions (12) or by chemical transformation that occurs simultaneously with transport (5). It is not yet known why this should be so, but can reasonably be guessed that charge transfer is a fundamental mechanism of active transport, not just an accidental consequence of the underlying chemical reactions. It is appropriate, therefore, that in recent years electrogenesis by active transport systems has been intensely studied from physical, physiological, and biochemical points of view. Most rapid progress has been made with the light-driven active transport of protons by membranes of chloroplasts and bacterial chromatophores. There, a combination of spectrochemical analysis, redox potentiometry, and voltage reporting by intrinsic electrochromic probes has revealed discrete, stepwise charge transfer through the membranes, and has yielded tentative biochemical assignments for the steps (6,14).

With transport systems that have not been so well characterized chemically, the strategy has evolved of developing detailed kinetic descriptions of transport, which can serve to limit or designate classes of molecular models. The approach is similar to that long used in the study of ordinary enzymatic reactions (27) and is particularly appropriate for active transport systems, where chemical catalysis [e.g., adenosine triphosphate (ATP) hydrolysis] as well as transmembrane transfer is brought about by the transport system. The usual parameters of such kinetic descriptions are transport velocity, binding capacity, or hydrolytic rate versus ion or substrate concentration. And, for electrogenic transport systems, the effects of the membrane electric field must also be included. Although a few laboratories have begun to apply field-pulse relaxation methods (9,26), both practical and theoretical considerations have limited most electrical-kinetic studies on active transport to steady-state conditions, analogous to those obtained in experiments with step changes of substrate or ion concentration.

Presented below is an outline of recent experiments relating the velocity of proton extrusion—through the plasma membrane of the fungus *Neurospora*—to the magnitude of electric potential difference imposed across the membrane. A major advantage of the particular organism (rather than conventional excitable cells) for studies on the electrogenicity of active transport is that the proton pump in *Neurospora* gives large electrical signals under long-term, steady-state conditions. The normal resting membrane potential of the organism ranges between -150 and -250 mV, of which only 0 to 40 mV can be attributed to ion diffusion, whereas the rest arises directly because of the proton pump. Both sustained electrogenicity and large membrane potentials make economic sense in organisms that have a continuous supply of ions (i.e., protons) from catabolism and which can use an enclosing wall (rather than an ion or salt pump) to prevent osmotic lysis. The proton pump of *Neurospora*, therefore, may perhaps be regarded as a physiological example for ion pumps in the envelope membranes of most nonanimal cells.

METHODS AND THEORY

Methods for maintaining cultures of *Neurospora* and preparing them for electrophysiological experiments have been described in detail elsewhere (20). Of interest for present purposes is the fact that the recording medium normally used is slightly acidic and ionically rather dilute, containing 25 mM K^+, or 25 mM ($K^+ + Na^+$), 1 mM free Ca^{2+}, and various anions or combinations of anions: Cl^-, SO_4^{2-}, phosphate, citrate, or dimethylglutarate, adjusted to pH 5.8.

Proton efflux through the electrogenic pump is assumed equal to the pump current, which is calculated from total membrane current. That, in turn, is determined by the voltage-clamp procedure illustrated in Fig. 1. Part A shows a *Neurospora* hypha with two microelectrodes inserted: one to record membrane potential (V_0) and a second to inject current (i) measured by the feedback amplifier. In practice, a second voltage electrode (V_1) is also inserted, at some distance from the two shown, in order to estimate current spread and voltage decrement in the long hyphal filaments of the organism. To obtain a systematic scan of the membrane current–voltage relationship, membrane potential is swept through a cascade of alternating (+ and −) pulses, from the resting value to 0 mV and to -300 mV, as shown in Fig. 1, B and C. For each pulse, V_0 (clamped) and the stable values of i and V_1 are measured, and a quasilinear theory for lumped cables (7) is used to calculate the membrane current (I_m) density at V_0. Plots of I_m versus V_0 constitute the membrane current–voltage relationship (*I–V* curve).

Theoretical analysis of membrane *I–V* curves is carried out by means of the simple kinetic model illustrated in Fig. 2. Only two states of the membrane "carrier" are specified, outside [$XH^+_{(o)}$] and inside [$P\sim XH^+_{(i)}$]; and these are connected by two pathways, one (k) for transferring charge across the membrane

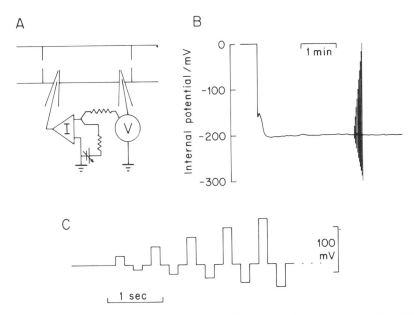

FIG. 1. Diagram of arrangements for voltage-clamping *Neurospora* plasma membrane, to obtain current–voltage curves. **A:** Clamp diagram; voltage (V_o) is monitored and set by the microelectrode at V, and required current (*i*) is delivered by the amplifier I. A second recording of voltage (V_1) is also made in an adjacent cytoplasmic compartment. Pore resistance through the cross-walls is high enough that each compartment can be considered isopotential. **B:** Sample trace for V_o, showing voltage jump at puncture, creep of 50 mV during sealing in of the electrode, and double staircase of clamped pulses. **C:** Diagrammatic expansion of initial clamp sequence.

and the other (κ) for recycling the carrier. Somewhat arbitrarily, voltage dependence for the charge-transfer pathway is introduced by means of a symmetric Eyring barrier (17), so that the individual reaction constants can be written

$$k_{io} = k_{io}^0 \exp(F\Delta\psi/2RT) \quad \text{and} \quad k_{oi} = k_{oi}^0 \exp(-F\Delta\psi/2RT), \quad (1a,b)$$

in which $\Delta\psi$ is the membrane potential (equivalent to V_o), the superscript 0 designates reaction constants for $\Delta\psi = 0$, and R, T, and F have their usual meanings. Association of the dephosphorylation reaction with the charge transfer step need not be taken literally but is drawn to indicate that chemical-bond energy is released into gradient energy at this step (see below and ref. 8). The carrier-recycling pathway subsumes all voltage-independent reaction steps: ion release at the external surface of the membrane, phosphorylation of the carrier protein, and ion rebinding at the internal surface. Therefore, the κ constants are referred to as lumped or condensed reaction constants. Finally, it has proven satisfactory to collect together all passive transport processes—including simple diffusion and any ion-coupled substrate transport that may occur under the conditions of these experiments—into a common leak pathway, with the linear conductance designated G_L in Fig. 2.

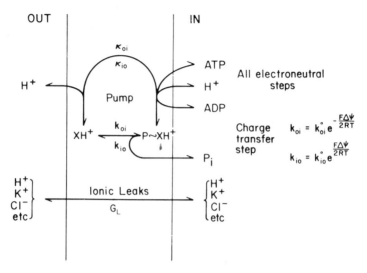

FIG. 2. Two-state carrier model for an electrogenic proton pump, drawn in parallel with the membrane leakage pathway(s). [From Sanders et al. (19), with permission.]

From this model, the total membrane I–V curve is given by

$$i = zFN \left[\frac{\kappa_{oi}k_{io} - \kappa_{io}k_{oi}}{k_{io} + k_{oi} + \kappa_{io} + \kappa_{oi}} \right] + G_L(\Delta\psi - E_L) \quad (2)$$

in which N is physically the carrier density in the membrane (but functions mathematically to scale the I–V curve and the reaction constants), z is the number of charges transferred across the membrane with each turn of the pump cycle, E_L is the equilibrium potential for the passive processes, and all other terms are defined in Fig. 2 or in the previous paragraph. Explicit voltage dependence is obtained by substituting Eq. (1a,b) into Eq. (2).

The reason for choosing a two-state model (rather than a more realistic five-state or higher n-state model) to interpret I–V curves is that steady-state I–V data alone are completely described by the concentrations and reaction constants for those two states of the carrier, which interconvert to move charges across the membrane; and, conversely, I–V data give no direct information about other carrier states. In other words, provided an electrogenic pump contains only one pathway for charge transit of the membrane, any current-voltage relationships for that pump can be described by a cycle containing only two pairs of reaction constants [i.e., Eq. (2) without the leakage terms].

The obvious limitations of this fact can be somewhat mitigated by introducing two reserve factors that incorporate reaction constants linking the inaccessible, lumped, carrier states, however many there may be. Calling these reserve factors r_o and r_i, we can rewrite the reaction constants in Eq. (2) as

$$k_{io} = \frac{k_{12}}{r_i}, \quad k_{oi} = \frac{k_{21}}{r_o}, \quad \kappa_{io} = \frac{\kappa_{12}}{r_i}, \quad \kappa_{oi} = \frac{\kappa_{21}}{r_o}, \quad (3a\text{–}d)$$

in which the subscripts 12 and 21 designate the i–o and o–i transitions in a fully expanded (n-state) kinetic model. It is apparent, thus, that concealed carrier states do influence the magnitude of the simple voltage-dependent reaction constants as determined from an I–V curve. However, that influence is quantitatively limited, between factors of, say, 0.1 and 10, except in one special circumstance that will be discussed again below. The explicit expansion of the reserve factors is algebraically straightforward, but depends, of course, on the number and arrangement of inaccessible carrier states. A detailed discussion of the nature and utility of reserve factors in kinetic models has been given previously (10).

RESULTS

The objective of the kinetic model described above, and of the experimental analysis to be presented below, is to obtain physically meaningful estimates of reaction parameters for an electrogenic proton pump. We should be able to approach this objective by fitting Eq. (2) to experimental I–V curves. Unfortunately, however, even though Fig. 2 and Eq. (2) represent major kinetic simplifications, steady-state I–V curves rarely contain sufficient information to specify all of the required parameters, either because the range of realizable voltages is too narrow or because the curves themselves are too nearly linear. We have therefore adopted two accessory procedures, one designed to simplify the equations further, and one to increase the amount of applicable experimental data.

The first procedure is to assign values to 3 of the 8 parameters in Eq. (2): $N = 10^{-12}$ moles/cm² (6,000 sites/μm²), which is roughly compatible with chemical data from vesicles of the *Neurospora* plasma membrane (2); $z = 1$, which is the only integral stoichiometry allowed by energetic calculations based either on maximal observed membrane potentials (−325 mV) or on extrapolated reversal potentials for the proton pump (−300 to −400 mV, under most conditions); and $E_L = 0$ mV, because experimentally estimated values of E_L do not differ systematically from 0. With these three parameters assigned, only five must be calculated for each I–V curve: k_{io}, k_{oi}, κ_{io}, κ_{oi}, and G_L. In principle, all can be extracted from the specific geometry of a single I–V curve (10). However, because the data resolution from one curve is often inadequate, we usually fit Eq. (2) jointly to several I–V curves, each of which represents a separate time point obtained during progressive action of a site-specific reagent. The fitting strategy, then, is to seek common values (satisfying all curves in a particular set) for as many as possible of the five parameters above, and to allow as few parameters as possible to vary from curve to curve. Optimization is carried out by varying each of the five in turn, and selecting the overall best fit by a least-squares method (18).

Response of the Proton Pump to Changes of Extracellular pH

It was known 15 years ago that the resting membrane potential of *Neurospora* is sensitive to extracellular pH (pH$_o$), having a slope of more than 30 mV/pH unit at pH_o values of 3 to 5 (20). This was assumed to reflect passive permeability of the *Neurospora* membrane, an interpretation supported at that time by related observations on the giant alga *Nitella clavata* (15). However, recent current-voltage analysis suggests a much different interpretation, namely, that the proton pump decreases its current by several $\mu A/cm^2$ for each unit lowering of pH_o, but passive permeability of the *Neurospora* plasma membrane is insensitive to pH_o.

Pertinent data are presented in Fig. 3. The inset shows the time course of

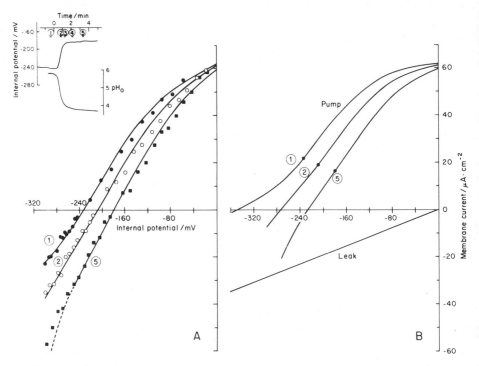

FIG. 3. Response of membrane and pump *I–V* relationships to extracellular acidification. *Inset:* Time course of membrane potential (*upper trace, left-hand ordinate scale*) and external pH (*lower trace, right-hand ordinate scale*) on shifting from pH 5.8 buffer to pH 3.6 buffer. *Numbered dashes* indicate intervals in which *I–V* scans were taken. **A:** Plots of *I–V* data for three of the five scans. Smooth curves were drawn by fitting text Eq. (2) to the plotted points, using a nonlinear least-squares algorithm (18). **B:** Component pump and leak curves from **A** and text Eq. (2). Optimal fitting required adjustment of k_{oi}^0 and κ_{io}, but no change of leak conductance (G_L). Table of parameter values:

Curve No.	k_{io}^0 (sec^{-1})	k_{oi}^0 (sec^{-1})	κ_{oi} (sec^{-1})	κ_{io} (sec^{-1})	G_L ($\mu S/cm^2$)	E_r (mV)
1, control	3.7×10^5	3.5×10^{-2}	6.5×10^2	5.7×10^3	95.4	−350
2	3.7×10^5	39.4×10^{-2}	6.5×10^2	10.9×10^3	95.4	−274
5	3.7×10^5	124×10^{-2}	6.5×10^2	22.3×10^3	95.4	−227

Buffer solutions contained 20 mM dimethylglutaric acid, titrated to pH 5.8 with 25 mM KOH or to pH 3.6 with KOH and supplemented to 25 mM K$^+$ with KCl; 1 mM CaCl$_2$; and 1% glucose.

of membrane potential (upper trace) and extracellular pH (lower trace, right-hand ordinate scale) when the normal recording buffer was replaced by low pH buffer. The bars labeled 1 to 5 indicate periods during which I–V scans were taken; and the actual data for three of these: 1 (control, $pH_o = 5.8$), 2 (transient), and 5 (stable end value, $pH_o = 3.7$) are plotted in Fig. 3A. All five curves were fitted jointly by adjusting two parameters, k_{oi} and κ_{io}, for each curve. Visually satisfactory fits were also obtained by adjusting κ_{io} and G_L. But this alternative was rejected both because the standard error of the fit was 30% larger and, more importantly, because separate experiments (D. Sanders, *unpublished data*) showed that passive conductance was constant for pH_o between 9.0 and 3.6.

Since, in any expanded kinetic model, lowered pH_o or, rather, elevated $[H^+]_o$ should increase the effective reaction constant for H^+ binding at the membrane outer surface, the major change at low pH was expected to be in κ_{io}. That it occurred instead in k^o_{oi} can be described as a parallel positive transfer effect (due to the reserve factors; see ref. 10, Table 2) and implies that proton dissociation at the membrane outer surface is fast relative to proton binding and relative to most other steps in the transport cycle.

The smooth curves drawn in Fig. 3A show best-fit membrane I–V curves for the three sets of data plotted, and the curves in Fig. 3B show the corresponding leak and pump I–V curves calculated from Eq. (2). Two features of these pump I–V curves should be especially noted: (a) the saturating pump current (at strong depolarization) is essentially unaffected by lowered pH_o, despite the fact that pump current at the resting membrane potentials (designated by dots on the curves) diminishes from 22.5 to 17 $\mu A/cm^2$ between pH 5.8 and pH 3.7; and (b) the apparent reversal potential of the pump shifts 123 mV, from -350 mV at pH 5.8 to -227 mV at pH 3.7, coinciding almost exactly with expectation for a pH shift of 2.1 units. It must be emphasized that neither of these results was assumed *a priori;* both emerged simply by fitting Eq. (2) simultaneously to all five sets of I–V data obtained during the time course of shifting pH_o.

Consequences of Lowered Intracellular pH

Although intracellular pH (pH_i) of *Neurospora* is well regulated against extracellular shifts [pH_i remains within 0.2 unit of normal (7.2) for $pH_o = 3.6$ to 9.0], cytoplasmic acidification can be forced by adding permeant weak acids to the extracellular solution. When this is done, both the proton pump and passive permeability respond, but the latter response dominates.

An I–V sequence conducted during treatment of *Neurospora* with 5 mM sodium butyrate ($pH_o = 5.8$) is shown in Fig. 4. The plotted I–V curves differ conspicuously from those during extracellular acidification, in that curve 4, obtained at low pH_i (6.69), is very much steeper than the control curve. They also differ from the curves of Fig. 3 in that satisfactory fitting of all five curves

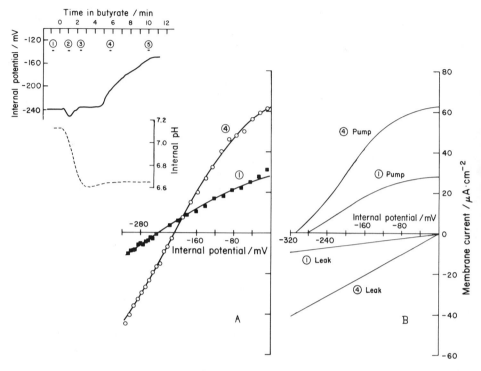

FIG. 4. Response of membrane, pump, and leak *I–V* curves to intracellular acidification. Conditions and general description of the experiments similar to Fig. 3. Internal pH was lowered by adding 5 mM neutralized sodium butyrate to the pH 5.8 buffer. *Dashed trace* in the *inset* designates the fact that pH$_i$ was measured, using an intracellular pH microelectrode (19), in a separate experiment. Note large change in slope of membrane *I–V* curve, from 1 (control) to 4. Table of parameter values:

Curve No.	k^0_{io} (sec^{-1})	k^0_{oi} (sec^{-1})	κ_{oi} (sec^{-1})	κ_{io} (sec^{-1})	G_L (μS/cm²)	E_r (mV)
1, control	7.5 × 10⁴	4.9 × 10⁻³	10	1.8 × 10³	25	−285
4	8.3 × 10⁴	4.9 × 10⁻³	25	1.8 × 10³	122	−309

(2, 3, 5 not shown) required adjustment of three of the five free parameters in Eq. (2): G_L, κ_{oi}, and k^0_{io}, although the change in k^0_{io} from curve 1 to curve 5 was less than 15%.

Inspection of the separated pump and leak curves, drawn in Fig. 4B, shows that the steepening slope of membrane *I–V* curves from 1 to 4 results from a small increase of pump conductance and a large increase of leak conductance. More detailed analysis (19) has shown, indeed, that, following a delay of about 1 min from the onset of butyrate treatment, G_L increases linearly with time, reaching about 10-fold the control value after 10 min. Since intracellular pH stabilizes in 2 to 3 min, the prolonged time course of changing G_L must arise from secondary effects of pH$_i$, perhaps mediated by some kind of control system.

The proton pump responds to lowered pH$_i$ much more rapidly, showing a detectable change by scan 2 (1 min) and near completion by scan 3 (2.5 min). The main parameter affected is κ_{oi} (which includes intracellular binding of H^+ ions), as would be expected from Fig. 2; and this is reflected in a nearly 2.5-fold increase of the saturating pump current. The calculated shift of reversal potential is 24 mV, from -285 mV for curve 1 to -309 mV for curve 4, again almost identical with expectation for cytoplasmic acidification from pH 7.12 to 6.69.

Turning back, briefly, to the inset of Fig. 4, the peculiar time course of voltage change can readily be interpreted. Since the initial, rapid effect of cytoplasmic acidification is to increase pump current, the membrane initially *hyperpolarizes*. But, as the steadily increasing leakage conductance becomes dominant, the membrane must depolarize.

Localization of the Energetic Transition and the Effect of ATP Withdrawal

In both sets of cases discussed above, fitting Eq. (2) to the pump $I-V$ curves yields a major asymmetry of reaction constants for the charge-transfer pathway. That is, $k_{io}^0/k_{oi}^0 \cong 10^7$. This circumstance suggests that energy is transferred from chemical bonds into the electrochemical gradient coincidentally with charge transfer across the membrane, although it is also consistent with very rapid charge dissociation after transfer (see parallel positive transfer effect, above).

In order to explore the reproducibility of this siting of the major energetic transition in the proton pump, and to ensure that it is not a peculiar consequence of manipulating H^+ concentrations, we have also examined $I-V$ relationships in *Neurospora* during metabolic inhibition, for example during treatment of this (obligately aerobic) organism with cyanide. Some representative results are shown in Fig. 5. In short-term experiments with cyanide, it is possible to fit sets of curves by manipulating only a single reaction constant in Eq. (2), and that—as would be expected from Fig. 2—is κ_{oi}, whose change can be seen graphically by the decrease of saturating pump current in Fig. 5B.

Although cytoplasmic ATP and adenosine diphosphate (ADP) concentrations were not measured in this experiment, previous experience (21,24) indicates that ATP should go from 3.0 to 0.9 mM between curves 1 and 2, and ADP should rise transiently from 0.8 to 1.4 mM and then recede again to 0.9 to 1.0 mM, by the action of adenylate kinase. Net hydrolysis would have little effect on free phosphate, since the normal cytoplasmic concentration of inorganic phosphate is approximately 10 mM (11). Therefore, the free energy available to the pump from ATP hydrolysis should fall only in accordance with the threefold drop in ATP concentration, which means about 30 mV. It is gratifying, therefore, that fitting of Eq. (2) to the $I-V$ plots in Fig. 5A yields a 25-mV lowering of reversal potential for the pump (Fig. 5B).

Again, as in the experiments of Figs. 3 and 4, the major energetic transition in the proton pump appears at the charge-transfer step, with $k_{io}^0/k_{oi}^0 > 10^7$. It will be shown, furthermore, that this same condition holds in experiments with

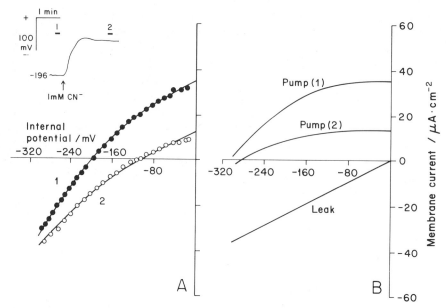

FIG. 5. Response of membrane and pump *I–V* curves to respiratory inhibition by 1 mM KCN. General description as in Fig. 3. [From Slayman (22), with permission.] Table of parameters:

Curve No.	k^0_{io} (sec^{-1})	k^0_{oi} (sec^{-1})	κ_{oi} (sec^{-1})	κ_{io} (sec^{-1})	G_L (μS/cm²)	E_r (mV)
1, control	2.8×10^6	1.8×10^{-1}	3.5×10^2	2.4×10^4	120	−308
2	2.8×10^6	1.8×10^{-1}	1.3×10^2	2.4×10^4	120	−283

direct pump inhibitors (see Fig. 6), as well as with respiratory inhibitors, so that it does appear to be a genuine physical property of the pump.

Another constant feature of the *I–V* analysis of the *Neurospora* proton pump, which is brought out by comparison of Figs. 3 to 5, is a small reaction asymmetry in the voltage-independent pathway. Under normal circumstances, $\kappa_{io}/\kappa_{oi} \cong 50$. It has previously been suggested, from comparison with results on the marine alga, *Acetabularia* (8), that this asymmetry results from the normal pH difference between *Neurospora* cytoplasm and the external medium. The present experiments cannot be said to test this question, in part because of the kinetic transfer effect seen with altered external pH (see above, Fig. 3), but the fact that the asymmetry diminishes with forced cytoplasmic acidification (Fig. 4) is at least consistent with the earlier suggestion.

However, a more interesting aspect of the asymmetry is brought out by the cyanide experiment. Since the change in free energy of ATP hydrolysis—on cyanide treatment, for example—is only about 30 mV, a persistent puzzle in metabolic experiments concerns accountability for the drop of ≥ 100 mV often observed in the membrane potential. That effect can now be seen as a kinetic feature, rather than a thermodynamic feature, of the pump. A threefold drop of ATP concentration yields roughly a threefold drop of saturating current

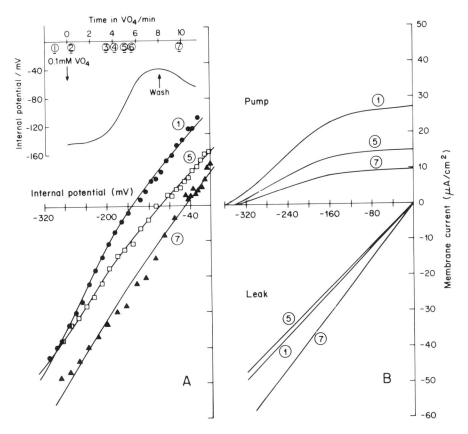

FIG. 6. Response of membrane, pump, and leak *I–V* curves to pump blockade by 100 μM orthovanadate. General description as in Fig. 3. Table of parameters:

Curve No.	k_{io}^0 (sec^{-1})	k_{oi}^0 (sec^{-1})	κ_{oi} (sec^{-1})	κ_{io} (sec^{-1})	G_L (μS/cm^2)	E_r (mV)
1, control	4.3×10^5	2.2×10^{-3}	2.7×10	3.1×10^3	156	−358
5	4.3×10^5	2.2×10^{-3}	1.5×10	3.1×10^3	148	−343
7	4.3×10^5	2.2×10^{-3}	0.95×10	3.1×10^3	191	−332

(with a not-quite-proportional change of pump current at all practical voltages); and the potential change so produced across the membrane leakage conductance (120 μS/cm^2, in Fig. 5) is about 100 mV. Very clearly, such a result cannot be described by ordinary linear equivalent circuit approaches to electrogenic pumps.

Kinetic Effects of a Pump-Specific Inhibitor

In addition to its primary effect of withdrawing ATP from the proton pump, a respiratory inhibitor like cyanide can be expected to have secondary effects as well, because of the multiple roles that ATP plays in metabolism. In this connection, we must note that the simple kinetic result presented in Fig. 5 persists for only a few minutes of cyanide treatment. Thereafter, membrane resistance and cytoplasmic ATP concentration rise (C. L. Slayman, *unpublished results*), and the membrane potential commences slow oscillation (23). Because of their obvious complexity, these phenomena have not yet been subjected to current–voltage analysis. Although it is probably unreasonable to expect any inhibitor to be really specific in its action, and devoid of secondary effects on the pump, it seemed important to check the results in Fig. 5 against the behavior of both *Neurospora* and the two-state kinetic model in experiments with a direct inhibitor of the proton pump. The best such inhibitor presently available is orthovanadate, which has been shown by Cantley and co-workers (3,13), on Na^+/K^+-ATPase, to inhibit the *Neurospora* membrane ATPase with an *in vitro* $K_{1/2}$ of about 1 μm (1,16).

The results of one I–V experiment with orthovanadate are displayed in Fig. 6. Again, the inset shows the time course of membrane potential, following addition of 0.1 mM orthovanadate to the recording medium. After a delay of approximately 2 min, the cell depolarized rather slowly to about −40 mV, at which point it was essentially insensitive to further depolarization by respiratory inhibitors. The delay and slow course of depolarization, as well as the requirement for a high concentration of vanadate, result from interaction—probably competition for transport and for intracellular binding—between vanadate and phosphate. Vanadate has practically no effect on intact *Neurospora* unless the cells are first starved of phosphate (16).

Seven I–V curves in all were generated during the vanadate treatment, and three of these are shown in Fig. 6A. Despite some obvious noise in the data (particularly in curve 7), fitting Eq. (2) to the whole set of I–V data yields the results of Fig. 6B, with two parameters changing from curve to curve: κ_{oi} and G_L. Because the change in G_L, unlike that in Fig. 4, is small and irregular, we regard it as spurious. Clearly, the major effect is the threefold decrease of κ_{oi}, so that when viewed by the two-state kinetic model (Fig. 2), vanadate has qualitatively the same effect on the *Neurospora* proton pump as does cyanide. This coincidence is indeed correct, since either ATP withdrawal or carrier withdrawal by a competing reaction (25) retards the lumped reloading step in the overall pump cycle.

DISCUSSION

It is helpful, in interpreting changes of kinetic parameters during ion pump manipulation, to consider how the parameters are related to geometric features of the I–V curves. For the two-state model there are two simple relationships, namely, that the lumped voltage-independent reaction constants are directly proportional to the positive and negative saturation currents:

$$\kappa_{oi} = i_{sat+}/zFN \quad \text{and} \quad \kappa_{io} = i_{sat-}/zFN. \quad (4a,b)$$

Since most I–V scans with the *Neurospora* proton pump extend far (ca. 300 mV) in the positive direction from E_r (the pump reversal potential), and since many different kinds of experiments have indicated that E_L (the net equilibrium potential for passive transport systems in the membrane) is near to 0 mV, i_{sat+} is closely approximated by the membrane short-circuit current. This means that any chemical change, such as of ATP concentration or pH_i, which affects the composite reaction constant κ_{oi} will be directly reflected in the membrane short-circuit current, and, conversely, that the curve-fitting procedure used above will interpret changes of short-circuit current mainly as changes of κ_{oi}.

Although this arrangement gives some confidence in evaluating κ_{oi}, there is one caveat; that is, changes of N (the concentration of functioning pump in the membrane) will—by altering the short-circuit current—mimic genuine changes of κ_{oi}. Thus, our initial decision to fix the value of N was somewhat arbitrary, and must be buttressed in each experimental case by other information about the mode of action of the particular test agent.

It should be added that there is actually a practical benefit to the intrinsic confusion of κ_{oi} and N. N is a simple scaling factor in Eq. (2), so that any inhibitor that diminishes N (say, by irreversible binding and inactivation of the carrier) acts as a switch, turning off a portion of the proton pump without affecting its kinetics. In this circumstance the I–V relationship for the switched portion of the pump can be obtained as the I–V difference curve before and after addition of the inhibitor. Now, the fact that changes of N mimic changes of κ_{oi} (and vice versa) means that pump I–V curves can be estimated by the difference method using inhibitors (i.e., cyanide) that only affect κ_{oi}, provided that effects on κ_{oi} are not too great (say, less than 10-fold, so changes in E_r are negligible). This argument provides theoretical justification for the empirical I–V difference method previously used to estimate current–voltage curves of the *Neurospora* proton pump (7).

There are two other almost simple relationships between the reaction constants of Eq. (2) and measurable features of the I–V curves:

$$E_r = \frac{RT}{zF} \ln \frac{k_{oi}^0}{k_{io}^0} \cdot \frac{\kappa_{io}}{\kappa_{oi}} \quad \text{and} \quad i_{sc} = zFN \frac{\kappa_{oi} k_{io}^0 + \kappa_{io} k_{oi}^0}{k_{io}^0 + k_{oi}^0 + \kappa_{io} + \kappa_{oi}} \quad (5a,b)$$

in which E_r is the pump reversal potential and i_{sc} is the pump short-circuit current. Unfortunately, k_{io}^0 and k_{oi}^0 are complicated functions of i_{sat+}, i_{sat-}, i_{sc}, and E_r, so that we must, in general, rely on the computer-fitting procedure to give estimates of these two reaction constants.

The same is true of κ_{io}, the reverse reaction constant in the voltage-independent pathway. Despite the simplicity of Eq. (4b), the saturating reversal current for the pump lies in the inaccessible range of extreme hyperpolarization, so it is only the distributed effect of κ_{io} on the experimental membrane I–V curve that impinges on the parameter calculations. There is, however, an important qualitative statement to be made about κ_{io}. As can be seen from Fig. 2, the most direct way to alter κ_{io} is to change the intracellular ADP concentration. Any inhibitor that causes ADP to be elevated must change i_{sat-} in rough proportion (Eq. 4b) and, if it happens to affect ATP at the same time, will have complicated effects on the overall pump I–V curve. Indeed, it is the effect on ADP levels, in computer simulations of the sodium pump, which led to the conclusion of Chapman et al. (4) that metabolic inhibitors could not be used to extract pump I–V curves by the difference method. It was our good fortune, for empirical I–V studies on the *Neurospora* proton pump (7), that adenylate kinase tends to stabilize cytoplasmic ADP near 1 mM, in the face of 10-fold (or larger) changes of ATP.

The major results presented in this chapter are, therefore, a mutual exploration of two systems: the electrogenic proton pump in the plasma membrane of *Neurospora*, and a rigorously abbreviated, two-state kinetic model for electrogenic pumps. The principal utility of the model is that its reaction parameters can be extracted—either by direct computation or by statistical methods—from actual current–voltage data on real electrogenic pumps. It is the physical significance of those parameters which needs to be explored further.

At a qualitative level, all four experimental treatments described above affect the model parameters in a sensible pattern, although with two surprises: lowered pH_i increases G_L, the membrane leakage conductance; and lowered pH_o increases k_{oi}^0, the back-reaction constant for transmembrane charge transfer. Both of these surprises can be rationalized. The first physiologically, by supposing that correction of pH_i by net proton extrusion requires new, non-H^+ leakage channels; and the second physically, by supposing that proton dissociation at the membrane outer surface is very rapid (see discussion of Fig. 3). Neither of these insights could emerge from a purely intuitive or empirical analysis of the membrane current–voltage relationships. At a quantitative level, cyanide inhibition, lowered pH_o, and lowered pH_i all produce correct changes in pump reversal potential, which can be regarded as a measure of the reliability for the two-state model.

We have already mentioned that the reaction constants in the two-state model are not elementary rate constants (see Eq. 3). Apart from inclusion of ligand concentrations (e.g., $[H^+]_i$, $[ATP]_i$, etc.) in certain of them, all reaction constants contain terms (reserve factors) describing the sequestering of carrier into experimentally inaccessible forms. Although reserve factors must modify the values of elementary rate constants, there is no reason why they should vary randomly from one experiment to another. This leaves unexplained the 50-fold variation of all corresponding parameter values among the four control curves in Figs. 3 to 6. Since each set of I–V curves is from a separate experiment, on separate cells, a substantial part of the parameter variation may be biological. Clearly, homogeneous data will be necessary to test the model further, but the question will still arise of whether the specific kinetic results from the four different treatments are mutually consistent. That question probably cannot be answered without expansion of the kinetic model.

ACKNOWLEDGMENTS

The authors are indebted to Dr. J. Warncke (Physiologisches Institut II, Universität des Saarlandes, Homburg-Saar, FRG) for the data in Figs. 5 and 6; to Dr. U.-P. Hansen (Institut für Angewandte Physik, Universität Kiel, FRG) for computer algorithms and analysis of the data in Figs. 5 and 6; and to Dr. D. Gradmann (Max-Planck-Institut für Biochemie, Martinsried, FRG) for many helpful discussions.

The authors are also indebted to Dr. J. B. Chapman (Department of Physiology, Monash University, Australia) for loan of the manuscript of ref. 4 prior to its publication.

The work was supported by research grant GM-15858 from the National Institute of General Medical Sciences.

REFERENCES

1. Bowman, B. J., and Slayman, C. W. (1979): The effects of vanadate on the plasma membrane ATPase of *Neurospora crassa. J. Biol. Chem.*, 254:2928–2934.
2. Bowman, E. J., Bowman, B. J., and Slayman, C. W. (1981): Isolation and characterization of plasma membranes from wild type *Neurospora crassa. J. Biol. Chem.*, 256:12336–12342.
3. Cantley, L. C., Jr., Josephson, L., Warner, R., Yanagisawa, M., Lechene, C., and Guidotti, G. (1977): Vanadate is a potent (Na,K)-ATPase inhibitor found in ATP derived from muscle. *J. Biol. Chem.*, 252:7421–7423.
4. Chapman, J. B., Johnson, E. A., and Kootsey, J. M. (1983): Electrical and biochemical properties of an enzyme model of the sodium pump. *J. Membr. Biol.*, 74:139–153.
5. Dills, S. S., Apperson, A., Schmidt, M. R., and Saier, M. H., Jr. (1980): Carbohydrate transport in bacteria. *Microbiol. Rev.*, 44:385–418.
6. Dutton, P. L., Mueller, P., O'Keefe. D. P., Packham, N. K., Prince, R. C., and Tiede, D. M. (1982): Electrogenic reactions of the photochemical reaction center and the ubiquinone-cytochrome b/c_2 oxidoreductase. In: *Electrogenic Ion Pumps*, edited by C. L. Slayman, pp. 323–343. Academic Press, New York.
7. Gradmann, D., Hansen, U.-P., Long, W. S., Slayman, C. L., and Warncke, J. (1978): Current-voltage relationships for the plasma membrane and its principal electrogenic pump in *Neurospora crassa*. I. Steady-state conditions. *J. Membr. Biol.*, 39:333–367.
8. Gradmann, D., Hansen, U.-P., and Slayman, C. L. (1982): Reaction-kinetic analysis of current-voltage relationships for electrogenic pumps in *Neurospora* and *Acetabularia*. In: *Electrogenic Ion Pumps*, edited by C. L. Slayman, pp. 257–276. Academic Press, New York.
9. Gräber, P. (1982): Phosphorylation in chloroplasts: ATP synthesis driven by $\Delta\psi$ and ΔpH of artificial or light-generated origin. In: *Electrogenic Ion Pumps*, edited by C. L. Slayman, pp. 215–245. Academic Press, New York.
10. Hansen, U.-P., Gradmann, D., Sanders, D., and Slayman, C. L. (1981): Interpretation of current-voltage relationships for "active" ion transport systems. I. Steady-state reaction-kinetic analysis of class-I mechanisms. *J. Membr. Biol.*, 63:165–190.
11. Harold, F. M. (1962): Depletion and replenishment of the inorganic polyphosphate pools in *Neurospora crassa. J. Bacteriol.*, 83:1047–1057.
12. Hoffman, J. F., editor (1978): *Membrane Transport Processes, vol. 1.* Raven Press, New York.
13. Josephson, L., and Cantley, L. C., Jr. (1977): Isolation of a potent (Na-K)ATPase inhibitor from striated muscle. *Biochemistry*, 16:4572–4578.
14. Junge, W. (1982): Electrogenic reactions and proton pumping in green plant photosynthesis. In: *Electrogenic Ion Pumps*, edited by C. L. Slayman, pp. 431–465. Academic Press, New York.
15. Kitasato, H. (1968): The influence of H^+ on the membrane potential and ion fluxes of *Nitella. J. Gen. Physiol.*, 52:60–87.

16. Kuroda, H., Warncke, J., Sanders, D., Hansen, U.-P., Allen, K. E., and Bowman, B. J. (1980): Effects of vanadate on the electrogenic proton pump in *Neurospora*. In: *Plant Membrane Transport: Current Conceptual Issues,* edited by R. M. Spanswick, W. J. Lucas, and J. Dainty, pp. 507–508. Elsevier, Amsterdam.
17. Läuger, P., and Stark, G. (1970): Kinetics of carrier-mediated ion transport across lipid bilayer membranes. *Biochim. Biophys. Acta,* 211:458–466.
18. Marquardt, D. W. (1963): An algorithm for least-squares estimation of nonlinear parameters. *J. Soc. Ind. Appl. Math.,* 11:431–441.
19. Sanders, D., Hansen, U.-P., and Slayman, C. L. (1981): Role of the plasma membrane proton pump in pH regulation in non-animal cells. *Proc. Natl. Acad. Sci. USA,* 78:5903–5907.
20. Slayman, C. L. (1965): Electrical properties of *Neurospora crassa*: Effects of external cations on the intracellular potential. *J. Gen. Physiol.,* 49:69–92.
21. Slayman, C. L. (1973): Adenine nucleotide levels in *Neurospora,* as influenced by conditions of growth and by metabolic inhibitors. *J. Bacteriol.,* 114:752–766.
22. Slayman, C. L. (1982): Charge-transport characteristics of a plasma membrane proton pump. In: *Membranes and Transport, vol. 1,* edited by A. N. Martonosi, pp. 485–490. Plenum, New York.
23. Slayman, C. L., Long, W. S., and Gradmann, D. (1976): "Action potentials" in *Neurospora crassa,* a mycelial fungus. *Biochim. Biophys. Acta,* 426:732–744.
24. Slayman, C. L., Long, W. S., and Lu, C.Y.-H. (1973): The relationship between ATP and an electrogenic pump in the plasma membrane of *Neurospora crassa. J. Membr. Biol.,* 14:305–338.
25. Smith, R. L., Zinn, K., and Cantley, L. C. (1980): A study of the vanadate-trapped state of the (Na,K)-ATPase. Evidence against interacting nucleotide site models. *J. Biol. Chem.,* 255:9852–9859.
26. Tittor, J., Hansen, U.-P., and Gradmann, D. (1983): Impedance of the electrogenic Cl^- pump in *Acetabularia*: Electrical frequency entrainments, voltage-sensitivity and reaction kinetic interpretation. *J. Membr. Biol.,* 75:129–139.
27. Wilbrandt, W., and Rosenberg, T. (1961): The concept of carrier transport and its corollaries in pharmacology. *Pharmacol. Rev.,* 13:109–183.

Electrogenic Transport: Fundamental Principles and Physiological Implications, edited by Mordecai P. Blaustein and Melvyn Lieberman. Raven Press, New York © 1984.

The *lac* Carrier Protein from *Escherichia coli*

H. Ronald Kaback, Nancy Carrasco, David Foster, Maria Luisa Garcia, Tzipora Goldkorn, Lekha Patel, and Paul Viitanen

Laboratory of Membrane Biochemistry, Roche Institute of Molecular Biology, Nutley, New Jersey 07110

Although the phenomenon of active transport (i.e., concentration of solute against a gradient at the expense of metabolic energy) has been recognized for some time, insight into the biochemistry of the reactions involved has begun to occur only recently. The primary reasons for the advances are essentially threefold: (a) Formulation of the chemiosmotic theory by Mitchell (19–22), which postulates that the immediate driving force for many processes in energy-coupling membranes is an electrochemical gradient of hydrogen ion ($\Delta\bar{\mu}_{H^+}$). (b) Development of techniques that allow detection and quantitation of electrochemical ion gradients in systems that are not amenable to a direct electrophysiological approach. (c) The availability of experimental model systems in which appropriate questions can be asked at a biochemical level. The purpose of this chapter is not a general review but rather to discuss the β-galactoside transport system in *Escherichia coli* as a representative example of current developments. However, as this is a rapidly evolving area, it should be emphasized that some of the notions expressed already may have been modified or even discarded.

BACKGROUND

Transport of β-galactosides in *E. coli* is catalyzed by the product of the *lac y* gene (4), the *lac* carrier protein or *lac* permease, which translocates these sugars with protons in a symport (co-transport) reaction as postulated in 1963 (20). Accordingly, it has been demonstrated virtually unequivocally that in the presence of $\Delta\bar{\mu}_{H^+}$, hydrogen ion moves down its electrochemical gradient and drives the uphill translocation of β-galactosides; conversely, downhill translocation of substrates drives uphill transport of hydrogen ion with generation of $\Delta\bar{\mu}_{H^+}$ (see ref. 12 for a review). The *lac* carrier protein was identified as a membrane protein chemically in 1965 (8) and functionally in 1970 (1), and it is one of a large class of substrate-specific polypeptides in bacteria that couple the downhill movement of a cation to the uphill accumulation of solute in response to a transmembrane electrochemical ion gradient.

In 1978 (32), the *lac y* gene was cloned into a recombinant plasmid, allowing the elucidation of its nucleotide sequence and the amino acid sequence of the *lac* carrier protein (2) as well as the amplification of the carrier (31) and its synthesis *in vitro* (5). Shortly thereafter, Newman and Wilson (23) successfully solubilized the carrier in octyl-β-D-glucopyranoside (octylglucoside) and reconstituted lactose transport activity in proteoliposomes, thus providing an experimental breakthrough that had been attempted unsuccessfully for some time (28). Almost simultaneously, it was demonstrated that *p*-nitrophenyl-α-D-galactopyranoside (NPG) is a highly specific photoaffinity label for the *lac* carrier protein (14), and the use of these techniques in concert culminated recently in the purification of a single polypeptide species in a functional state and its identification as the product of the *lac y* gene (7,24). The purified protein reconstituted into proteoliposomes catalyzes all of the known reactions of the *lac* transport system with excellent efficiency (i.e., $\Delta\bar{\mu}_{H^+}$-driven lactose accumulation, facilitated influx and efflux, lactose-induced proton movements, counterflow, and exchange). Therefore, it seems highly likely that the product of the *lac y* gene is the only polypeptide in the cytoplasmic membrane of *E. coli* required for lactose transport.

MECHANISTIC STUDIES

Employing an approach similar to that used previously with isolated membrane vesicles (13,15), carrier-mediated lactose efflux down a concentration gradient was studied in order to probe the translocation mechanism in proteoliposomes reconstituted with purified *lac* carrier protein.

The maximal rate of efflux is pH-dependent, increasing more than 100-fold from pH 5.5 to pH 9.5 in a sigmoidal fashion with a midpoint at approximately pH 8.3. In contrast, experiments performed under identical conditions with equimolar lactose in the external medium (i.e., under exchange conditions) demonstrate that the exchange reaction is insensitive to pH and very fast relative to efflux, particularly at relatively acid pH values ($<$ pH 7.5). Proton symport occurs during lactose efflux, resulting in the transient formation of a membrane potential ($\Delta\psi$, interior negative) as shown by efflux-dependent accumulation of rubidium in the presence of valinomycin. Comparison of the efflux and exchange reactions suggests that the rate-determining step for efflux involves a reaction corresponding to the return of the unloaded carrier to the inner surface of the membrane and that either loss of the symported proton from the carrier or translocation of the unloaded carrier may be limiting. Counterflow experiments conducted at various pH values reveal that external lactose affects proton loss from the carrier. When external lactose is present at concentrations below the apparent K_m of the carrier, counterflow is pH-dependent and decreases from pH 5.5 to 7.5, indicating that deprotonation of the carrier occurs frequently under these conditions to limit the counterflow process. When external lactose is saturating, however, counterflow is unaffected by pH. Moreover, transient formation of Δψ observed during lactose efflux is abolished under these conditions. The observations suggest an ordered mechanism for efflux, whereby lactose is released first, followed by loss of a proton, and they indicate that the loaded carrier recycles in the protonated form during counterflow and exchange (Fig. 1).

These conclusions are supported by experiments in which the effects of deuterium oxide on the reconstituted system were investigated. At equivalent pH and pD (i.e., pD = pH + 0.4), the rate of lactose-facilitated diffusion (influx as well as efflux) is approximately 3 to 4 times slower in deuterated medium (with over 95% of the protium replaced with deuterium) relative to control conditions in protium, whereas the rate of exchange is identical in the presence of deuterium and protium. Furthermore, during counterflow with the external lactose concentration below the apparent high-affinity K_m of the carrier, the magnitude of the overshoot is 2 to 3 times greater in deuterium relative to protium. Finally, and remarkably, Δψ-driven lactose accumulation exhibits essentially no solvent deuterium isotope effect.

On the basis of these observations, it is reasonable to suggest that under conditions where carrier turnover is driven by a lactose concentration gradient, the rate of translocation is determined by a step(s) involving protonation, deprotonation, or a subsequent step (i.e., a reaction corresponding to the return of the unloaded carrier). In contrast, when there is a driving force on the protons (i.e., in the presence of $\Delta\bar{\mu}_{H^+}$), this step(s) is no longer rate-determining. As a cautionary note, it should be emphasized that the solvent deuterium isotope effects described cannot be attributed definitively to a true kinetic isotope effect as opposed to a pK_a effect (i.e., deuterium increases the pK_a values of various functional groups from 0.4 to 0.7 pH units), because the isotope effect on efflux disappears at ≥ pH 9.0. On the other hand, at these alkaline pHs, the rate of efflux approaches the rate of exchange, suggesting that the rate-determining

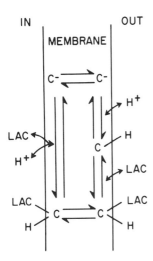

FIG. 1. Schematic representation of reactions involved in lactose efflux. C represents the *lac* carrier protein. The order of substrate binding at the inner surface of the membrane is not implied.

step for efflux may change at high pH. In any event, it seems evident that different steps are rate determining when carrier turnover is driven by $\Delta\bar{\mu}_{H^+}$ or by a solute concentration gradient.

STRUCTURE/FUNCTION CONSIDERATIONS

Proteolysis experiments with isolated cytoplasmic membrane vesicles, in which the *lac* carrier was photolabeled specifically with NPG, demonstrate that the protein spans the bilayer and that the binding site is contained within a transmembrane segment of the protein (9). In these experiments, topologically sealed right-side-out (25–27) and inverted (29) vesicles, previously photolabeled with NPG, were incubated with papain, chymotrypsin, or trypsin for various periods of time, solubilized and subjected to sodium dodecylsulfate-polyacrylamide gel electrophoresis (SDS-PAGE). In untreated control samples, essentially all of the radioactivity is associated with a band that migrates at an M_r of about 33,000 daltons[1], and it is apparent from experiments with vesicles devoid of *lac* carrier, vesicles containing amplified levels of *lac* carrier, and the purified protein itself that this band represents the product of the *lac y* gene. With both right-side-out and inverted vesicles, radioactivity appearing at 33,000 daltons disappears with time and is replaced by a radioactive band at about 20,000 daltons, thus demonstrating directly that the *lac* carrier is accessible to proteases from either side of the membrane. In contrast, when the vesicles are dissolved in SDS and then subjected to proteolysis, the *lac* carrier is fragmented into very small pieces that electrophorese at the front of the gels. It is apparent, therefore, that the proteolysis patterns observed with intact membrane vesicles are due to structural constraints imposed by the membrane.

Detailed kinetic studies of lactose transport in isolated membrane vesicles demonstrate that, in addition to acting thermodynamically as the driving force for active transport, $\Delta\bar{\mu}_{H^+}$ alters the distribution of the *lac* carrier between two markedly different kinetic states (30). In the absence of $\Delta\bar{\mu}_{H^+}$, transport exhibits an apparent K_m of about 20 mM for lactose, and when a $\Delta\psi$ (interior negative) or a ΔpH (interior alkaline) is applied, the apparent K_m decreases by about 100-fold to 0.2 mM. Furthermore, the distribution of the carrier between these two pathways varies as the square of $\Delta\psi$ or ΔpH. On the basis of these observations, it was suggested very tentatively that the *lac* carrier protein might exist in two forms, monomer and dimer, that the monomer catalyzes facilitated diffusion (high apparent K_m) and the dimer active transport (low apparent K_m), and finally, that $\Delta\bar{\mu}_{H^+}$ causes a monomer-dimer transition.

Interestingly, application of radiation inactivation analysis (16) to the system provides support for this notion. In these experiments, vesicles containing the *lac* carrier were frozen in liquid nitrogen before or after energization and sub-

---

[1] Although the molecular weight of the *lac* carrier protein is 46,504, for unknown reasons, the protein migrates with an apparent M_r of about 33,000 daltons during SDS-Page (12%).

jected to a high-intensity electron beam for various periods of time. Since the vesicles become very permeable after short periods of irradiation, it is necessary to extract and reconstitute the *lac* carrier in order to assay activity. Thus, after irradiation, the samples were extracted with octylglucoside, reconstituted into proteoliposomes and tested for counterflow activity (10).

Under all conditions tested, the decrease in activity exhibits pseudo-first-order kinetics as a function of radiation dosage, allowing straightforward application of target theory for determination of functional molecular mass. When *lac* carrier activity solubilized from nonenergized vesicles was assayed under these conditions, the results obtained are consistent with a functional molecular weight of 45,000 to 50,000 daltons, a value similar to the molecular weight of the *lac* carrier protein as determined from the sequence of the *lac y* gene (2) or from amino acid analysis of the purified protein (24). Importantly, moreover, similar values were obtained when the octylglucoside extract was irradiated, and target volumes observed for D-lactate dehydrogenase (D-LDH) and dicyclohexylcarbodimide-sensitive adenosine triphosphatase (ATPase) activity in the same vesicle preparations are comparable to the known molecular weights of D-LDH and the F_1 portion of the H^+-ATPase. Strikingly, when the same procedures were carried out with vesicles that had been energized with appropriate electron donors prior to freezing and irradiation, a functional molecular weight of 85,000 to 90,000 daltons was obtained for the *lac* carrier with no change in the target mass of D-LDH. In contrast, when the vesicles were energized in the presence of the protonophore carbonylcyanide-*m*-chlorophenylhydrazone, the target mass of the *lac* carrier returned to 45,000 to 50,000 daltons.

In addition to these observations, genetic studies demonstrating that certain *lac y* gene mutations are dominant (18) also support the idea that oligomeric structure may be important for *lac* carrier function.

MONOCLONAL ANTIBODIES AGAINST THE *lac* CARRIER PROTEIN

Recently, monoclonal antibodies directed against purified *lac* carrier protein were prepared by somatic cell fusion of mouse myeloma cells with splenocytes from an immunized mouse (3). Several clones produce antibodies that react with the purified protein as demonstrated by solid-phase radioimmunoassay (SP-RIA) and by immunoblotting experiments, and culture supernatants from one clone (4B1) inhibits active transport in isolated membrane vesicles. Five stable clones were selected for expansion, formal cloning and production of ascites fluid, and the antibodies secreted *in vivo* by 4B1 also were found to inhibit lactose transport. Antibody from hybridoma 4B1, an IgG2a immunoglobulin, inhibits active transport of lactose with purified *lac* carrier in proteoliposomes and in right-side-out membrane vesicles. In contrast, the antibody has no effect on the generation of $\Delta\bar{\mu}_{H^+}$ by membrane vesicles, nor does it interfere with the ability of vesicles containing the *lac* carrier to bind NPG. In order to achieve 50% inhibition of transport activity, a two- to threefold molar excess of antibody to *lac* carrier is required, regardless of the amount of *lac* carrier in the membrane. Thus, the concentration of antibody needed for a given degree of inhibition is proportional to the amount of *lac* carrier in the membrane. Notably, antibody-induced inhibition occurs within seconds, suggesting that the epitope is accessible on the surface of the membrane, and Fab fragments prepared from antibody 4B1 inhibit transport as effectively as the intact antibody.

In proteoliposomes reconstituted with purified *lac* carrier, 4B1 markedly inhibits $\Delta\psi$-driven lactose accumulation with no effect on the exchange reaction and no effect on NPG binding. Given these observations, the finding that NPG binding is not inhibited by antibody, and evidence suggesting that $\Delta\bar{\mu}_{H^+}$ may induce a major conformational alteration in the *lac* carrier (e.g., dimerization), it is tempting to speculate that the antibody might inhibit active transport by preventing this conformational transition. In any event, initial efforts to localize the epitopes within the *lac* carrier are promising. When the purified protein is fragmented with cyanogen bromide, 4B1 epitope activity is retained, and a number of cyanogen bromide fragments can be identified by high-performance liquid chromatography (HPLC). In addition, preliminary use of monoclonal affinity columns prepared with antibodies 4B1 and 4A10R appears to be encouraging with regard to epitope isolation and identification. Clearly, the early indications suggest that further characterization of these antibodies combined with detailed structure-function studies may provide important insight into the structure and mechanism of action of the *lac* carrier protein.

A STRUCTURAL MODEL FOR THE *lac* CARRIER PROTEIN

In order to pursue some of these observations and proposals, a testable secondary structure model of the *lac* carrier protein would be particularly useful (7a). The *lac* carrier is a very hydrophobic polypeptide of molecular weight 46,504 daltons (10,16), containing 417 amino acid residues of known sequence (2). Circular dichroic measurements on the purified protein solubilized in octylglucoside or reconstituted into proteoliposomes demonstrate that 85 ± 5% of the amino acid residues are arranged in helical secondary structures. Analysis of the sequential hydropathic character of the protein (17) indicates that the protein is composed of 12 to 13 hydrophobic segments with a mean length of 24 ± 4 residues per segment (Fig. 2). Approximately 70% of the 417 amino acids in the *lac* carrier are found in these domains. The hydropathic profile, together with the circular dichroic measurements, suggests that the 12 to 13 hydrophobic segments are largely in a helical conformation. If the segments are assumed to be α-helical, the mean length of each domain approximates the thickness of the hydrophobic portion of the lipid bilayer. Based on these considerations, it is proposed that the *lac* carrier protein consists of 12 to 13 α-helical segments that traverse the membrane in a zig-zag fashion as suggested for bacteriorhodopsin (6,11). Obviously, the model makes certain predictions that are immediately applicable to some of the proteolysis and monoclonal antibody experiments

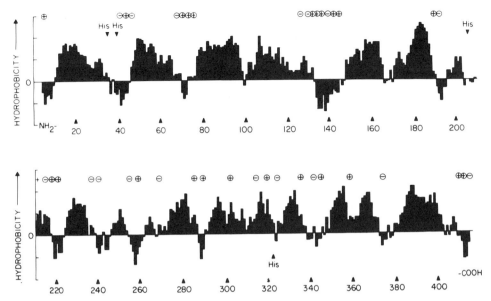

FIG. 2. Hydropathic profile of the *lac* carrier protein. A modified version of the SOAP program described by Kyte and Doolittle (17) was used. The program assigns values to amino acids on the basis of their hydrophobic character. Using a moving segment approach that continuously determines the average hydropathy of a segment of 7 residues as it advances through the sequence, consecutive scores are plotted at the middle of each segment from the amino to the carboxy terminals. The positions of all histidines (His) and charged residues are noted. The hydropathic profile was also determined with a 19-residue moving segment. Generally, the results were similar to the profile given, but the resolution between segments was considerably reduced.

discussed above. Thus, evidence regarding its validity (or the lack thereof) should be forthcoming within the near future.

REFERENCES

1. Barnes, E. M., Jr., and Kaback, H. R. (1970): β-Galactoside transport in bacterial membrane preparations: energy coupling via membrane-bound D-lactic dehydrogenase. *Proc. Natl. Acad. Sci. USA*, 66:1190–1198.
2. Büchel, D. E., Gronenborn, B., and Müller-Hill, B. (1980): Sequence of the lactose permease gene. *Nature*, 283:541–545.
3. Carrasco, N., Tahara, S. M., Patel, L., Goldkorn, T., and Kaback, H. R. (1982): Preparation, characterization and properties of monoclonal antibodies against the *lac* carrier protein from *Escherichia coli*. *Proc. Natl. Acad. Sci. USA*, 79:6894–6898.
4. Cohen, G. N., and Monod, J. (1957): Bacterial permeases. *Bacteriol. Rev.*, 21:169–194.
5. Ehring, R., Beyreuther, K., Wright, J. K., and Overath, P. (1980): *In vitro* and *in vivo* products of *E. coli* lactose permease gene are identical. *Nature*, 283:537–540.
6. Engelman, D. M., Henderson, R., McLachlan, A. D., and Wallace, B. A. (1980): Path of the polypeptide in bacteriorhodopsin. *Proc. Natl. Acad. Sci. USA*, 77:2023–2027.
7. Foster, D. L., Garcia, M. L., Newman, M. J., Patel, L., and Kaback, H. R. (1982): Lactose proton symport by purified *lac* carrier protein. *Biochemistry*, 21:5634–5638.
7a. Foster, D. L., Boublik, M., and Kaback, H. R. (1983): Structure of the *lac* carrier protein of *Escherichia coli*. *J. Biol. Chem.*, 258:31–34.
8. Fox, C. F., and Kennedy, E. P. (1965): Specific labeling and partial purification of the m protein, a component of the β-galactoside transport system of *Escherichia coli*. *Proc. Natl. Acad. Sci. USA*, 54:891.

9. Goldkorn, T., Rimon, G., and Kaback, H. R. (1981): Topology of the *lac* carrier protein in the membrane of *Escherichia coli. Proc. Natl. Acad. Sci. USA,* 80:3322–3326.
10. Goldkorn, T., Rimon, G., Kempner, E. S., and Kaback, H. R. (1982): Functional molecular weight of the *lac* carrier protein. *Fed. Proc.,* 41:1415 (Abstr. No. 6692).
11. Henderson, R., and Unwin, P. N. T. (1975): Three-dimensional model of purple membrane obtained by electron microscopy. *Nature,* 257:28–32.
12. Kaback, H. R. (1981): Mechanism of proton lactose support in *Escherichia coli.* In: *Chemiosmotic Proton Circuits in Biological Membranes,* edited by V. P. Skulachev and P. C. Hinkle, pp. 525–536. Addison-Wesley, Reading, MA.
13. Kaczorowski, G. J., and Kaback, H. R. (1979): Mechanism of lactose translocation in membrane vesicles from *Escherichia coli.* I. Effect of pH on efflux, exchange, and counterflow. *Biochemistry,* 18:3691–3697.
14. Kaczorowski, G. J., LeBlanc, G., and Kaback, H. R. (1980): Specific labeling of the *lac* carrier protein in membrane vesicles of *Escherichia coli* by a photoaffinity reagent. *Proc. Natl. Acad. Sci. USA,* 77:6319–6323.
15. Kaczorowski, G. J., Robertson, D. E., and Kaback, H. R. (1979): Mechanism of lactose translocation in membrane vesicles from *Escherichia coli.* 2. Effect of imposed $\Delta\psi$, ΔpH and $\Delta\bar{\mu}_{H^+}$. *Biochemistry,* 18:3697–3704.
16. Kempner, E. J., and Schlegel, W. (1979): Size determination of enzymes by radiation inactivation. *Anal. Biochem.,* 92:2–10.
17. Kyte, J., and Doolittle, R. F. (1982): A simple method for displaying the hydropathic character of a protein. *J. Mol. Biol.,* 157:105–132.
18. Mieschendahl, M., Büchel, D., Bocklage, H., and Müller-Hill, B. (1981): Mutations in the *lac y* gene of *Escherichia coli* define functional organization of lactose permease. *Proc. Natl. Acad. Sci USA,* 78:7652–7656.
19. Mitchell, P. (1961): Coupling of phosphorylation to electron and hydrogen transfer by a chemiosmotic type of mechanism. *Nature,* 191:144–148.
20. Mitchell, P. (1963): Molecule, group, and electron translocation through natural membranes. *Biochem. Soc. Symp.,* 22:142–169.
21. Mitchell, P. D. (1966): Chemiosmotic Coupling in Oxidative and Photophosphorylation. Glynn Research Ltd., Bodmin, England.
22. Mitchell, P. D. (1968): Chemiosmotic Coupling and Energy Transduction. Glynn Research Ltd., Bodmin, England.
23. Newman, M. J., and Wilson, T. H. (1980): Solubilization and reconstitution of the lactose transport system from *Escherichia coli. J. Biol. Chem.,* 255:10583–10586.
24. Newman, M. J., Foster, D. L., Wilson, T. H., and Kaback, H. R. (1981): Purification and reconstitution of functional lactose carrier from *Escherichia coli. J. Biol. Chem.,* 256:11804–11808.
25. Owen, P., and Kaback, H. R. (1978): Molecular structure of membrane vesicles from *Escherichia coli. Proc. Natl. Acad. Sci. USA,* 75:3148–3152.
26. Owen, P., and Kaback, H. R. (1979): Immunochemical analysis of membrane vesicles from *Escherichia coli. Biochemistry,* 18:1413–1422.
27. Owen, P., and Kaback, H. R. (1979): Antigenic architecture of membrane vesicles from *Escherichia coli. Biochemistry,* 18:1422–1426.
28. Padan, E., Schuldiner, S., and Kaback, H. R. (1979): Reconstitution of *lac* carrier function in cholate-extracted membranes from *Escherichia coli. Biochem. Biophys. Res. Commun.,* 91:854–861.
29. Reenstra, W. W., Patel, L., Rottenberg, H., and Kaback, H. R. (1980): Electrochemical proton gradient in inverted membrane vesicles from *Escherichia coli. Biochemistry,* 19:1–9.
30. Robertson, D. E., Kaczorowski, G. J., Garcia, M.-L., and Kaback, H. R. (1980): Active transport in membrane vesicles from *Escherichia coli:* The electrochemical proton gradient alters the distribution of the *lac* carrier between two different kinetic states. *Biochemistry,* 19:5692–5702.
31. Teather, R. M., Bramhall, J., Riede, I., Wright, J. K., Fürst, M., Aichele, G., Wilhelm, U., and Overath, P. (1980): Lactose carrier protein of *Escherichia coli.* Structure and function of plasmids carrying the y gene of the *lac* operon. *Eur. J. Biochem.,* 108:223–231.
32. Teather, R. M., Müller-Hill, B., Abrutsch, U., Aichele, G., and Overath, P. (1978): Amplification of lactose carrier protein in *Escherichia coli* using a plasmid vector. *Mol. Gen. Genet.,* 159:239–248.

Electrogenic Transport: Fundamental Principles and Physiological Implications,
edited by Mordecai P. Blaustein and Melvyn Lieberman. Raven Press, New York © 1984.

Electrogenic Ion Transport in Higher Plants

Roger M. Spanswick and Alan B. Bennett

Section of Plant Biology, Division of Biological Sciences, Cornell University, Ithaca, New York 14853

Compared to animal cells, plant cells have large membrane potentials, the magnitude of which appears to result mainly from the presence of electrogenic ion pumps (44). Some of the effects we now attribute to electrogenic pumps were observed by Blinks and co-workers (5,6) in the 1930s. However, explicit recognition of the presence of electrogenic pumps followed the adoption of modern microelectrode techniques. In particular, Slayman's work (36,37) on the fungus *Neurospora crassa* has been very influential, partly because the membrane potential in this species is too large to be accounted for by the passive diffusion of the major ions across the membrane. Also, the membrane potential was found to be sensitive to respiratory inhibitors and was closely correlated with the cellular adenosine triphosphate (ATP) concentration (39). Similar work on freshwater algae soon followed (21,33–35,41,43).

Although early work with inhibitors (15) indicated the presence of electrogenic pumps in higher plants, the membrane potential was within the range of possible diffusion potentials. Extensive flux measurements were necessary to demonstrate unambiguously that the effects could not be attributed to changes in the diffusion potential brought about by the effects of the inhibitors on the permeability coefficients for the major ions (20). There are, however, a number of higher plant tissues, such as red beet storage tissue (30) and the leaves of the aquatic plants *Lemna* (23,26) and *Elodea* (42), in which the membrane potentials are comparable to that of *Neurospora* and well beyond the negative limit of possible diffusion potentials. In *Elodea canadensis* the membrane potential in the light is −257 mV, a value that might at one time have been considered beyond the limit for dielectric breakdown.

The hypothesis that, in nonmarine plants, H^+ is the ion pumped electrogenically (21,37,42) is now almost universally accepted (31,44). Adoption of this hypothesis has been fostered by the successful application of Mitchell's chemiosmotic hypothesis to plant cell membranes (31,38), particularly in the demonstration of proton co-transport systems for sugars and amino acids (16,22,26,40), and by the growing acceptance of the acid-growth hypothesis for the action of the plant growth substance indole acetic acid (9,19). However, it has been difficult to test this hypothesis directly using intact systems, because of the impossibility of measuring unidirectional H^+ fluxes, and much of the evidence in its favor is indirect (44,45).

The necessity of using pH changes to estimate net H^+ fluxes makes it advantageous to use isolated membrane vesicles to avoid the complications caused by the buffering capacity of cell walls and the CO_2 exchanges in photosynthesis and respiration. There is, however, one further complication in working with plant cells that causes difficulties: the plant cell has two membranes. One, the plasmalemma (plasma membrane), corresponds to the plasma membrane of animal cells, whereas the other, the tonoplast, surrounds the central vacuole. The magnitude and direction of the electrochemical gradient for H^+ across the tonoplast suggests that this membrane must also contain a proton pump directed from the cytoplasm to the vacuole (43,46).

In this chapter we review the progress that has been made and the problems that have arisen as a result of using isolated cell membranes to characterize the electrogenic proton pumps of higher plants.

THE ISOLATION OF SEALED MEMBRANE VESICLES

There are two basic requirements for the demonstration of electrogenic ion transport in a vesicular system. First, the catalytic site for ATP hydrolysis must be on the external surface of the vesicles and, second, the passive leakage of the ion across the membrane must be sufficiently small to permit detection of the concentration gradient and electrical potential resulting from pump activity.

The initial problems encountered in demonstrating ATP-dependent transport in microsomal preparations from plant roots (52) led to attempts to reduce the passive permeability of the microsomal vesicles. In the case of H^+ transport, it is possible to estimate the passive permeability by imposing a pH gradient across the vesicle membranes and monitoring the relaxation of the gradient by observing the recovery of fluorescence quench of weak bases such as quinacrine or 9-aminoacridine (27,29). Corn root plasma membrane vesicles rapidly equilibrated imposed pH gradients and, although the passive permeability to H^+ could be reduced by the addition of Ca^{2+} or phospholipid, the vesicles did not demonstrate ATP-dependent transport.

An alternative approach was introduced by Sze (50), who applied a technique to separate sealed erythrocyte vesicles from unsealed vesicles to membranes from tobacco callus tissue. A microsomal fraction suspended in 250 mM sucrose was centrifuged onto a 10% dextran T70 cushion. The ATPase activity of the membranes collected from the sucrose/dextran interface was shown to be stimulated by uncouplers, implying that the vesicles were sealed and that the ATPase was involved in H^+ transport (50). In contrast, a corn root microsomal fraction centrifuged onto a discontinuous Ficoll gradient (5% and 10% Ficoll) yielded an interface fraction enriched in Mg^{2+}-dependent ATPase activity but lacking

ATP-dependent H$^+$ transport (29). However, a number of reports have recently appeared demonstrating both ATP-dependent H$^+$ transport (11,17,25,48) and the formation of an electrical gradient (4,32,51) in microsomal preparations from a variety of plant tissues using a number of techniques. In most preparations, however, there is considerable uncertainty as to the origin of the transport-exhibiting vesicles.

H$^+$ TRANSPORT IN MICROSOMAL VESICLES FROM CORN ROOTS

A preparation of corn root membranes was eventually obtained in which it was possible to demonstrate ATP-dependent H$^+$ transport. Essentially, the method was a modification of Sze's (50) dextran-cushion technique. Factors contributing to the observation of transport included the use of root tips rather than whole seedling roots, substitution of sorbitol for sucrose in the extraction medium, and the inclusion of protectants such as bovine serum albumin and β-mercaptoethanol, the use of Bis-Tris-propane/Mes buffer in place of Tris/Mes, and the substitution of quinacrine for 9-aminoacridine in the assay to give greater sensitivity. The effect of these changes was cumulative, and none of them alone was critical.

The initial characterization of the H$^+$ transport (11) revealed, as expected, that it was specific for MgATP, stimulated by KCl, and inhibited by diethylstilbestrol (DES) but not by oligomycin. The pH gradient was rapidly abolished by the uncoupler gramicidin and removal of the ATP with glucose and hexokinase after establishment of the pH gradient revealed the passive leak. There was, however, one feature of the transport that was not expected. It was specifically stimulated by Cl$^-$ and not by K$^+$. This contrasts with the ATPase activity of membranes purified from corn leaf mesophyll protoplasts (28), which is stimulated by K$^+$ and is indifferent to the accompanying anion. It was also found that, unlike the plasma membrane ATPase (13,28), the H$^+$-ATPase associated with vesicles isolated by this method was insensitive to vanadate.

IDENTIFICATION OF VESICLE ORIGIN

Dissimilarity between the ATPase of purified plasma membranes and the ATPase and transport characteristics of the vesicles obtained by Sze's method prompted us to characterize the vesicles exhibiting transport in greater detail, with a view to determining their origin (12,14).

The main characteristics of the ATPase and the associated H$^+$ transport are summarized in Table 1 and compared with the properties of the ATPase associated with corn leaf plasma membranes (28) and with the ATPase of isolated vacuoles (1,53). There is both a clear contrast with the plasma membrane ATPase and a distinct similarity of the transport ATPase to the tonoplast ATPase. A general parallel also exists with the ATPase and H$^+$ transport properties of the vacuo-lysosomal membranes of lutoids from the latex of *Hevea brasiliensis* (10,24).

TABLE 1. *Comparison of the effects of ions and inhibitors on the microsomal vesicle ATPase and H^+ transport characteristics with the ATPase activities of the plasma membrane and tonoplast*

	Plasma membrane ATPase	Microsomal vesicles		Tonoplast ATPase	Ref.
		ATPase	H^+ transport		
K^+ stimulation	Yes	No	No	No	1,28,53
Cl^- stimulation	No	Yes	Yes	Yes	1,28,53
NO_3^- inhibition	No	Yes	Yes	Yes	28,53
Vanadate inhibition	Yes	—	No	No	28,53
DCCD inhibition	Yes	Yes	Yes	Yes	1,28,53
DES inhibition	Yes	Yes	Yes	Yes	28,53

DCCD = N,N'-dicyclohexylcarbodiimide.
DES = diethylstilbestrol.

Examination of transport activity distribution on linear sucrose gradients showed the distribution to be broad (12). However, the maximum activity at 1.10 to 1.12 g/cm^3 was clearly separable from mitochondrial markers and from β-glucan synthetase II, a marker for the plasma membrane. In fact, the transport activity coincides closely with the distribution of tonoplast membrane vesicles obtained from tobacco suspension cells (Fig. 1; ref. 7).

Thus, current evidence strongly points to the tonoplast rather than the plasma membrane as the source of the vesicles active in H^+ transport.

ELECTROGENIC PROPERTIES OF H^+ TRANSPORT IN CORN ROOT VESICLES

Stimulation of ATP-dependent H^+ transport by anions suggested that the H^+-ATPase catalyzed an electrogenic influx of protons and that the formation of a pH gradient was impeded by a positive interior membrane potential in the absence of permeant ions. In fact, chloride stimulates H^+ transport by dissipating a large membrane potential associated with activity of the H^+ pump but, in addition, directly stimulates the rate of ATP hydrolysis by direct activation of the H^+-ATPase. These two mechanisms of Cl^- stimulation of ATP-dependent H^+ transport share common kinetic properties that suggest that the H^+-ATPase may be associated with an anion channel and that at physiological Cl^- concentrations, H^+ transport is closely coupled to Cl^- influx.

Evidence supporting this model of coupling of Cl^- and H^+ fluxes comes from a kinetic analysis of the effects of Cl^- on the ATP-dependent formation of pH and electric potential gradients as well as on ATPase activity associated with corn root membrane vesicles (4). Initially it was necessary to determine that relative initial rates of H^+ influx could be measured reliably. The fluorescence probes of ΔpH, 9-aminoacridine and quinacrine were used, since internalization of the probe could be monitored continuously. Quenching of 9-aminoacridine

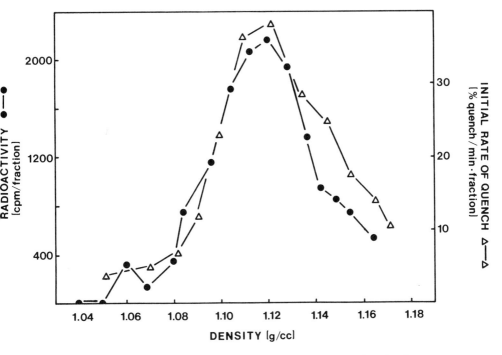

FIG. 1. Comparison of the distribution of ATP-dependent quenching of quinacrine fluorescence (△) and ^{14}C-choline-labeled tonoplast membranes (●) on continuous sucrose gradients. Distribution of ^{14}C-choline-labeled tonoplast membranes is replotted from Fig. 1 of Briskin and Leonard (7) with permission.

fluorescence was shown to respond in a linear fashion to imposed pH gradients across corn root vesicle membranes. It was also possible to derive a theoretical relationship demonstrating a direct proportionality between the initial rate of fluorescence quenching and H^+ flux through the pump (4). Formation of positive interior membrane potentials was monitored by measuring the shift in absorbance of Oxonol VI (610 to 580 nm) associated with redistribution of the dye in response to membrane potential (3). The response of this dye was calibrated by the imposition of H^+ diffusion potentials and found to be linear to approximately 100 mV, which was the magnitude of the ATP-induced membrane potential when measured in the absence of monovalent permeant ions.

Measurement of initial rates of ATP-dependent H^+ transport as a function of increasing KCl concentration indicated a biphasic response of the pump to Cl^-, which could be resolved into a linear and saturable component (Fig. 2). As discussed above, all Cl^- salts were equally effective in stimulating H^+ transport, ruling out a role for the accompanying cation. Measurements of ATP-dependent formation of membrane potential indicated that Cl^- discharged the membrane potential with similar biphasic kinetics (Fig. 3). These results clearly

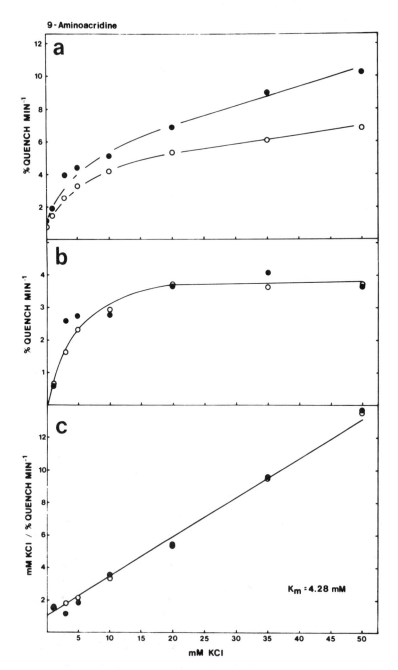

FIG. 2. Kinetics of KCl stimulation of the initial rate of fluorescence quenching of 9-aminoacridine. Fluorescence quench assayed in the presence of 5 mM $MgSO_4$ and 0 to 50 mM KCl (●), or 5 mM $MgSO_4$, 20 mM K_2SO_4, and 0 to 50 mM KCl (○). **(b):** Saturable component of Cl^--stimulated rates of fluorescence quench after subtraction of the linear component as described by Bennett and Spanswick (4). **(c):** Hanes-Woolf plot of **(b)**. [From Bennett and Spanswick (4), with permission.]

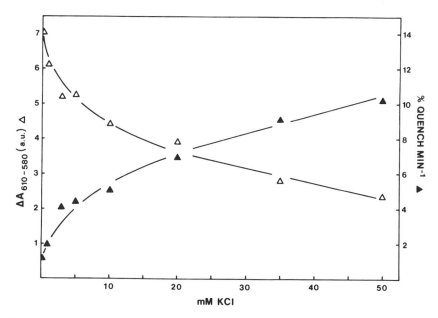

FIG. 3. Effect of KCl on the ATP-dependent initial rates of fluorescence quenching of 9-aminoacridine (▲) and absorbance changes of Oxonol VI (△). [From Bennett and Spanswick (4), with permission.]

show that Cl⁻ acted as a permeant anion at all concentrations tested, stimulating H⁺ transport by alleviating an electrical limitation on the rate of H⁺ transport.

The biphasic nature of Cl⁻ permeation further suggested two pathways of permeation with the saturable component attributable to permeation through an anion channel and the nonsaturable component attributable to permeation through the lipid bilayer. This contention is supported by the effects of the anion channel blockers 4-acetamido-4'-isothiocyano-2,2'-stilbenedisulfonic acid (SITS) and 4,4'-diisothiocyano-2,2'-stilbenedisulfonic acid (DIDS), which abolished the saturable component but not the linear component of Cl⁻-stimulated H⁺ transport (4).

In addition to the role played by Cl⁻ in relieving the electrical limitation imposed on the rate of H⁺ transport, it became apparent that Cl⁻ also directly stimulated the catalytic activity of the H⁺-ATPase. This effect is best demonstrated by the stimulation of ATPase activity by Cl⁻, even in the presence of gramicidin, conditions where the activity of the H⁺-ATPase would be unencumbered by the formation of any electric potential or ion gradients. Table 2 shows the stimulation of ATPase activity by several Cl⁻ salts both in the absence and presence of gramicidin. All activities in the presence of gramicidin are higher, due to an apparent limitation of the ATPase activity by the formation of $\Delta\bar{\mu}_{H^+}$, but the stimulation of ATPase activity by Cl⁻ was proportionally

TABLE 2. *Salt stimulation of ATPase activity in the absence and presence of gramicidin*

Additions	ATPase activity (μmol P_i/mg/hr)			% Salt stimulation
	Mg^{2+}-dependent	Salt stimulation	Gramicidin stimulation	
With gramicidin				
No salt	1.70	—	—	—
KCl	2.42	0.712	—	41.8
NaCl	2.36	0.66	—	38.7
LiCl	2.51	0.81	—	47.5
RbCl	2.61	0.91	—	53.4
Without gramicidin				
No salt	3.21	—	1.51	—
KCl	4.69	1.47	2.27	45.9
NaCl	4.77	1.55	2.40	48.3
LiCl	4.86	1.65	2.35	51.3
RbCl	4.96	1.74	2.34	54.2

Adenosine triphosphatase assayed at 25°C with the indicated salts added at a concentration of 50 mM. Mg^{2+}-dependent activity was calculated as the difference in ATPase activity measured in the presence and absence of 5 mM $MgSO_4$.
From Bennett and Spanswick (4), with permission.

similar in the absence and presence of gramicidin. This result indicated that Cl^- directly activated the H^+-ATPase and, furthermore, indicated that this direct effect was entirely responsible for the stimulation of ATPase activity, since, if Cl^- stimulated ATPase activity by both relieving membrane potential and direct activation of the H^+-ATPase, one would expect a proportionately greater stimulation of ATPase activity in the absence of gramicidin than in its presence. This can be rationalized in that the ATPase activity is limited by $\Delta \bar{\mu}_{H^+}$ and, although Cl^- may relieve the membrane potential, a pH gradient readily forms so that Cl^- addition alone would only transiently reduce $\Delta \bar{\mu}_{H^+}$ and only transiently relieve the limitation on ATPase activity.

The kinetics of Cl^- stimulation of ATPase activity were hyperbolic with a K_m nearly identical to the K_m for the saturable component of Cl^- stimulation of ATP-dependent H^+ transport. This similarity in kinetic constants suggested a common rate-limiting process in both the saturable component of Cl^- permeation of the vesicle membrane and direct activation of ATPase activity. This common reaction may involve entry of Cl^- to an anion channel that is physically associated with the H^+-ATPase. A model suggesting such an interaction is shown in the upper panel of Fig. 4. Alternatively, the anion channel responsible for the saturable component of Cl^- permeation of the membrane may be physically separate from the H^+-ATPase (lower panel Fig. 4) and the similarities in kinetic constants coincidental. The close coupling of Cl^- and H^+ transport in tonoplast vesicle is consistent with the small positive potential across the tonoplast *in vivo* (47).

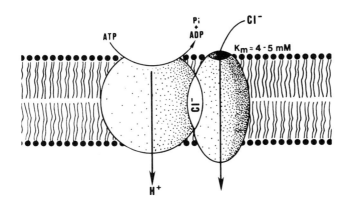

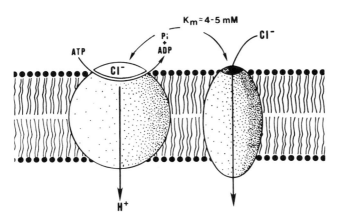

FIG. 4. Proposed models of the interaction between the H$^+$-ATPase and an anion channel. The *upper model* accounts for similarities in kinetic constants for Cl$^-$ stimulation of H$^+$-ATPase activity and Cl$^-$ stimulation of H$^+$ transport by a physical association of the H$^+$-ATPase with a Cl$^-$ channel. The *lower model* accounts for the observed similarities by coincidence. [From Bennett and Spanswick (4), with permission.]

SOLUBILIZATION AND RECONSTITUTION OF THE H$^+$-ATPase

As discussed above, the properties of the tonoplast H$^+$-ATPase differ markedly from the plasma membrane ATPase of plant cells with respect to salt and inhibitor sensitivity. The plasma membrane ATPase of corn roots has recently been shown to form a phosphorylated intermediate, indicating its similarity to the Na$^+$/K$^+$-ATPase and other plasma membrane ion-translocating ATPases (8). In contrast, the tonoplast H$^+$-ATPase is more similar to the chromaffin granule or lysosomal H$^+$-ATPases that have not been structurally characterized to the extent of the cation-stimulated, vanadate-sensitive, ion-translocating

ATPases. It has been suggested, however, that the chromaffin granule H^+-ATPase is structurally similar to the mitochondrial F_1/F_o-ATPase (2). In order to examine the structural details of the corn root H^+-ATPase, it was necessary to begin purification of the enzyme, relying on reconstitution as an assay for purification.

Solubilization of the H^+-ATPase was achieved with either deoxycholate or the zwitterionic sulfobetaine detergent, Zwittergent 3-14. The activity of the ATPase following solubilization was strongly dependent on the presence of glycerol during solubilization, and full activity was dependent on the addition of phospholipid to the deoxycholate-solubilized enzyme. Interestingly, although glycerol was required for the maintenance of ATPase activity during solubilization with Zwittergent 3-14, added phospholipids were not required for full activity of the solubilized ATPase. This suggested that Zwittergent 3-14 may have been less effective than deoxycholate in delipidating the solubilized ATPase, or alternatively that Zwittergent 3-14 actually substituted for phospholipids in maintaining a hydrophobic environment favorable for full activity of the enzyme.

Solubilized H^+-ATPase was reconstituted into phospholipid vesicles by removal of deoxycholate by gel filtration of the solubilized H^+-ATPase through Sephadex G-200. Reformed vesicles were collected in the column void volume and found to catalyze ATP-dependent quenching of acridine orange fluorescence

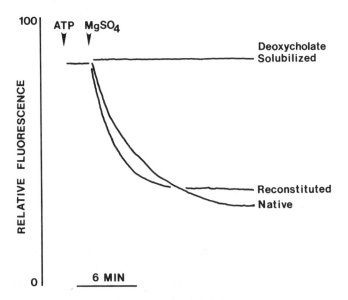

FIG. 5. ATP-dependent quenching of acridine orange fluorescence with native, 0.4% deoxycholate-solubilized, and reconstituted membranes. Equal amounts of protein (110 μg) of either native, deoxycholate-solubilized, or reconstituted membranes were assayed. Fluorescence quenching of acridine orange was initiated by the sequential addition of ATP and MgSO$_4$ to final concentrations of 5 mM at the indicated *arrows*.

at rates comparable to that of native vesicles (Fig. 5). This result demonstrated the feasibility of using reconstituted deoxycholate-solubilized H^+-ATPase both as an assay for purification of the ATPase as well as a means to study interactions of the ATPase with the lipid membrane environment. Further attempts at purification have, to date, been unsuccessful because of the instability of the solubilized H^+-ATPase, which, at ice temperature, loses activity within 24 hr after solubilization. It is hoped that further work will overcome this difficulty, and will lead to the preparation of a purified H^+-ATPase that will be suitable for structural analysis.

DISCUSSION

We have presented evidence for an electrogenic membrane-bound H^+-ATPase in corn root tips and have shown that it probably derives from the tonoplast. Furthermore, it has been possible to solubilize the ATPase with detergents and reconstitute the transport function in artificial lipid vesicles.

As noted above, there are now several reports in the literature of H^+ transport in membrane vesicles isolated from plant cells. Except for Mettler et al. (25), who also suggest a tonoplast origin, and Hager et al. (17) who suggest the Golgi or endoplasmic reticulum, although later suggest that the vesicles may be prevacuoles (18), most authors assume that the vesicles derive from the plasma membrane (32,48,51). However, the most recent reports (18,49) indicate that these preparations also contain a Cl^--stimulated ATPase. Thus, it is possible to suggest that, to date, H^+ transport has been demonstrated only in tonoplast vesicles. There appears to be no literature that unequivocally demonstrates H^+ transport in plasma membrane vesicles. The reasons for this are not clear, although it is possible that sealed plasma membrane vesicles are oriented with the catalytic site for ATP hydrolysis on the inside.

The use of isolated vacuoles and tonoplast vesicles may well provide more information about the tonoplast than it is possible to obtain by experiments on intact tissues, since in higher plants it is very difficult to insert a microelectrode into the thin layer of cytoplasm to measure the potential relative to a microelectrode in the vacuole. Also, measurement of ion fluxes requires lengthy compartmental analysis, and measurement of H^+ fluxes is not feasible.

Although this work has provided useful information about the tonoplast, a major objective for the future must be to obtain an equivalent preparation that will permit us to characterize the relationship between H^+ transport and the membrane potential in the plasma membrane of higher plants.

ACKNOWLEDGMENTS

The authors' work reported in this chapter was supported by NSF grants PCM 78-12119 and PCM 81-11007.

REFERENCES

1. Admon, A., Jacoby, B., and Goldschmidt, E. E. (1981): Some characteristics of the Mg-ATPase of isolated red beet vacuoles. *Plant Sci. Lett.*, 22:89–96.
2. Apps, D. K., and Schatz, G. (1979): An adenosine triphosphatase isolated from chromaffin granule membranes is closely similar to F_1-adenosine triphosphatase of mitochondria. *Eur. J. Biochem.*, 100:411–419.
3. Bashford, C. L., and Thayer, W. S. (1977): Thermodynamics of the electrochemical proton gradient in bovine heart submitochondrial particles. *J. Biol. Chem.*, 252:8459–8463.
4. Bennett, A. B., and Spanswick, R. M. (1982): Optical measurements of ΔpH and $\Delta\psi$ in corn root membrane vesicles. Kinetic analysis of Cl^- effects on a proton-translocating ATPase. *J Membr. Biol.*, 71:95–107.
5. Blinks, L. R. (1935): Protoplasmic potentials in *Halicystis*. IV. Vacuolar perfusion with artificial sap and sea water. *J. Gen. Physiol.*, 18:409–420.
6. Blinks, L. R., Darsie, M. L., and Skow, R. K. (1938): Bioelectric potentials in *Halicystis*. VII. The effects of low oxygen tension. *J. Gen. Physiol.*, 22:255–279.
7. Briskin, D. P., and Leonard, R. T. (1980): Isolation of tonoplast vesicles from tobacco protoplasts. *Plant Physiol.*, 66:684–687.
8. Briskin, D. P., and Leonard, R. T. (1982): Partial characterization of a phosphorylated intermediate associated with the plasma membrane ATPase of corn roots. *Proc. Natl. Acad. Sci. USA*, 79:6922–6926.
9. Cleland, R. (1971): Cell wall extension. *Ann. Rev. Plant Physiol.*, 22:197–222.
10. Cretin, H. (1982): The proton gradient across the vacuo-lysosomal membrane of lutoids from the latex of *Hevea brasiliensis*. I. Further evidence for a proton-translocating ATPase on the vacuo-lysosomal membrane of intact lutoids. *J. Membr. Biol.*, 65:175–184.
11. DuPont, F. M., Bennett, A. B., and Spanswick, R. M. (1982): Proton transport in microsomal vesicles from corn roots. In: *Plasmalemma and Tonoplast: Their Functions in the Plant Cell*, edited by D. Marmé, E. Marrè, and R. Hertel, pp. 409–416. Elsevier Biomedical Press, Amsterdam.
12. DuPont, F. M., Bennett, A. B., and Spanswick, R. M. (1982): Localization of a proton-translocating ATPase on sucrose gradients. *Plant Physiol.*, 70:1115–1119.
13. DuPont, F. M., Burke, L. L., and Spanswick, R. M. (1981): Characterization of a partially purified adenosine triphosphatase from a corn root plasma membrane fraction. *Plant Physiol.*, 67:59–63.
14. DuPont, F. M., Giorgi, D. L., and Spanswick, R. M. (1983): Characterization of a proton-translocating ATPase in microsomal vesicles from corn roots. *Plant Physiol.*, 70:1694–1699.
15. Etherton, B., and Higinbotham, N. (1960): Transmembrane potential measurements of cells of higher plants. *Science*, 131:409–410.
16. Etherton, B., and Rubinstein, B. (1978): Evidence for amino acid H^+ co-transport in oat coleoptiles. *Plant Physiol.*, 61:933–937.
17. Hager, A., Frenzel, R., and Laible, D. (1980): ATP-dependent proton transport into vesicles of microsomal membranes of *Zea mays* coleoptiles. *Z. Naturforsch.*, 35C:783–793.
18. Hager, A., and Helmle, M. (1981): Properties of an ATP-fueled, Cl^--dependent proton pump localized in membranes of microsomal vesicles from maize coleoptiles. *Z. Naturforsch.*, 36:997–1008.
19. Hager, A., Menzel, H., and Krauss, A. (1971): Versuche and Hypothese zur Primärwirkung des Auxins beim Streckungswachstum. *Planta*, 100:47–75.
20. Higinbotham, N., Graves, J. S., and Davis, R. F. (1970): Evidence for an electrogenic ion transport pump in cells of higher plants. *J. Membr. Biol.*, 3:210–222.
21. Kitasato, H. (1968): The influence of H^+ on the membrane potential and ion fluxes of *Nitella*. *J. Gen. Physiol.*, 52:60–87.
22. Lichtner, F. T., and Spanswick, R. M. (1981): Electrogenic sucrose transport in developing soybean cotyledons. *Plant Physiol.*, 67:869–874.
23. Löppert, H. (1979): Evidence for electrogenic proton extrusion by subepidermal cells of *Lemna paucicostata*, 6746. *Planta*, 144:311–315.
24. Marin, B., Marin-Lanza, M., and Komor, E. (1981): The protonmotive potential difference across the vacuo-lysosomal membrane of *Hevea brasiliensis* (rubber tree) and its modification by a membrane-bound adenosine triphosphatase. *Biochem. J.*, 198:365–372.
25. Mettler, I. J., Mandola, S., and Taiz, L. (1982): Proton gradients produced *in vitro* by microsomal

vesicles from corn coleoptiles. Tonoplast origin? In: *Plasmalemma and Tonoplast of Plant Cells: Their Functions in the Plant Cell,* edited by D. Marmé, E. Marrè, and R. Hertel, pp. 395–400. Elsevier Biomedical Press, Amsterdam.
26. Novacky, A., Ullrich-Eberius, C. I., and Lüttge, U. (1978): Membrane potential changes during transport of hexoses in *Lemna gibba* Gl. *Planta,* 138:263–270.
27. Perlin, D. S., and Spanswick, R. M. (1980): Proton transport in plasma membrane vesicles. In: *Plant Membrane Transport: Current Conceptual Issues,* edited by R. M. Spanswick, W. J. Lucas, and J. Dainty, pp. 529–530. Elsevier North-Holland Biomedical Press, Amsterdam.
28. Perlin, D. S., and Spanswick, R. M. (1981): Characterization of ATPase activity associated with corn leaf plasma membranes. *Plant Physiol.,* 68:521–526.
29. Perlin, D. S., and Spanswick, R. M. (1982): Isolation and assay of corn root membrane vesicles with reduced proton permeability. *Biochim. Biophys. Acta,* 690:178–186.
30. Poole, R. J. (1966): The influence of the intracellular potential on potassium uptake by beetroot tissue. *J. Gen. Physiol.,* 49:551–563.
31. Poole, R. J. (1978): Energy coupling for membrane transport. *Annu. Rev. Plant Physiol.,* 29:437–460.
32. Rasi-Caldogno, F., De Michelis, M. I., and Pugliarello, M. C. (1981): Evidence for an electrogenic ATPase in microsomal vesicles from pea internodes. *Biochim. Biophys. Acta,* 642:37–45.
33. Richards, J. L., and Hope, A. B. (1974): The role of protons in determining membrane electrical characteristics in *Chara corallina. J. Membr. Biol.,* 16:121–144.
34. Saito, K., and Senda, M. (1973): The light-dependent effect of external pH on the membrane potential of *Nitella. Plant Cell Physiol.,* 14:147–156.
35. Saito, K., Senda, M. (1974): The electrogenic ion pump revealed by the external pH effect on the membrane potential of *Nitella.* Influences of external ions and electric current on the pH effect. *Plant Cell Physiol.,* 15:1007–1016.
36. Slayman, C. L. (1965): Electrical properties of *Neurospora crassa.* Effects of external cations on the intracellular potential. *J. Gen. Physiol.,* 49:69–92.
37. Slayman, C. L. (1965): Electrical properties of *Neurospora crassa.* Respiration and the intracellular potential. *J. Gen. Physiol.,* 49:93–116.
38. Slayman, C. L. (1974): Proton pumping and generalized energetics of transport: A review. In: *Membrane Transport in Plants,* edited by U. Zimmermann and J. Dainty, pp. 107–119. Springer-Verlag, New York.
39. Slayman, C. L., Long, W. S., and Lu, C. Y-H. (1973): The relationship between ATP and an electrogenic pump in the plasma membrane of *Neurospora crassa. J. Membr. Biol.,* 14:305–338.
40. Slayman, C. L., and Slayman, C. W. (1974): Depolarization of the plasma membrane of *Neurospora* during active transport of glucose: evidence for a proton-dependent cotransport system. *Proc. Natl. Acad. Sci. USA,* 71:1935–1939.
41. Spanswick, R. M. (1972): Evidence for an electrogenic ion pump in *Nitella translucens.* I. The effects of pH, K^+, Na^+, light and temperature on the membrane potential and resistance. *Biochim. Biophys. Acta,* 288:73–89.
42. Spanswick, R. M. (1972): Electrical coupling between the cells of higher plants: A direct demonstration of intercellular transport. *Planta,* 102:215–227.
43. Spanswick, R. M. (1974): Hydrogen ion transport in giant algal cells. *Can. J. Bot.,* 52:1029–1034.
44. Spanswick, R. M. (1981): Electrogenic ion pumps. *Annu. Rev. Plant Physiol.,* 32:267–289
45. Spanswick, R. M. (1982): The electrogenic pump in the plasma membrane of *Nitella. Curr. Topics Membr. and Transport,* 16:35–47.
46. Spanswick, R. M., and Miller, A. G. (1977): Measurement of the cytoplasmic pH in *Nitella translucens. Plant Physiol.,* 59:664–666.
47. Spanswick, R. M., and Williams, E. J. (1964): Electrical potentials and Na, K, and Cl concentrations in the vacuole and cytoplasm of *Nitella translucens. J. Exp. Bot.,* 15:422–427.
48. Stout, R. G., and Cleland, R. E. (1982): MgATP-generated electrochemical proton gradient in oat root membrane vesicles: In: *Plasmalemma and Tonoplast: Their Functions in the Plant Cell,* edited by D. Marmé, E. Marrè, and R. Hertel, pp. 401–407. Elsevier Biomedical Press, Amsterdam.
49. Stout, R. G., and Cleland, R. E. (1982): Evidence for a Cl^--stimulated Mg-ATPase proton pump in oat root membranes. *Plant Physiol.,* 69:798–803.
50. Sze, H. (1980): Nigericin-stimulated ATPase activity in microsomal vesicles of tobacco callus. *Proc. Natl. Acad. Sci. USA,* 77:5904–5908.

51. Sze, H., and Churchill, K. A. (1981): Mg^{2+}/KCl-ATPase of plant plasma membranes is an electrogenic pump. *Proc. Natl. Acad. Sci. USA,* 78:5578–5582.
52. Sze, H., and Hodges, T. K. (1976): Characterization of passive ion transport in plasma membrane vesicles of oat roots. *Plant Physiol.,* 58:304–308.
53. Walker, R. R., and Leigh, R. A. (1981): Characterization of a salt-stimulated ATPase activity associated with vacuoles isolated from storage roots of red beet (*Beta vulgaris* L.). *Planta,* 153:140–149.

Electrogenic Transport: Fundamental Principles and Physiological Implications, edited by Mordecai P. Blaustein and Melvyn Lieberman. Raven Press, New York © 1984.

Electrogenic Sodium Pumping in *Xenopus* Blastomeres: Apparent Pump Conductance and Reversal Potential

*Luca Turin

Department of Anatomy and Embryology, University College, London WC1E 6BT, England

The electrogenic sodium pump performs both electrical and chemical work by using a fixed source of energy. Assuming that pump stoichiometry is not grossly altered by the electric field across the membrane, there must be a (negative) voltage at which the sum of all the work it must perform in a pump cycle equals that provided by adenosine triphosphate (ATP) hydrolysis. Chapman and Johnson (1) have calculated that this voltage should be in the range of −100 to −200 mV. By analogy with pump reversal by chemical means (3), this zero-current potential might even be expected to be a true reversal potential, past which the pump dissipates electrochemical energy and makes ATP. Gradmann et al. (4) have shown that near this point, the pump should behave like a voltage source in series with a conductance, whereas further away it is expected to saturate and behave like an ideal current source.

Evidence to date (2) has not revealed any striking voltage dependence of electrogenic pumping. When some is seen, the data are difficult to obtain (10) or can be explained by the activation of an unrelated conductance by sodium loading (6).

Three years ago, while studying the electric properties of single, isolated *Xenopus* blastomeres, I chanced on a way to reduce their membrane conductance by lowering extracellular pH. The important finding, as far as the present chapter is concerned, is that at an external pH of 6.5, membrane conductance is often so low that resting membrane potential is largely governed by resting, unstimulated pump current. Electrogenic pumping can then be measured under favorable conditions: steady-state pump current; low, near-linear background conductance; and unrestricted access to the cell membrane. Simple current-clamp measurements of membrane potential and conductance are then sufficient to show that the electrogenic pump deviates from ideal current source behavior even at modest values of membrane potential. These are only preliminary results, but they sug-

---

* Present address: E. R. Biologie du Developpement, Station Zoologique, La Darse, F-06230 Villefranche/Mer, France.

gest that *Xenopus* blastomeres may be an ideal preparation on which to study sodium-pump electrophysiology.

METHODS

Fertilized eggs were obtained by injecting sexually mature pairs of *Xenopus laevis* with human chorionic gonadotrophin (Organon Labs) and leaving them to mate and lay eggs overnight. The blastula is a hollow sphere of cells organized in an epithelium-like fashion. The central cavity, or blastocoel, contains a fluid approximating frog Ringer (9). The pH of the central cavity, measured with pH-sensitive microelectrodes, is 8.5 (11). This blastocoel fluid constitutes the real extracellular medium of the blastomeres, since the membrane facing the outside of the blastula is highly impermeable (intact embryos can be reared in distilled water). The isolation method was essentially that of Ito and Loewenstein (5). Using fine forceps, the vitelline membrane was opened and the blastula dissociated mechanically with glass needles. This was done in a medium containing 120 mM Na, 2.4 mM K, 4 mM Ca, and 2 mM Mg all as their chloride salts and 5 mM of tris (hydroxymethyl) methylaminopropanesulfonic acid (TAPS) (Sigma) titrated to pH 8.5 with NaOH. Acid Ringer contains the same ions, but it is buffered to pH 6.5 with 5 mM 2-(2-amino-2-oxoethyl) aminoethanesulfonic acid (ACES) (Sigma). Strophanthidin was obtained from Boehringer. Many cells break when the blastula is opened, and I found that the best compromise between cell size and number surviving the isolation procedure was at stages 6½ to 7 of Nieuwkoop and Faber (7). At this stage, blastomeres range from approximately 100 to 300 μm in diameter, increasing in size as one goes from the animal (pigmented) pole to the vegetal pole.

All experiments described below were performed on the large vegetal pole blastomeres because of their inherently larger size, surface area, and electrical resistance. Using a fine pipette, the blastomeres were transferred to the experiment bath, onto a piece of blue Millipore filter to give increased contrast. Within 1 or 2 min they flatten slightly and adhere solidly to the filter, which enables fast solution flow. Bath volume was 2.5 ml, and a change of solution consisted of 20 ml made to flow through the bath in 15 sec. The cell was impaled with two microelectrodes of 10 to 30 MΩ resistance, filled with 3 M KCl. One electrode was used for current passing, the other to record membrane potential. Voltages were displayed on an oscilloscope and chart recorder.

RESULTS

In physiological medium, the membrane of *Xenopus* blastomeres responds to raised extracellular K^+ in the range 2.4 to 120 mM in a near-Nernstian fashion, and membrane conductance is linear (8).

Figure 1 shows a typical record of a vegetal blastomere in response to an extracellular pH change. The record begins with the blastomere in medium at pH 8.5. Constant-current pulses of 0.68 nA (1 sec long, every 5 sec) are passed

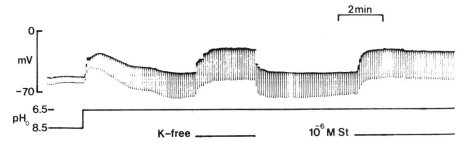

FIG. 1. Influence of electrogenic sodium pumping on membrane potential and conductance of a single blastomere after background conductance is reduced by lowering extracellular pH. Membrane potential first falls then rises again as conductance reaches a steady state. Potassium-free medium or 1 μM strophanthidin solution causes a rapid and simultaneous fall in membrane potential and conductance. Current pulses: 0.68 nA.

through the current electrode, and membrane potential with the voltage deflections is measured through a second electrode. Membrane potential and conductance are stable at −52 mV and 117 nS, respectively. The pH of the medium is then lowered to pH 6.5. Membrane potential falls immediately, as does conductance. After approximately 1 min, membrane potential begins to rise again, while conductance falls gradually; 4 min after the pH change, a new steady state has been reached, with membrane potential and conductance at −48 mV and 22 nS, respectively. The superfusing solution is then changed to one containing no potassium. Membrane potential falls within 1 min by 30 mV, while conductance simultaneously falls to 17 nS, after which a new steady state is reached. Readmission of normal potassium-containing medium to the bath rapidly restores potential and conductance to their previous values. A solution containing micromolar strophanthidin is then admitted to the bath. Potential and conductance fall again with a time course similar to that obtained in K-free solution. A steady state is reached within 30 sec, at −20 mV and 18 nS. Strophanthidin is then removed (not shown on trace), but membrane potential and conductance show no further changes in the next 5 min. The changes in membrane potential in this experiment are most easily explained by supposing that electrogenic sodium pumping provides a membrane current, the influence of which makes itself increasingly felt as the membrane conductance falls in response to acid pH_o. Inhibition of sodium pumping either by potassium removal or cardiac glycosides restores membrane potential to its true level, i.e., governed by passive conductances only. Occasionally, the background conductances appear to be entirely leakage components. In one experiment (not shown), no difference in potential was seen when extracellular potential was raised to 60 mM K, suggesting that all potassium conductance had been removed by exposure to acid medium. It is difficult to say whether electrogenic pumping is already present in alkaline medium, or whether it arises in response to external acidification.

The conductance changes observed during pump inhibition cannot be due

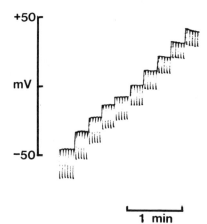

FIG. 2. Effect of voltage on membrane conductance of a vegetal blastomere in pH 6.5 medium. Current pulses of 1.4 nA are superimposed on a steady current. The membrane rectifies in the outward direction.

to the change in membrane potential. Figure 2 shows a trace obtained from another blastomere in medium at pH 6.5 by superimposing a variable DC current to pulses of 1.4 nA. The membrane clearly exhibits outward rectification. This is an invariable feature of blastomeres in acid media; the true conductance decrease in Fig. 1 is thus likely to be slightly larger.

Using a 10-fold higher concentration of strophanthidin, potential and conductance changes are much more rapid, but still occur simultaneously. Sections taken from four records at the time of superfusion with 10 μM strophanthidin are shown in Fig. 3. The legend to Fig. 3 gives the values for membrane potential and conductance before and after exposure to strophanthidin. The effect of the drug begins immediately after the solution change is started. It is clear that in each case, potential and conductance changes occur simultaneously, and are complete within 2 to 3 pulse intervals, or 15 sec.

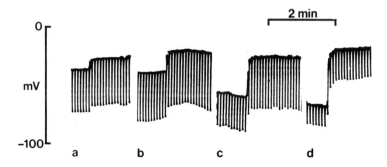

FIG. 3. Segments of records taken from 4 experiments at the time of sodium-pump inhibition by 10 μM strophanthidin. Extracellular pH is 6.5 throughout. Membrane potential and conductance values before and after strophanthidin are: **(a)** −37 mV, 18 nS; −27 mV, 15 nS. **(b)** −40 mV, 17 nS; −22 mV, 13 nS. **(c)** −58 mV, 24 nS; −27 mV, 16 nS. **(d)** −65 mV, 25 nS; −15 mV, 20 nS.

The fact that both potential and conductance rapidly reach a steady state, and the large size and low conductance of the cells make it unlikely that ionic rearrangements are responsible for the conductance change.

This close association between the fall in membrane potential and conductance immediately suggests treating the electrogenic pump as a linear voltage source, in series with a conductance equal in magnitude to the change observed when the pump is inhibited. The membrane equivalent circuit is then assumed to consist of two limbs, one containing the pump, with an unknown electromotive force (emf) in series with a known conductance, the other containing the passive membrane emf and conductance (E_m and g_m after pump inhibition), assumed to be unaffected by strophanthidin. Kirchoff's law then gives values of the pump emf for Fig. 1 and the four traces in Fig. 3 of −138, −96, −160, and −164 mV, respectively. This treatment assumes linearity in all the conductances, which is known not to be the case for background conductance and is unlikely to be true of pump conductance. (This point is dealt with in greater detail in the chapter by De Weer, *this volume*.) They should therefore be taken only as rough estimates.

If the electrogenic sodium pump behaves as a voltage source in this range, and its emf is around 100 mV, as these records suggest, it should be possible to take membrane potential past this level by applying enough current, at which point the pump should be driven backwards, and strophanthidin should have a hyperpolarizing effect. Figures 4a and 4b show that this is indeed the case. In Fig. 4a the trace begins with the cell at its normal resting potential of −35 mV. Constant current pulses of 4 nA are applied every 4 sec. A constant current is then superimposed on the pulses in order to hyperpolarize E_m to −100 mV. Conductance immediately falls due to the nonlinearity of membrane resistance. After 1 min at this new voltage level, 5 μM strophanthidin is admitted to the bath. Within the next minute, E_m hyperpolarizes by 30 mV. Peak membrane

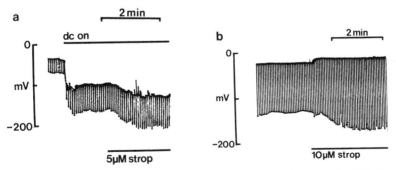

FIG. 4a: Hyperpolarizing effect of 5 μM strophanthidin when membrane potential is held at −100 mV by a steady hyperpolarizing current. *Pulse current:* 4 nA. **b:** Differing effects of 10 μM strophanthidin on resting membrane potential and a hyperpolarized potential caused by the injection of 3.8-nA current pulses. Resting potential falls from −35 mV, while peak potential during the pulses rises from −135 mV.

potential is then approximately −200 mV, and the small spikes visible in this section of the trace are most likely due to transient membrane breakdown. The membrane current–voltage relationship in this voltage range is too nonlinear to be able to calculate the emf of the pump as was done for Figs. 1 and 2. All one can say is that at the initial holding voltage of −100 mV, the electrogenic pump was having a depolarizing effect and that its emf must, therefore, be lower. Figure 4b shows a similar experiment on another cell. In this case no DC current is used, and instead pulses of 3.8 nA take the membrane potential from its resting level of −32 mV to −135 mV. When 10 μM strophanthidin is admitted to the bath, the two potential levels immediately begin to diverge. The resting potential falls by 12 mV, while the potential during the current pulse rises to −170 mV during the following minute. The trace becomes slightly irregular at these high voltages, presumably again due to the excessive electrical stress on the membrane. Figure 4b clearly illustrates that the membrane current-voltage relationships before and after pump inhibition cross each other at some point between −32 and −135 mV.

DISCUSSION

The conductance changes accompanying pump inhibition, and the apparent reversal of electrogenic pumping at large negative voltages suggest that the electrogenic pump in *Xenopus* blastomeres does not behave as an ideal current source, and that current can be made to flow through a ouabain-sensitive membrane entity in both directions. Several possibilities exist to account for these results.

The simplest view is that they reflect genuine changes in pump current in response to varying workloads, in which case a complete pump current-voltage relationship obtained under voltage-clamp conditions could yield information about pump kinetics (4).

The second, no less attractive view is that at some point during its cycle the pump is open and behaves like an ion channel proper. If this channel state were highly selective for potassium, a reversal potential of about −100 mV would be expected, which is not inconsistent with these results. In this view, pump-current source and conductance, although closely associated, are in parallel, and the true potential at which pump current actually stops is not being measured.

The third possibility is that pump stoichiometry, rather than pumping rate, is altered by voltage. In view of the complex and cooperative nature of the sodium-pumping mechanism, it would be surprising if all its rate constants were voltage independent. If stoichiometry changes markedly with voltage, any current-voltage relationship can in principle be obtained. The results described here would then simply be explained by the fact that above a certain voltage the Na:K ratio becomes, say, 2:3, rather than 3:2, while ATP is still being consumed in the normal fashion. Finally, the last possibility is that strophanthidin is exerting an unrelated effect on a small but highly selective potassium conductance. This seems unlikely in view of the high specificity of cardiac glycoside action in other systems, and of the small doses used.

What experiments would distinguish between these possibilities? The crucial point seems to be whether ATP is being synthesized from adenosine diphosphate (ADP) and inorganic phosphate when inward current flows through the pump. Measurements performed under intracellular dialysis may answer this question, if leakage conductances can be kept sufficiently low to still benefit from the favorable conditions of measurement afforded by blastomeres.

ACKNOWLEDGMENTS

This work is dedicated to S.M.R. I am pleased to acknowledge the help, advice, and hospitality of Dr. Anne Warner, in whose laboratory all the experiments were performed. I also thank Robert Purves, Stuart Bevan, and David Armstrong for helpful discussions.

This work was supported by the Medical Research Council and the Wellcome Trust.

REFERENCES

1. Chapman, J. B., and Johnson, E. A. (1978): The reversal potential for an electrogenic sodium pump. *J. Gen. Physiol.*, 72:403–408.
2. Eisner, D. A., and Lederer, W. J. (1980): Characterization of the electrogenic sodium pump in cardiac Purkinje fibers. *J. Physiol. (Lond.)*, 303:441–474.
3. Garrahan, P. J., and Glynn, I. M. (1967): The incorporation of inorganic phosphate into adenosine triphosphate by reversal of the sodium pump. *J. Physiol. (Lond.)*, 192:237–256.
4. Gradmann, D., Hansen, U.-P., and Slayman, C. P. (1981): Reaction-kinetic analysis of current-voltage relationships for electrogenic pumps in Neurospora and Acetabularia. *Curr. Topics Membr. Transport*, 16:258–276.
5. Ito, S., and Loewenstein, W. R. (1969): Ionic communication between early embryonic cells. *Dev. Biol.*, 19:228–243.
6. Kononenko, N. I., and Kostyuk, P. G. (1976): Further studies on the potential-dependence of the sodium-induced membrane current in snail neurones. *J. Physiol. (Lond.)*, 256:601–615.
7. Nieuwkoop, P. D., and Faber, J. (1956): *Normal Table of Xenopus laevis*. North-Holland, Amsterdam.
8. Slack, C., and Warner, A. E. (1975): Properties of surface and junctional membranes of embryonic cells isolated form blastula stages of Xenopus laevis. *J. Physiol. (Lond.)*, 248:97–120.
9. Slack, C., Warner, A. E., and Warren, R. L. (1973): The distribution of sodium and potassium in amphibian embryos during early development. *J. Physiol., (Lond.)*, 232:297–312.
10. Thomas, R. C. (1981): Electrophysiology of the sodium pump in a snail neuron. *Curr. Topics Membr. Transport*, 16:3–16.
11. Turin, L., and Warner, A. E. (1980): Intracellular pH in Xenopus embryos: Its effect on ionic communication between blastomeres. *J. Physiol. (Lond.)*, 300:489–504.

Electrogenic Transport: Fundamental Principles and Physiological Implications, edited by Mordecai P. Blaustein and Melvyn Lieberman. Raven Press, New York © 1984.

Electrogenic Sodium Pump Current Associated with Recovery from Intracellular Acidification of Snail Neurons

R. C. Thomas

Department of Physiology, Bristol University, Bristol BS8 1TD, England

The intracellular pH (pH_i) of snail neurons, squid axons, and other animal cells is kept near 7.2 (3), although a passive distribution of H^+, OH^-, and HCO_3^- ions would give a pH_i of < 7. When it is decreased by acid addition the pH_i rapidly recovers, as long as bicarbonate is available (1,7). Experiments on snail neurons (involving ionic substitution and direct measurement of intracellular Na and Cl) suggest that the pH_i is kept alkaline by a specific transport system in the cell membrane. This appears to have a mechanism that uses the energy of the Na gradient across the cell membrane to extrude Cl^- and probably H^+ ions in exchange for Na^+ and HCO_3^- ions (11). A similar mechanism operates in squid axons and barnacle muscle, although not in other cells investigated. [For details see review by Roos and Boron (3).]

The proposed ionic movements are illustrated in Fig. 1, left. The various ion gradients in snail neurons are such that the entry of one Na^+ ion would provide enough energy for the exit of two H^+ ion equivalents (i.e., the exit of one H^+ and the entry of one HCO_3^- ion). Russell and Boron (4) have been able to investigate the stoichiometry of pH_i regulation in squid axons using radioactive isotopes. Snail neurons are too small for this, however, and thus far the only attempt to measure the stoichiometry in neurons has involved measurements of changes in intracellular Na activity, which are fraught with uncertainties (7).

This chapter describes a preliminary exploration of an alternative approach. In this I have assumed that the Na that enters when the acid is extruded is subsequently pumped out by the electrogenic Na pump, as shown in Fig. 1.

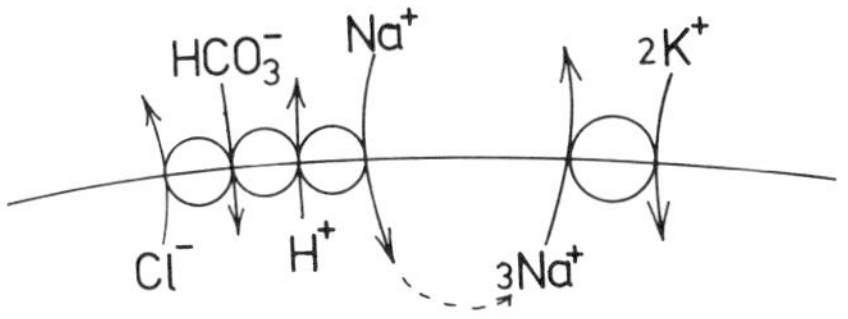

FIG. 1. Diagram of how Na influx via the pH_i regulating system stimulates the sodium pump.

Can the quantity of Na pumped out be estimated by measuring the charge generated by the Na pump (5)? The preliminary answer is yes, but with reservations.

The approach I adopted was to inhibit the Na pump by removing the external K and then to inject HCl. Once pH_i had almost recovered, external K was replaced, so that the Na pump then extruded the Na that had entered via the pH_i regulating system. Thus, the measurement of the charge generated by the Na pump was not obscured by the effects of a large intracellular acidification on membrane permeability.

METHODS

Experiments were done on exposed large neurons at the rear of the visceral and right pleural ganglia of the isolated brain of the snail *Helix aspersa*. The procedures used were modified from those originally used some 14 years earlier in a study of the Na pump itself (5). Indeed, so numerous are the modifications that only the voltage clamp and current amplifiers are virtually the same. The general setup is illustrated in Fig. 2. Further details are available elsewhere (8,10,11).

The Preparation

The isolated brain was mounted on a plastic block specially shaped to give a minimum bath volume when placed in an acrylic chamber (2). The connective

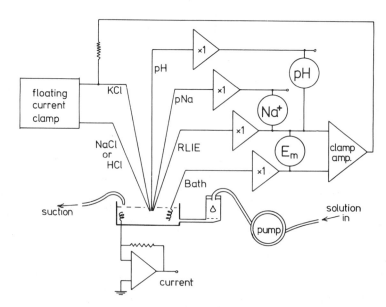

FIG. 2. Schematic diagram of experimental arrangement. See text and previous descriptions for details.

tissue was cut away and then torn to expose the neurons. Most experiments were done on the largest cell at the rear of the right pleural ganglion.

Physiological Solutions

The preparation was continually superfused with either normal snail Ringer containing 4 mM KCl, 80 mM NaCl, 7 mM $CaCl_2$, 5 mM $MgCl_2$, and 20 mM $NaHCO_3$, or a K-free variation. Both were equilibrated with 2.5% CO_2 in air to give a pH close to 7.5. A constant superfusion rate was maintained by a pump next to the experimental chamber. The pump tubing, however, allowed excessive loss of CO_2 by the pumped solution.

Measurement of Membrane Potential

To minimize problems of drift and sudden changes in tip potential I used silanized microelectrodes filled with a reference liquid ion-exchanger (RLIE) to measure membrane potential (E_m) (12). Since I was only interested in slow potential changes, their high resistance was no problem.

Other Microelectrodes

Recessed-tip glass microelectrodes for measuring pH_i and free internal Na concentration were made, calibrated, and used as previously described (8,10). The microelectrodes for current and ion injection were made from filamented aluminosilicate glass and filled with 2 M HCl, 2 M NaCl, 3 M KCl, or, in one experiment, 3 M CsCl. As filled, the resistances were 20 to 30 MΩ; before use the NaCl and KCl microelectrode tips were broken to reduce their resistances to about 10 MΩ.

Voltage Clamping and Iontophoretic Injections

Interbarrel iontophoretic injections of NaCl or HCl were made by a floating current clamp, current being passed between KCl and NaCl or HCl-filled microelectrodes. Between injections a backing current of 3 to 5 nA was passed to prevent diffusion of Na^+ or H^+ ions into the cell. The transport number for NaCl injection was taken to be 1.0 (9) and 0.93 for HCl (6).

To measure current generated by the Na pump the average E_m was held constant by a slow voltage-clamp circuit, as before (5). The only significant changes from this circuit were that unity-gain amplifiers were used instead of cathode followers, and the clamp amplifier output was fed back to the neuron via the KCl-filled iontophoretic electrode. The other injection microelectrode (NaCl or HCl) was connected to the current-measuring side of the current clamp.

The voltage-clamp current was measured using a current-to-voltage transducer between the bath and ground, as before. To ensure that there was no other pathway between the path and ground, the solution inflow was via a sealed drip system, and waste solution was collected in a bottle inside the Faraday cage.

Estimation of Pump-Generated Charge

Given a record of clamp current against time, the charge generated by the Na pump in extruding a small quantity of injected Na can be taken as the area between the clamp-current record associated with the extrusion and the extrapolated baseline current. The main difficulty is the objective measurement of such an area when the baseline is noisy. In the following calculations the baselines were drawn as conservatively as possible and the areas measured by weighing cut-out tracings.

Special Technical Problems

Before describing some of the results obtained, it may be helpful to mention the main difficulties encountered. One problem is that some of the large cells appear to have relatively weak Na pumps. When NaCl was injected into such cells it was extruded so slowly (as seen with a Na^+-sensitive microelectrode) that the current generated by the pump was too small to be easily measured. A similarly small pump current was seen in other cells when they were so damaged by microelectrode penetration that internal Na had risen well above its normal level. In this case, the Na pump was already so loaded by the high internal Na, that NaCl injection has little effect on the pump rate. Accordingly, most experiments were done on the largest cell at the rear of the right pleural ganglion, which has proven capable of surviving multiple penetrations and to have a very active Na pump.

The second and perhaps most distressing and time-consuming problem is that on many occasions this cell, and others with similar properties, were subject to random and powerful synaptic inputs. When clamped near their resting potentials, the clamp currents were so unstable and noisy that currents generated by the Na pump were partly obscured. Attempts to isolate living large cells by ligaturing their axons were unsuccessful, as were attempts to minimize synaptic transmission by lowering Ca and raising Mg, adding Mn, or applying pharmacologic blockers.

A third problem that often ended experiments prematurely was the tendency of the injection microelectrodes to block irreversibly. The experiment could sometimes be saved by removing the blocked electrode from the neuron, breaking its tip, and reinserting it; but more often, this damaged the cell beyond recovery. Perhaps, if the current clamp could produce more than 15 V it might be possible to unblock electrodes without removing them from the cell.

RESULTS

Before observing the effects of acid injection, I did a number of experiments in which I injected NaCl. Most of these injections were made into clamped cells bathed in normal Ringer. I checked that the apparent charge generated by the Na pump was not changed by the many minor changes in experimental procedure since the original experiments. For example, 14 years ago Ringer was buffered with Tris, and I did not use a backing current to prevent Na^+ leakage between injections. I also needed to check the effects of external K removal and replacement.

NaCl Injection Effects on Clamp Current, pH_i, and Na_i

Figure 3 shows an experiment of this type. After a period of equilibration, I started the experiment by penetrating a neuron with Na^+-sensitive, pH-sensitive, and reference microelectrodes. Six minutes later I thrust a KCl-filled microelectrode into the cell and used it to pass a brief hyperpolarizing current of 2 nA across the cell membrane. This was both to check the current measurement system and to see that the first three microelectrodes all measured the same change in E_m. The absence of a deflection on the pH_i and Na_i records indicates complete cell penetration.

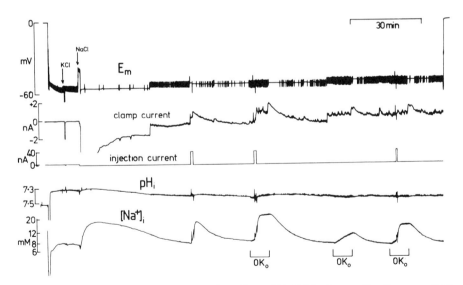

FIG. 3. Pen recording of experiment showing effect of NaCl injection and/or the removal of external K on the clamp current, pH_i, and internal Na^+ concentration of a snail neuron whose membrane potential (E_m) was loosely held at 56 and then 52 mV. Spontaneous action potentials were reduced to only a few millivolts by the slow response of the pen recorder. *Arrows* above E_m trace indicate when KCl and NaCl electrodes were inserted.

Next, the NaCl-filled microelectrode was inserted, rather roughly. This caused a 16-mV depolarization and a large increase in internal Na. A 4-nA backing current was then switched on to prevent NaCl diffusion from the electrode, and the voltage clamp circuit was activated. Initially, the holding potential was set at 56 mV, but after 30 min it was reduced to 52 mV. Seventeen minutes later NaCl was injected by a current of 40 nA for 1 min.

This injection increased Na_i by over 10 mM, and caused the clamp current to change by 1.2 nA in the outward direction. After about 20 min both clamp current and Na_i had largely recovered. I then switched to a 0 K Ringer, and 2 min later made a second NaCl injection. The 0 K Ringer blocked the Na pump sufficiently to cause a continual increase in Na_i.

After 8 min, external K was replaced, allowing Na_i to recover. As the K reached the cell, the clamp current increased to about 2 nA beyond its previous level in normal Ringer, and then declined as the Na_i decreased. Once Na_i had stabilized, and the holding potential had been adjusted, the cell was exposed twice more to 0 K solution. The first exposure was without NaCl injection, and gives an idea of the normal Na influx rate, assuming that 0 K has no effect other than that of blocking the Na pump.

The charge generated by the clamp in keeping E_m constant during and after the first NaCl injection was 0.57 μC, or 24% of the charge used to inject the Na^+ ions. Allowing for the effect of 0 K alone, the charge generated when external K was replaced after the second injection was 26% of the injection charge.

Effect of Loading the Neuron with Cesium

It seemed possible that replacing some of the normal internal K with Cs might facilitate measurements of the charge generated by the Na pump. Figure 4 shows the last part of such an experiment, in which the KCl microelectrode was replaced with one filled with 3 M CsCl. Thus the 5-nA backing current would carry Cs^+ ions into the cell except when NaCl was injected. An unexpected effect of this Cs loading was to increase the sensitivity of the Na^+-sensitive microelectrode. The glass NAS 11-18 is presumably more sensitive to K^+ than to Cs^+, so K^+ ions interfere with Na^+ measurements more than Cs^+ ions.

As seen in Fig. 4, the effects of Cs on the clamp current were minimal, but the Na pump appeared to be insensitive to the removal of external K. (This effect was even more marked in an experiment where Rb was injected.) The experiment does demonstrate very well, however, an interesting difference between the time course of the clamp-current decline, as seen after an injection and after a period in 0 K. The current declines more rapidly after a brief than after a long exposure to 0 K, and more rapidly after 0 K than after an injection in normal Ringer. Perhaps this might be due to differences in adenosine triphosphate (ATP) levels.

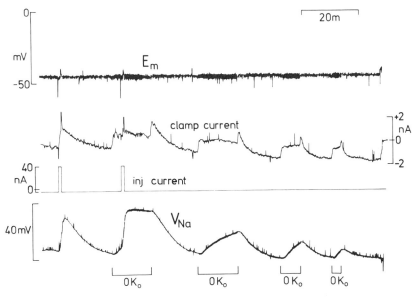

FIG. 4. Responses of clamp current and Na⁺ sensitive electrode potential (V_{Na}) to NaCl injection and/or removal of external K. Neuron injected with CsCl for 90 min before the record begins.

Effect of HCl Injection on pH_i and Na_i

Figure 5 shows an experiment in which it was possible to compare directly the effects on Na_i of HCl and NaCl injections. The clamp current was too noisy to allow estimation of the charge generated by the sodium pump, but the Na_i record allows comparison of the increase caused by pH_i recovery with that caused by direct NaCl injection.

The second 0 K exposure caused an increase in Na_i of 5.1 mM. Assuming that all the periods in 0 K solution caused a similar Na_i increase due to the normal Na influx, the extra Na influx due to the (admittedly incomplete) pH_i recovery after two HCl injections can be estimated. For the first the extra Na_i was 3.1 mM, and for the second, 2.4 mM. The two injection charges were 3 and 3.6 μC, respectively.

The third injection of NaCl can be used to estimate the ratio of increase in Na_i to injection charge: it was about 4.8 mM for an injection charge of 1.8 μC, that is, 2.7 mM/μC. This compares with 1 and 0.7 for the two acid injections. These numbers are roughly as expected if one Na⁺ enters for each two H⁺ extruded or neutralized, especially as the pH_i recovery is incomplete by the time external K is replaced.

Effect of HCl Injection on Clamp Current

In approximately 20 attempts to measure the clamp charge associated with the restoration of external K after an acid injection, only 3 experiments produced

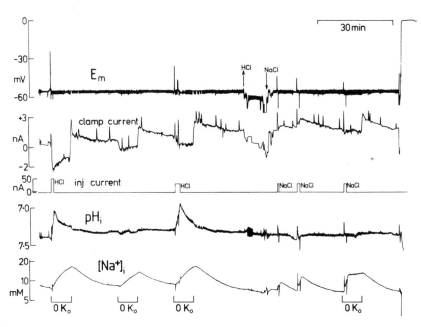

FIG. 5. The effects of HCl and NaCl injections, with and without external K, on the clamp current, pH_i, and Na_i of a large snail neuron. *Arrows* above the E_m trace indicate where HCl electrode was withdrawn and where NaCl electrode was inserted.

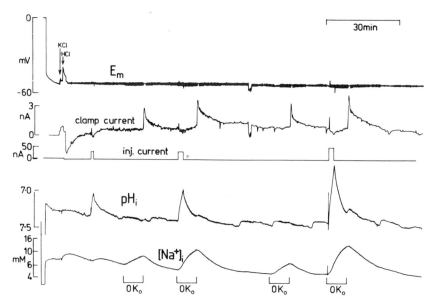

FIG. 6. The effects of HCl injection on the clamp current, pH_i, and Na_i. Where indicated, cell was superfused with K-free solution. *Arrows* above E_m record indicate where KCl and HCl electrodes were inserted.

reasonably quiet clamp-current records. Figure 6 shows the best of these experiments.

External K was removed for four 8-min periods. The first and third served as controls, and during the second and fourth HCl was injected. Each time the K was replaced, there was a rapid shift in the clamp current, followed by a decline as the Na_i recovered. Allowing for the clamp charge component due to resting Na influx, the clamp charge associated with the first HCl injection was 0.19 μC, and with the second, 0.76 μC. The charges used for the two injections were 3.6 and 6 μC, respectively.

Average Charge Ratios

In the complete series of experiments, the charge generated by the Na pump could be estimated in response to 24 injections of NaCl (only 3 made in the absence of external K) and 6 injections of HCl. The results are summarized in Table 1. They suggest that HCl injection is nearly half as effective as NaCl injection in subsequently stimulating the Na pump.

DISCUSSION

Coupling Ratio of the Na Pump

A detailed discussion of the implications of the charge ratio for NaCl injection is beyond the scope of this chapter. The charge ratio of 0.27 implies that a quarter or more of the injected Na is pumped out uncoupled to K influx, and is in good agreement with the earlier work (5). This ratio can be reconciled with the 0.33 expected for a pump-coupling ratio of 3 Na:2 K by suggesting that the transport number was probably < 1.0, that some of the injected Na may have left other than via the Na pump, and indeed that some may not have left at all if the cell volume increased. [For further discussion see a recent paper by Thomas (9).] A particular problem with the present work was the probable underestimation of the charge generated by the pump.

TABLE 1. *Average charge ratios*

Injection	Ratio of clamp to injection charge (mean ± SEM)
NaCl with normal K	0.27 ± 0.06 (N = 21)
NaCl during 0 K	0.21 ± 0.07 (N = 3)
HCl during 0 K	0.09 ± 0.03 (N = 6)

Collected data from experiments on 15 snail neurons. Charge generated by voltage clamp in keeping E_m constant during period of Na pump hyperactivity divided by charge used to inject NaCl or HCl.

Stoichiometry of the pH_i Regulating System

In their study of pH_i regulation in squid axons, Russell and Boron (4) reported that one Na^+ ion entered the axon for each two H^+ ion equivalents extruded or neutralized, as expected from the scheme originally proposed for snail neurons (7), and shown in Fig. 1. In the present experiments, the charge generated by the pump on return to normal external K after HCl injection was almost half that generated after an equivalent NaCl injection, also as expected from the proposed scheme. But the results can only be taken as preliminary.

Technical Considerations

There are several difficulties in the interpretation of the results of this type of experiment. Before one can be fully convinced that the charge ratios can be relied on, some way must be found to reduce the noise on the clamp current. More measurements must be made with small injections less likely to cause changes in cell volume or loss of Na by routes other than the Na pump, and new measurements of the transport number must be made.

ACKNOWLEDGMENTS

I am grateful to the Medical Research Council for financial support, to Michael Rickard and Avril Lear for technical and secretarial services, and to Robert Meech for improving an early draft of this chapter.

REFERENCES

1. Boron, W. P., and De Weer, P. (1976): Active proton transport stimulated by CO_2/HCO_3^-, blocked by cyanide. *Nature*, 259:240–241.
2. Evans, M. G., and Thomas, R. C. (1982): Minimising the volume of an experimental bath for intracellular recording from a superfused snail brain preparation. *J. Physiol. (Lond.)*, 327:18P–19P.
3. Roos, A., and Boron, W. P. (1981): Intracellular pH. *Physiol. Rev.*, 61:296–434.
4. Russell, J. M., and Boron W. P. (1982): Intracellular pH regulation in squid giant axons. In: *Intracellular pH: Its Measurement, Regulation, and Utilization in Cellular Functions*, edited by R. Nuccitelli and D. W. Deamer, pp. 221–237. Alan R. Liss, New York.
5. Thomas, R. C. (1969): Membrane current and intracellular sodium changes in a snail neurone during extrusion of injected sodium. *J. Physiol. (Lond.)*, 201:495–514.
6. Thomas, R. C. (1976): The effect of carbon dioxide on the intracellular pH and buffering power of snail neurones. *J. Physiol. (Lond.)*, 255:715–735.
7. Thomas, R. C. (1977): The role of bicarbonate, chloride and sodium ions in the regulation of intracellular pH in snail neurones. *J. Physiol. (Lond.)*, 273:317–338.
8. Thomas, R. C. (1978): *Ion-Sensitive Intracellular Microelectrodes*. Academic Press, London.
9. Thomas, R. C. (1982): Electrophysiology of the sodium pump in a snail neuron. In: *Current Topics in Membranes and Transport, Vol. 16: Electrogenic Ion Pumps*, edited by C. L. Slayman, pp. 3–16. Academic Press, New York.
10. Thomas, R. C. (1982): Ion-sensitive microelectrodes. In: *Techniques in Cellular Physiology*, Vol. P125, pp. 1–12. Elsevier/North-Holland, Amsterdam.
11. Thomas, R. C. (1982): Snail neuron intracellular pH regulation. In: *Intracellular pH: Its Measure-*

ment, Regulation and Utilization in Cellular Functions, edited by R. Nuccitelli and D. W. Deamer, pp. 189–204. Alan R. Liss, New York.
12. Thomas, R. C., and Cohen, C. J. (1981): A liquid ion-exchanger alternative to KCl for filling intracellular reference microelectrodes. *Pfluegers Arch.*, 390:96–98.

Electrogenic Transport: Fundamental
Principles and Physiological Implications,
edited by Mordecai P. Blaustein and Melvyn
Lieberman. Raven Press, New York © 1984.

Sodium-Dependent Calcium Efflux and Sodium-Dependent Current in Perfused Barnacle Muscle Single Cells

*M. T. Nelson[1] and †W. J. Lederer

*Fakultät für Biologie, Universität Konstanz, D-7750 Konstanz, West Germany; and
†Department of Physiology, University of Maryland School of Medicine,
Baltimore, Maryland 21201

Evidence of a Na-Ca exchange mechanism has been obtained in many biological tissues (2). Although the physiological role of this transport system has not been unequivocally established (1,5), much evidence exists suggesting that it is important in keeping intracellular calcium at a low level (Blaustein, *this volume;* 2,4). Fundamental to understanding the Na-Ca exchange mechanism is the determination of its stoichiometry, the number of Na^+ ions exchanged for each Ca^{2+}. Investigations of the stoichiometry of this exchange mechanism suggest that between 2 and 6 Na^+ ions are exchanged for each Ca^{2+} ion (Blaustein, *this volume,* Mullins, *this volume;* 2,4,17). Conservative interpretation of these results is warranted, however, because the experimental maneuvers used to obtain these results are known to affect processes other than the Na-Ca exchange mechanism. For example, when extracellular sodium (Na_o) is altered, changes in Na_i or pH_i may occur independently of Na-Ca exchange (Lederer et al., *this volume*). Similarly, changes of Na_o may affect a cell's membrane potential without altering Na-Ca exchange (3,6–8,12,19). We have attempted to eliminate such uncontrolled factors by perfusing the intracellular compartment (13–15) and controlling the membrane potential of single muscle cells from barnacles (9–11,16). Using this preparation, Na_o-dependent membrane current and Na_o-dependent Ca efflux were measured simultaneously. We find that Na_o-dependent Ca efflux is associated with large changes in membrane current. We conclude that, even under these highly controlled conditions, the Na_o-dependent membrane current and Ca efflux cannot be easily attributed to a simple and unique Na_o-dependent process such as the Na-Ca exchange mechanism.

METHODS

Single muscle cells from the giant barnacle, *Balanus nubilus,* were perfused through a single cut end (13). Membrane potential was controlled by passing

---

[1] Dr. Nelson is presently at the University of Maryland address.

current through a platinized-platinum axial wire (11,16). The internal perfusate flowed at 5 μl/min and contained 189 mM K glutamate, 10 mM Na glutamate, 38 mM KCl, 350 mM sucrose, 60 mM Hepes, 7.9 mM $CaCl_2$ (50 to 100 μCi/ml ^{45}Ca), 8 mM EGTA (about 10 μM free calcium), 8 mM $MgCl_2$, apyrase (3 units/ml), 0.2 mM phenol red, pH = 7.3. To minimize any ATP-dependent process from modifying the measured Ca efflux or the measured membrane current, apryrase [an adenosine triphosphate (ATP) and adenosine diphosphate (ADP) phosphatase] was in the perfusion fluid. The superfusate flowed at about 9.1 ml/min and contained 456 mM NaCl, 6 mM Tris-HCl, 10 mM KCl, 25 mM $MgCl_2$, pH = 7.8, temperature = 16°C. After passing over the barnacle muscle cell, the superfusing solution was pumped through an on-line scintillation counter (Beckman Beta-Mate II). All experiments were performed in Ca_o-free medium in order to prevent Ca_o-dependent Ca efflux from contributing to our Ca-efflux measurements (cf. 11,14,15).

RESULTS AND DISCUSSION

The effects of extracellular sodium (Na_o) on Ca efflux and membrane current were examined in cells that contained low and high $[Ca_i^{2+}]$. When $[Ca_i^{2+}]$ is low (about 0.1 μM), Ca efflux is also low (about 0.2 pmol/sec cm²) and relatively insensitive to changes in Na_o (using Li_o as a replacement cation). With low $[Ca_i^{2+}]$, reducing Na_o to 0 mM causes a slight depolarization or increase in inward current. Increasing $[Ca_i^{2+}]$ causes Ca efflux to rise and the membrane to depolarize. Figure 1 illustrates the effects of changing Na_o on Ca efflux and membrane current in a cell perfused with high $[Ca_i^{2+}]$. The membrane potential was clamped at −40 mV and 2.3 mV depolarizations were applied for 2 sec every minute. The effects of these depolarizations on membrane current are shown on the top trace and the calculated membrane conductance (g_m) is shown on the bottom of the figure. On increasing Na_o, there is a rapid and large increase in Ca efflux and in inward current. In this experiment, following the change in inward current and Ca efflux, there was a slow decrease in membrane conductance. Because membrane current changes before a change in membrane conductance (both on addition and removal of Na_o), one need not ascribe the change in membrane current to the observed change in conductance. In other experiments, changes in Na_o (with Li_o as the replacement) had no measurable effect on membrane conductance. Na_o-dependent Ca efflux and Na_o-dependent membrane current are maximal at about 200 mM Na_o (15) and are insensitive to extracellular lanthanum (5 mM) (10).

In Fig. 1, after changing Na_o, the delay of about 1 min in the change of Ca efflux represents the time it takes the superfusate to be pumped from the experimental chamber to the counter. Adjusting for this delay and avoiding the first minute after the solution change, we have plotted the relationship between Ca efflux and membrane current in Fig. 2. The slope of the line in Fig. 2 suggests that on altering Na_o, a change of approximately 1 pmol/sec of Ca

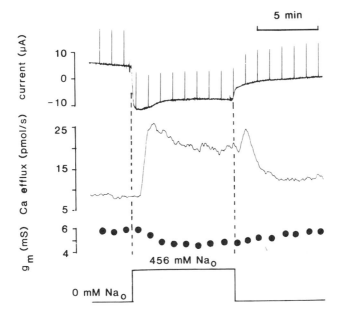

FIG. 1. Na_o-dependent Ca efflux and Na_o-dependent membrane current in voltage-clamped barnacle muscle single cells. The time courses of changes in membrane current, Ca efflux, and membrane conductance are shown. Original current and Ca-efflux records are shown. A time delay (about 1 min) is readily apparent on the record of Ca efflux. This delay reflects the time it takes superfusate to get to the scintillation counter. Holding potential is −40 mV. Depolarizations of 2.3 mV are made every minute in order to measure membrane conductance (g_m). $Na_o + Li_o = 456$ mM.

efflux is associated with a change of current of 1 µA (10 peq/sec). Thus, if both the Na_o-dependent Ca efflux and the Na_o-dependent current are mediated by a Na-Ca exchange mechanism, then the relative magnitudes of these changes would imply that approximately 12 Na^+ ions exchanged for 1 Ca^{2+} ion. This possible coupling ratio is clearly higher than those previously reported (between 2 and 6 Na^+ ions exchanged for each Ca^{2+} ion) (cf. Blaustein, *this volume;* Mullins, *this volume;* 4). For reasons stated below, we do not believe that the coupling ratio is as large as 12:1. We have inquired, nevertheless, whether there are theoretical or experimental reasons to exclude a particular coupling ratio. Figure 3 shows that even with very large coupling ratios, $E_{Na\text{-}Ca}$, the thermodynamic reversal potential of the transport system is well behaved. As r, the coupling ratio, varies from zero to infinity, $E_{Na\text{-}Ca}$ goes through a discontinuity at $r = 2$. At $r = 0$, $E_{Na\text{-}Ca} = E_{Ca}$ and at $r = $ infinity, $E_{Na\text{-}Ca} = E_{Na}$. However, the stoichiometry of transport has no necessary bearing on the voltage dependence of transport by the Na-Ca exchange. Such voltage dependence of transport is a function of the voltage dependence of the rate-limiting step of the transport mechanism (see DeWeer, *this volume*). Thus, if the rate-limiting

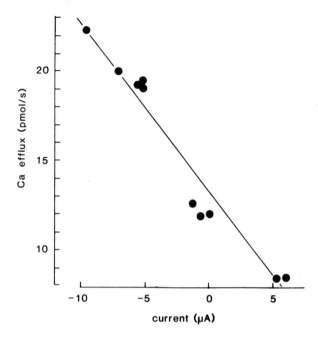

FIG. 2. Relationship between the Na_o-dependent Ca efflux and the Na_o-dependent current changes. Data from Fig. 1 have been plotted. The Ca-efflux data have been adjusted to account for the time delay. Data for the first minute after changing solution have not been plotted. The slope of the line (fit by eye) indicates that a change of Ca efflux of approximately 1 pmole/sec is associated with a change of membrane current of 1 μA (or about 10 peq/sec).

step is insensitive to voltage, then even with large coupling ratios, the transport system need not be strongly voltage dependent. Over limited ranges of membrane potential (−50 to −30 mV), the Na_o-dependent Ca efflux was not voltage dependent (W. J. Lederer and M. T. Nelson, *unpublished observation*). Clearly, at extreme potentials a charged carrier must become rate limiting as it gets trapped on one side of the membrane or on the other side. If the rate-limiting step in ion transport involved ion translocation, then, depending on the details (e.g., simultaneous or sequential, "ping-pong"), a large coupling ratio could be linked to significant voltage dependence of ion transport. We did not observe this, however.

If our results were interpreted solely in terms of a Na-Ca exchange mechanism, two contributing factors must be evaluated experimentally. The first concern is the possibility that Ca efflux, which could be attributed to the Na-Ca exchange mechanism, has been underestimated. The second point of inquiry is whether the membrane current that could be attributed to the Na-Ca exchange mechanism may be overestimated.

Calcium efflux may be underestimated when ^{45}Ca-specific activity is locally

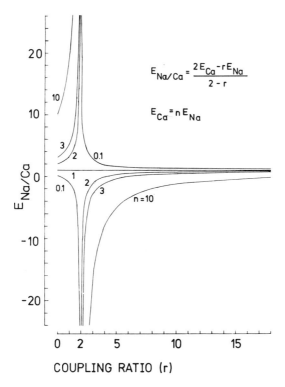

FIG. 3. Theoretical relationship between the coupling ratio (r) of the Na-Ca exchange mechanism and the reversal potential ($E_{\text{Na-Ca}}$) for the Na-Ca exchange mechanism. $E_{\text{Na-Ca}}$ is plotted in units of E_{Na}, the reversal potential for Na$^+$ across the cell membrane. The various curves indicate how the relationship changes as the relative values of E_{Ca} change with respect to E_{Na}. (Note that $n = E_{\text{Ca}}/E_{\text{Na}}$). For a given n, $E_{\text{Na-Ca}} = E_{\text{Na}}(2n - r)/(2 - r)$, and there is a clear singularity at $r = 2$ when $n \neq 1$ as is evident in the figure.

reduced near the Na-Ca exchange mechanism due to cold calcium leaking from intracellular Ca buffers. We have attempted to control for this in our preparation by buffering intracellular Ca with a large amount of EGTA and by removing ATP-dependent buffer systems (by perfusing with apyrase and no added ATP). Furthermore, the addition of extracellular EGTA does not materially affect our results and suggests that ^{45}Ca-recycling is not a major factor. Because there are no specific blockers of Na-Ca exchange yet identified, the only properties of this transport system available to us for identification of Na-Ca exchange current or flux are its [Ca$_i^{2+}$]-activation and Na$_o$-dependence. However, other [Ca$_i^{2+}$]-activated, Na-dependent current sources could exist and make it difficult for us to interpret our results in terms of only the Na-Ca exchange mechanism. One such current source has been reported in cardiac tissue (TI channels) and described as a Ca-activated, nonselective channel (3,6–8,12). TI channels may be activated in parallel with the Na-Ca exchange mechanism in cardiac Purkinje fibers and account for some of the Na_o-dependent current changes seen in that preparation (Lederer et al., *this volume*). The finding of TI-like channels in sheep (6,7; Lederer et al., *this volume*) and rat heart (3) and in neuroblastoma (19) suggests that similar channels may be found elsewhere, including in those preparations used to examine Na-Ca exchange stoichiometry (e.g., 17). If increasing Na_o augmented inward current or Na influx through TI channels (or equivalent Ca-activated conductances) (see ref. 18) while also activating the Na-Ca exchange mechanism, the identification of the specific contributions of each current source would be very difficult. Because, in barnacle muscle, the Na_o-dependent current saturates at about 200 mM Na_o and is La_o^{3+}-insensitive (5 mM) (W. J. Lederer and M. T. Nelson, *unpublished observation*), these Ca-activated channels would, presumably, have to have similar properties.

In summary, the results of our experiments raise one major issue. How confident can we be in any of the experimentally determined stoichiometries of the Na-Ca exchange mechanism? Numerous reports to date have discussed the physiological implications of particular stoichiometries of the Na-Ca exchange mechanism. Previous work has suggested that the stoichiometry may range from 2:1 to about 6:1 (Blaustein, *this volume;* Mullins, *this volume;* 2). If no parallel processes existed, our results would suggest that the stoichiometry is larger than this range of values. Higher stoichiometries are consistent with lower resting $[Ca_i^{2+}]$ if the Na-Ca exchange is at its electrochemical equilibrium. However, the exact level of $[Ca_i^{2+}]$ under normal physiological conditions has not been resolved satisfactorily and, therefore, does not provide enough support for a particular coupling ratio. Furthermore, the assumption that the Na-Ca exchange is at equilibrium requires that the Na-Ca exchange mechanism does not transport net Na or Ca either into or out of the intracellular compartment. The large stoichiometry that could be inferred from our data appears unlikely, because it implies greater complexity of the Na-Ca exchange mechanism than seems reasonable at present. We suspect that additional charge movement and/or ^{45}Ca buffering may exist in parallel with the Na-Ca exchange mechanism, as has been suggested elsewhere (Lederer et al., *this volume;* 18). This possibility could be tested directly when a specific blocker of the Na-Ca exchange mechanism becomes available.

ACKNOWLEDGMENTS

We would particularly like to thank M. P. Blaustein for his encouragement, support, and counsel during the course of our work on this project. We also thank D. A. Eisner, L. J. Mullins, R. A. Sjodin, and S-S. Sheu for valuable discussions; L. Goldman and L. Horn for loan of equipment; E. Santiago for laboratory assistance; and D. Darragh for preparation of the figures.

This work has been supported by the NSF and the Muscular Dystrophy Association of America (M. P. Blaustein); the Maryland Affiliate of the American Heart Association (MTN), the Alexander von Humboldt Foundation (MTN), an Established Investigatorship of the American Heart Association and its Maryland Affiliate (WJL), the NIH (HL25675) (WJL), and a Basil O'Connor starter grant from the March of Dimes Birth Defects Foundation (WJL).

REFERENCES

1. Barry, W., and Smith, T. (1982): Mechanisms of transmembrane calcium movement in cultured chick embryo ventricular cells. *J. Physiol.* (*Lond.*), 325:243–260.
2. Blaustein, M. P., and Nelson, M. T. (1982): Sodium-calcium exchange: Its role in the regulation of cell calcium. In: *Calcium Transport across Biological Membranes*. edited by E. Carafoli, pp. 217–236. Academic Press. New York.
3. Colquhoun, D., Neher, E., Reuter, H., and Stevens, C. F. (1981): Inward current channels activated by intracellular Ca in cultured cardiac cells. *Nature,* 294:752–754.
4. DiPolo, R., Requena, J., Brinley, F. J., Mullins, L. J., Scarpa, A., and Tiffert, T. (1976): Ionized calcium concentrations in squid axons. *J. Gen. Physiol.*, 67:433–467.
5. Droogmans, G., and Casteels, R. (1979): Sodium and calcium interactions in vascular smooth muscle cells of the rabbit ear artery. *J. Gen. Physiol.*, 74:57–70.
6. Eisner, D. A., and Lederer, W. J. (1979): Inotropic and arrhythmogenic effects of potassium depleted solutions on mammalian cardiac muscle. *J. Physiol.* (*Lond.*), 294:255–277.
7. Kass, R. S., Lederer, W. J., Tsien, R. W., and Weingart, R. (1978): Role of calcium ions in transient inward currents and aftercontractions induced by strophanthidin in cardiac Purkinje fibres. *J. Physiol.*, (*Lond.*), 281:187–208.
8. Kass, R. S., Tsien, R. W., and Weingart, R. (1978): Ionic basis of transient inward current induced by strophanthidin in cardiac Purkinje fibres. *J. Physiol.* (*Lond.*), 281:209–226.
9. Lederer, W. J., and Nelson, M. T. (1981): Electrogenic sodium pumping in the internally perfused barnacle muscle fibre: Simultaneous measurement of sodium efflux and membrane current. *J. Physiol.* (*Lond.*), 319:62P–63P.
10. Lederer, W. J., and Nelson, M. T. (1981): Current associated with Na_o-dependent Ca-efflux in barnacle muscle cells. *J. Physiol.* (*Lond.*), 320:119P–120P.
11. Lederer, W. J., Nelson, M. T., and Rasgado-Flores, H. (1982): Electroneutral Ca_o-dependent Ca-efflux from barnacle muscle single cells. *J. Physiol.* (*Lond.*), 325:34P–35P.
12. Lederer, W. J., and Tsien, R. W. (1976): Transient inward current underlying arrhythmogenic effects of cardiotonic steroids in Purkinje fibres. *J. Physiol.* (*Lond.*), 263:73–100.
13. Nelson, M. T., and Blaustein, M. P. (1980): Properties of sodium pumps in internally perfused barnacle muscle fibers. *J. Gen. Physiol.*, 75:183–206.
14. Nelson, M. T., and Blaustein, M. P. (1981): Effects of ATP and vanadate on calcium efflux from barnacle muscle fibres. *Nature,* 289:314–316.
15. Nelson, M. T., and Blaustein, M. P. (1981): Effect of Na_o-dependent calcium efflux on membrane potential of internally perfused barnacle muscle fibers. *Biophys. J.*, 33:61A.
16. Nelson, M. T., and Lederer, W. J. (1983): Stoichiometry of the electrogenic Na-pump in barnacle muscle: Simultaneous measurement of Na-efflux and membrane current. In: *Third International Meeting on the Na-K ATPase,* edited by J. F. Hoffman and W. B. Forbush, IIIrd. Academic Press, New York.
17. Reeves, J., and Sutko, J. (1980): Sodium-calcium exchange activity generates a current in cardiac membrane vesicles. *Science,* 208:1461–1464.
18. Sheu, S-S., Blaustein, M. P., and Santiago, E. M. (1982): Effects of extracellular and intracellular Ca^{2+} on membrane potential and ^{22}Na influx in internally perfused barnacle muscle. *J. Gen. Physiol.*, 80:22a–23a.
19. Yellen, G. (1982): Single Ca^{2+}-activated nonselective cation channels in neuroblastoma. *Nature,* 296:357–359.

Electrogenic Transport: Fundamental Principles and Physiological Implications, edited by Mordecai P. Blaustein and Melvyn Lieberman. Raven Press, New York © 1984.

The Effects of Na-Ca Exchange on Membrane Currents in Sheep Cardiac Purkinje Fibers

*W. J. Lederer, *S -S. Sheu, **R. D. Vaughan-Jones, and †D. A. Eisner

*Department of Physiology, University of Maryland School of Medicine, Baltimore, Maryland 21201; **Department of Pharmacology, Oxford University, South Parks Road, Oxford OX1 3QT, England; and †Department of Physiology, University College London, London WC1E 6BT, England

There is compelling evidence for the existence of a sarcolemmal Na-Ca exchange in cardiac muscle. Such a mechanism is important in determining the resting level of intracellular calcium ($[Ca^{2+}]_i$) and, hence, tension in the heart. It has recently been suggested that Na-Ca exchange may also contribute to the influx and efflux of Ca, which occurs during the cardiac cycle and which is important in producing a phasic contraction (see refs. 3 and 16 for recent reviews).

One of the factors that determines whether Na-Ca exchange produces a net efflux or influx of Ca ions under a given set of conditions is the stoichiometry of the exchange, i.e., the number of Na ions exchanged for each Ca ion. Unfortunately, there is, as yet, no generally agreed value for this stoichiometry. Experiments in squid axons are consistent with 2 to 6 Na ions being exchanged per Ca ion (1), but for many years, the Na-Ca exchange in cardiac muscle was thought to exchange only two Na ions per Ca ion (18). However, recent work on cardiac sarcolemmal vesicles suggests that not two but three Na ions are transported (17). If this is the case, then the mechanism in the heart will be electrogenic, since net positive charge will accompany the movement of Na through the exchange.

It has been suggested that the current generated by such an electrogenic mechanism could have significant effects on the electrophysiology of cardiac muscle (16). Indeed, recent experiments suggest that Na-Ca exchange can affect the resting membrane potential of cardiac muscle, and this effect has been attributed to the action of an electrogenic Na-Ca exchange current (5,13). In these experiments, the removal of external Na ions resulted in a transient increase of tension thought to be produced by a transient influx of Ca through a Na-Ca exchange. The transient increase of tension was accompanied by a transient hyperpolarization of the resting potential. Since, under these conditions, Na ions would be leaving the fiber in exchange for Ca entering, the hyperpolarization was taken as evidence that the exchange was electrogenic. If three or more Na ions left in exchange for one Ca ion, this would generate an outward current that would hyperpolarize the membrane potential. Furthermore, the effect would be transient since cardiac fibers become depleted of Na rapidly (10). However, it is equally possible that the hyperpolarization is caused by the rise of $[Ca^{2+}]_i$ acting on Ca-dependent membrane channels (4,14,15). It is therefore important to establish whether the changes of membrane potential produced by alterations of Na-Ca exchange activity can be attributed solely to an electrogenic Na-Ca exchange current, or whether they are secondary to changes of $[Ca^{2+}]_i$.

In an attempt to separate these possibilities, we have studied the changes of membrane current produced by removing external Na or raising external Ca, two procedures believed to alter the activity of the Na-Ca exchange mechanism (2,6,19,20). The experiments have been performed on voltage-clamped sheep cardiac Purkinje fibers. The results suggest that most of the current changes can be attributed to changes of membrane conductance rather than to an electrogenic Na-Ca exchange current. Although the results do not exclude the existence of such an electrogenic current, they nevertheless emphasize that it may easily be obscured by other membrane currents that are a secondary consequence of Na-Ca exchange.

METHODS

The experimental methods have been described in detail elsewhere (8). Isolated sheep cardiac Purkinje fibers were shortened to a length of 1 to 2 mm. They were mounted in the experimental chamber with one end attached to a sensitive tension transducer. A 2-microelectrode voltage-clamp technique was used to control the membrane potential and to measure membrane current. In some experiments, the intracellular Na activity (a_{Na}^i) was measured simultaneously using a liquid ion-exchanger, Na^+-sensitive microelectrode (20). Intracellular Na measurements have been expressed in terms of the activity (a_{Na}^i), assuming the activity coefficient of the external calibrating solution to be 0.75 (see refs. 8 and 20 for further details). Normal Tyrode solution consisted of 145 mM NaCl; 4.0 mM KCl; 2.0 mM $CaCl_2$; 1 mM $MgCl_2$; 10 mM Tris-HCl (pH 7.4 at 37°C); 10 mM glucose. When required, low Na or Na-free solutions were obtained by substituting equivalent amounts of Tris-HCl or LiCl for NaCl (pH maintained at 7.4).

RESULTS AND DISCUSSION

The experiment illustrated in Fig. 1 was designed to investigate the effects of Na removal on membrane current. The Na-K pump had been inhibited by removing external K and, at the time indicated, the external Na concentration was reduced to 36 mM (replaced by Tris). This produced an increase of resting tension which then decayed away. Throughout this period the membrane poten-

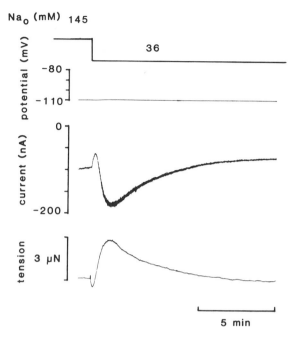

FIG. 1. The effects of Na reduction on membrane current and tension. Traces show **top:** membrane potential; **middle:** current; **bottom:** tension. The membrane potential was held at −110 mV throughout and, at the time shown, Na_o was reduced from 145 to 36 mM (using Tris as a replacement cation). In order to inhibit the Na pump all solutions were K-free throughout.

tial was held constant at −110 mV. Removing external Na produced a transient increase of inward current, which had a similar time course to the rise of tension. It should be noted that a Na-Ca exchange that carries more than 2 Na ions out of the cell in exchange for each Ca ion would be expected to produce a transient increase of outward current rather than the observed increase of inward current.

In order to investigate the origin of this transient increase in inward current, the effects of Na removal were examined at 3 different membrane potentials (Fig. 2). The Na-K pump had again been inhibited by exposing the preparation to the K-free solution. As described in the figure legend, the membrane potential was held at −55 mV and voltage-clamp pulses (5-sec duration) were applied to −38 and −75 mV. Sodium removal once again produced a transient contracture. However, the effects of Na removal on membrane current depend on the membrane potential. The transient change of current is outward at −38 mV but inward at −75 mV, whereas at −55 mV there is little effect of Na removal on membrane current. The principal effect of Na removal is therefore to produce a transient increase of membrane conductance. The apparent reversal potential for this conductance is close to −55 mV.

The Na-K pump had been inhibited before removing external Na in the

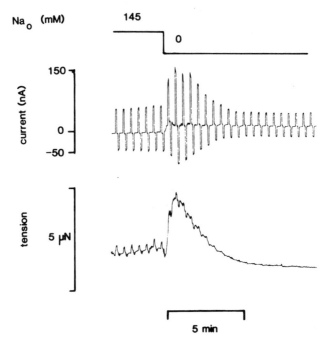

FIG. 2. The effects of membrane potential on the changes of current and tension produced by Na removal. Traces show **top:** current; **bottom:** tension. The membrane potential was held at −55 mV and 5-sec pulses were applied to −38 mV and −75 mV. Therefore, the *upper envelope* of the current trace shows the current at −38 mV and the *lower envelope,* the current at −75 mV. The *intermediate level* of current is −55 mV. In order to inhibit the Na pump, the solutions were K-free throughout. At the time shown external Na was removed and replaced by Tris.

experiments of Figs. 1 and 2. If the Na-K pump had not been inhibited, much smaller increases of resting tension were produced by Na removal (cf. 2 and 5). This is presumably because Na_i is lower when the Na-K pump is working, and therefore Na removal produces a smaller entry of Ca via Na-Ca exchange. Similarly, when Na_i was low, the transient increase of membrane conductance produced by Na removal was also much less. This observation is important since it indicates that the removal of external Na cannot, by itself, account for the observed changes of current and conductance. Instead, they are likely to be mediated via changes of the intracellular ionic composition.

Reducing external Na to intermediate levels produces smaller contractures than those produced by complete Na removal and correspondingly smaller transient increases of membrane conductance. These observations suggest that the magnitude of the transient increase of conductance may depend on the magnitude of the increase of $[Ca^{2+}]_i$ (as indicated by the increase of tension).

In a further series of experiments we examined the effects of another maneuver, which should affect Na/Ca exchange: increasing the external Ca concentration.

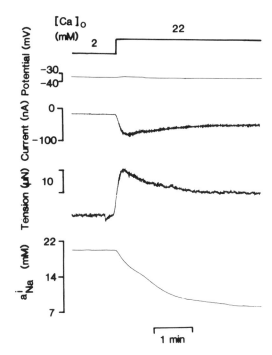

FIG. 3. The effects of an elevation of external calcium concentration on a_{Na}^i, current and tension. Traces show (from **top** to **bottom**): membrane potential, current, tension, and a_{Na}^i. In order to inhibit the Na pump, all solutions were K-free and contained 10^{-5} M strophanthidin. External calcium concentration was increased from 2 to 22 mM at the time indicated.

In the experiment of Fig. 3 the fiber had been exposed to a K-free solution containing 10^{-5} M strophanthidin to inhibit the Na-K pump. This produced an increase of the intracellular Na activity (a_{Na}^i) to about 20 mM. Increasing Ca_o from 2 to 22 mM then resulted in a fall of a_{Na}^i to 7 mM. This observation, coupled with the fact that in the Purkinje fiber a_{Na}^i does not rise above 20 to 30 mM following inhibition of the Na-K pump, has led to the suggestion that Na-Ca exchange under these conditions actively extrudes Na ions from the cell (6). Thus, the fall of a_{Na}^i seen on raising Ca_o is thought to be mediated via Na-Ca exchange. Elevation of Ca_o also produces a transient increase of resting tension, presumably reflecting a transient rise of $[Ca^{2+}]_i$ (Fig. 3). This can also be explained on a Na-Ca exchange model since, as a_{Na}^i falls, Ca entry through Na-Ca exchange will decline. If Na-Ca exchange is electrogenic, the fall of a_{Na}^i should be accompanied by a decrease of outward current, but, in fact, the opposite occurs. There is an *increase* of outward current.

This change of current was investigated more thoroughly in the experiment of Fig. 4. Here, depolarizing voltage-clamp pulses were applied in order to measure the membrane conductance while Ca_o was elevated from 0.5 to 5 mM. The increase of Ca_o produced a transient increase of both tension and membrane

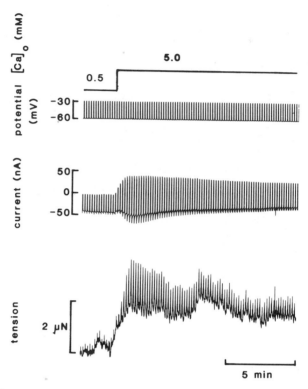

FIG. 4. The effects of membrane potential on the changes of membrane current produced by elevating Ca$_o$. Traces show **top:** membrane potential; **middle:** current; **bottom:** tension. The membrane potential was held at −55 mV and a 2-sec duration pulse to −30 mV was applied at 0.1 Hz. All solutions were K-free and, at the time shown, Ca$_o$ was elevated from 0.5 to 5 mM. On raising Ca$_o$ from 0.5 to 5 mM there was an increase of resting tension. Unfortunately, mechanical interference on the tension transducer obscures the subsequent relaxation. However, the depolarizing pulses produce tonic contractures that are visible as spikes on the tension record. It is evident that, following the elevation of Ca$_o$, the size of these contractures first increases and then decreases.

conductance (cf. 9). Therefore, much of the current change produced by elevation of Ca$_o$, like that due to removal of Na$_o$, may result from changes of membrane conductance and not from an electrogenic Na-Ca exchange. It is notable that at −60 mV there was a transient increase of inward current on raising Ca$_o$, whereas at −30 mV there was a transient increase of outward current. This is again consistent with a conductance mechanism with an apparent reversal potential between −30 and −60 mV.

The above results illustrate the difficulty of measuring the putative electrogenic Na-Ca exchange current in cardiac Purkinje fibers. The large changes of membrane conductance that occur during such measurements will clearly prevent the accurate recording of such currents. The present results therefore cannot tell us whether or not Na-Ca exchange in heart is electrogenic.

It is of interest to estimate the magnitude of the electrogenic current expected for a 3 Na$^+$ per 1 Ca^{2+} exchange. This can be calculated from the rate of fall of a_{Na}^i in experiments such as that of Fig. 3. If the fall of a_{Na}^i occurs via a Na-Ca exchange, which exchanges 3 Na$^+$ per Ca^{2+}, then, knowing the volume of the preparation we can calculate that the maximum electrogenic (outward) current should be 25 nA. Since the actual current observed in Fig. 3 is in the opposite direction (inward) the electrogenic current, if it exists, must have been obscured by the change of membrane conductance producing a larger increase in inward current.

Previous workers (5) concluded that the transient hyperpolarization produced by Na removal could not be accounted for by any simple conductance change. They rejected the possibility that the hyperpolarization could be a consequence of the transient rise of [Ca^{2+}]$_i$ acting on a Ca-activated membrane channel. This conclusion came from the observation that application of caffeine to produce a contracture of similar magnitude to that produced by Na removal did not produce a change of membrane potential. It was assumed that [Ca^{2+}]$_i$ rose to similar levels in both cases, so that the possibility of the hyperpolarization resulting from a Ca-activated conductance could be rejected. It should be noted, however, that (a) caffeine increases the apparent affinity of the contractile proteins for Ca$_i^{2+}$ (11); and (b) an intracellular acidification is known to occur during exposure to Na-free solutions (2,7), and this decreases the apparent affinity of the contractile proteins for Ca$_i^{2+}$ (12). It is therefore likely that [Ca^{2+}]$_i$ was much less during the application of caffeine. Consequently, contributions from Ca-activated conductances cannot be excluded.

Finally, we must consider the origin of the changes of membrane conductance seen in the present work. The apparent reversal potential for the current changes was about −55 mV in Na-free solution (Fig. 2) and it fell between −30 and −60 mV in high Ca$_o$, normal Na$_o$ solution (Fig. 4). These values are too positive to be attributed to a Ca-activated K channel. The nonspecific Ca-activated cation channels recently described for cardiac Purkinje fibers have a reversal potential of −20 to −40 mV in Na-free solutions, which would be consistent with the present results (15). However, these channels have a reversal potential of about 0 mV in normal, Na-containing solutions (4,15), which is less negative than that seen in Fig. 4. We must conclude therefore that the apparent reversal potentials observed in the present work cannot simply be described in terms of known Ca-activated ion conductances. It should be pointed out, however, that if Na-Ca exchange in cardiac muscle is electrogenic, it will contribute to the measured current and influence the apparent reversal potential. Once again, this emphasizes the difficulty of interpreting changes of membrane current measured when Na$_o$ or Ca$_o$ are varied.

In conclusion, measurements of membrane current in voltage-clamped sheep Purkinje fibers neither support nor deny the existence of an electrogenic Na-Ca exchange. It is clear, however, that current changes hitherto attributed to a Na-Ca exchange mechanism may now be attributed, at least partly, to changes of membrane conductance. Such conductance changes are probably a secondary consequence of variations in $[Ca^{2+}]_i$.

REFERENCES

1. Baker, P. F., Blaustein, M. P., Hodgkin, A. L., and Steinhardt, R. A. (1969): The influence of calcium on sodium efflux in squid giant axons. *J. Physiol. (Lond.)*, 200:431–458.
2. Bers, D. M., and Ellis, D. (1982): Intracellular calcium and sodium activity in sheep heart Purkinje fibres. Effects of changes of external sodium and intracellular pH. *Pfluegers Arch.*, 393:171–178.
3. Chapman, R. A. (1979): Excitation-contraction coupling in cardiac muscle. *Prog. Biophys. Mol. Biol.*, 35:1–52.
4. Colquhoun, D., Neher, E., Reuter, H., and Stevens, C. F. (1981): Inward current channels activated by intracellular Ca in cultured cardiac cells. *Nature*, 294:752–754.
5. Coraboeuf, E., Gautier, P., and Guiraudou, P. (1981): Potential and tension changes induced by sodium removal in dog Purkinje fibres: Role of an electrogenic sodium-calcium exchange. *J. Physiol. (Lond.)*, 311:605–622.
6. Deitmer, J. W., and Ellis, D. (1978): Changes in the intracellular sodium activity of sheep heart Purkinje fibres produced by calcium and other divalent cations. *J. Physiol. (Lond.)*, 277:437–453.
7. Deitmer, J. W., and Ellis, D. (1980): Interactions between the regulation of the intracellular pH and sodium activity of sheep cardiac Purkinje fibres. *J. Physiol. (Lond.)*, 304:471–488.
8. Eisner, D. A., Lederer, W. J., and Vaughan-Jones, R. D. (1981): The dependence of sodium pumping and tension on intracellular sodium activity in voltage-clamped sheep Purkinje fibres. *J. Physiol. (Lond.)*, 317:163–187.
9. Eisner, D. A., Lederer, W. J., and Vaughan-Jones, R. D. (1982): The effects of extracellular calcium on the intracellular pH and Na activity of sheep cardiac Purkinje fibres. *J. Physiol. (Lond.)* (in press).
10. Ellis, D. (1977): The effects of external cations and ouabain on the intracellular sodium activity of sheep heart Purkinje fibres. *J. Physiol. (Lond.)*, 273:211–240.
11. Endo, M., and Kitizawa, T. (1978): Excitation-contraction coupling in chemically skinned cardiac muscle. In: *Proceedings of the VIII World Congress of Cardiology*, edited by S. Hayase and S. Murao, pp. 800–803. Excerpta Medica, Amsterdam.
12. Fabiato, A., and Fabiato, F. (1978): Effects of pH on the myofilaments and the sarcoplasmic reticulum of skinned cells from cardiac and skeletal muscles. *J. Physiol. (Lond.)*, 276:233–255.
13. Horackova, M., and Vassort, G. (1979): Sodium-calcium exchange in regulation of cardiac contractility. Evidence for an electrogenic, voltage-dependent mechanism. *J. Gen. Physiol.*, 73:403–424.
14. Kass, R. S., Lederer, W. J., Tsien, R. W., and Weingart, R. (1978): Role of calcium ions in transient inward currents and aftercontractions induced by strophanthidin in cardiac Purkinje fibres. *J. Physiol. (Lond.)*, 281:187–208.
15. Kass, R. S., Tsien, R. W., and Weingart, R. (1978): Ionic basis of transient inward current induced by strophanthidin in cardiac Purkinje fibres. *J. Physiol. (Lond.)*, 281:209–226.
16. Mullins, L. J. (1981): *Ion Transport in Heart*. Raven Press, New York.
17. Pitts, B. J. R. (1979): Stoichiometry of sodium calcium exchange in cardiac sarcolemmal vesicles. *J. Biol. Chem.*, 254:6232–6235.
18. Reuter, H. (1974): Exchange of calcium ions in the mammalian myocardium. Mechanisms and physiological significance. *Circ. Res.*, 34:599–605.
19. Reuter, H., and Seitz, N. (1968): The dependence of calcium efflux from cardiac muscle on temperature and external ion composition. *J. Physiol. (Lond.)*, 195:451–470.
20. Sheu, S.-S., and Fozzard, H. A. (1982): Transmembrane Na^+ and Ca^{2+} electrochemical gradients in cardiac muscle and their relationship to force development. *J. Gen. Physiol.*, 80:325–351.

Electrogenic Transport: Fundamental Principles and Physiological Implications, edited by Mordecai P. Blaustein and Melvyn Lieberman. Raven Press, New York © 1984.

Photoelectric Properties of the Light-Driven Proton Pump Bacteriorhodopsin

*E. Bamberg,[1] *A. Fahr, and **G. Szabo

*Universität Konstanz, Fakultät für Biologie, D-755 Konstanz, West Germany; and **University of Texas Medical Branch Department of Physiology and Biophysics, Galveston, Texas 77550*

Bacteriorhodopsin, a membrane protein of *Halobacterium halobium*, acts as a light-driven proton pump (25). In the intact organism, it creates a light-induced transmembrane proton gradient that may be used for adenosine triphosphate (ATP) synthesis or oxidative phosphorylation. In bacterial membranes bacteriorhodopsin is present in the form of two-dimensional crystalline arrays, the so-called purple membrane. It is a chromoprotein in which the chromophore retinal is linked by a Schiff base to a lysine residue at position 216 (3,20). Considerable attempts have been made to elucidate the structure and photochemistry of the purple membrane (for reviews see refs. 13,28,33,34). Most of these studies considered the photochemical reactions of the photocycle of isolated purple membranes suspended in aqueous solution (see refs. 7,9,21,24–26). Although these studies reveal the existence of a number of intermediate steps in the overall photocycle, the connection between single-reaction steps and the proton transfer is not completely clear.

An alternative approach to this problem consists in orienting purple membranes or vesicles reconstituted with purple membranes at an interface and recording the photoelectric response that follows the excitation of the chromoprotein by light. Various systems and procedures have been used for this purpose, for example, orienting purple membrane sheets at the hydrocarbon–water interface (17) or binding lipid vesicles containing purple membranes to lipid-impregnated membrane filters (5,6) or to thick films (10,11). A number of studies have been carried out with planar lipid bilayer membranes (8,14,15,18,31).

The purpose of this study was to use bacteriorhodopsin as a model for electrogenic pumps. Different techniques of membrane electrical measurements, for example, stationary, kinetic, and fluctuation analysis of membrane currents, can be applied to this model system. Such studies are expected to reveal the mechanisms by which protons are translocated across purple membranes.

---

[1] Present address: Max-Planck Institut für Biophysik, Heinrich-Hoffmann Str. 7, D-6000 Frankfurt, West Germany.

MATERIALS AND METHODS

The experiments reported here were performed on a sandwich-like structure formed by purple membrane absorbed to planar lipid bilayers (1). In some experiments phospholipid vesicles incorporating bacteriorhodopsin were absorbed to planar lipid bilayers (2,14).

STATIONARY PHOTOCURRENTS

When a suspension of purple membranes is added to one of the aqueous compartments bathing a black lipid membrane, photosensitivity develops within minutes. Typical records of short-circuit current observed under these circumstances are shown in Figs. 1A and B. The data shown in Fig. 1A were obtained with an undoped lipid bilayer that has an extremely low conductance (of the order of 10 nS/cm^2) and it is practically proton impermeable. In all experiments the sign of the photocurrent was the same, upward deflections corresponding to proton transfer toward the bacteriorhodopsin-free side. In the second part of the experiment, the membrane was doped with gramicidin A, which forms channels permeable to alkali metal cations and protons. For the composition of the aqueous phase in these experiments (0.1 M MgCl$_2$; 0.5 mM Tris, pH 7.2), H$^+$ is the only permeant species. Figure 1B illustrates that in the presence of gramicidin A, a large stationary photocurrent is observed. Under steady illumination such a photocurrent may be observed for at least 20 min. That is to say, continuous, steady-state proton pumping is observed in these experiments.

Figure 2 compares the action spectrum of the stationary photocurrent I_∞ with the absorption spectrum of purple membrane. An excellent agreement is seen indicating that I_∞ is, indeed, generated by bacteriorhodopsin. A similar agreement between absorption spectrum and the action spectrum of the photovoltage has been reported by Dancshazy and Karvaly (8). Comparable results are obtained when lipid vesicles reconstituted with purple membrane are absorbed to a lipid bilayer (for details see ref. 2). That both initial photocurrent, observed after switching on the light, and the stationary photocurrent, I_∞, saturate with increasing light intensity is illustrated in Fig. 3, where the reciprocal values for I_0 and I_∞ are plotted as a function of the reciprocal light intensity, J. I_0 and I_∞ can be described by the simple relationship:

$$I = I^s \frac{J}{J + J^{1/2}} \qquad (1)$$

where I^s is the saturation current and $J^{1/2}$ is the half-saturating light intensity, usually about 5 mW/cm^2. The data of Figs. 1A and B are consistent with the notion that purple membrane is only attached to the underlying lipid bilayer, forming a sandwich-like structure. Under these circumstances stationary photocurrents can be obtained only by increasing the proton permeability of the

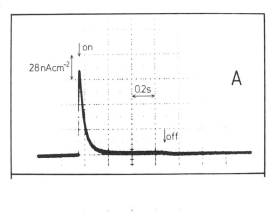

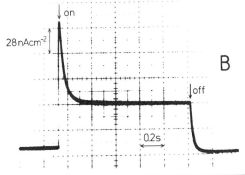

FIG. 1. Short-circuit photocurrent after addition of purple membrane fragments to one aqueous compartment. The aqueous phase contained 0.1 M $MgCl_2$ and 0.5 mM Tris, pH 7.0. The light was filtered through a K 4 filter (Balzers, λ_{max} = 550 nm), yielding a light intensity in the plane of the membrane of about 5.0 mW/cm². The area of the black film was 1.8 mm². **A:** Photocurrent in the absence of gramicidin. The sign of the photocurrent corresponds to a proton transfer toward the bacteriorhodopsin-free side. **B:** Photocurrent after addition of 30 nM gramicidin A to the bacteriorhodopsin-free compartment. After switching off the light the current falls to a slightly negative value and approaches zero level (*horizontal trace* at *left*) within a few seconds. The record was made about 30 min after the addition of gramicidin; the higher amplitude of the initial current I_o in record **B** as compared with record **A** is presumably due to the binding of additional purple membrane during this time (a similar increase of I_o is observed in the absence of gramicidin). The dark conductance of the membrane after addition of gramicidin was about 100 nS/cm².

supporting lipid bilayer. The proposed structure is shown schematically in Fig. 4A.

Figure 4B shows the electrical equivalent circuit diagram for this system. When the supporting bilayer has a very low conductance G_m, only transient photocurrents arise as a result of the loading of capacitors C_p and C_m. Stationary photocurrents may be induced by increasing G_m, for example, by the addition of a protonophore or gramicidin A. The properties of this transport system have been characterized in the following experiments.

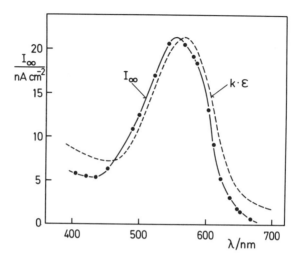

FIG. 2. Action spectrum of the stationary photocurrent I_∞, as measured with a series of narrow-band interference filters. The experimental conditions were similar to those of Fig. 1B. Correction factors accounting for the emission spectrum of the lamp and the transmission of the filters were obtained by calibration with a bolometer. The action spectrum was normalized to equal quantum flux density. *Dashed line* represents the extinction coefficient ϵ of the purple membrane (28), normalized to equal peak height by multiplication with a constant factor k.

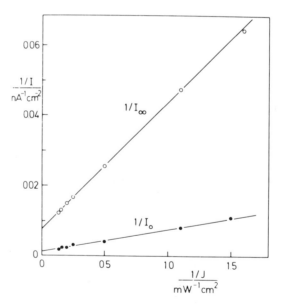

FIG. 3. Reciprocal values of the initial current I_o and of the stationary current I_∞ as a function of the reciprocal light intensity. White light was filtered by a K 4 filter (550 nm). The light intensity was varied by calibrated gray filters. Similar conditions as described for Figs. 1A and B.

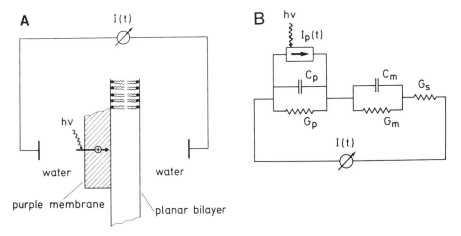

FIG. 4. A: Proposed structure of the photocurrent generator. **B:** Equivalent circuit diagram of the proposed photocurrent generator. G_p and C_P are the conductance and capacitance of the purple membrane, respectively. G_m and C_m are the corresponding values of the underlying part of the black film. Under short-circuit conditions the uncovered areas of the black film can be omitted from the equivalent circuit.

DETERMINATION OF THE ORIENTATION OF THE TRANSIENT MOMENT OF CHROMOPHORE RETINAL WITH RESPECT TO THE PLANE OF THE MEMBRANE

When the exciting light is linearly polarized, the rate of absorption is proportional to the average value of $\cos^2\psi$, where ψ is the angle between the vector of the radiation and the transition moment of the chromophore. Thus, at small light intensities (outside the saturation region) the photocurrent should be proportional to $\cos^2\psi$. The experimental situation is shown in Fig. 5. According to calculation of Steinemann et al. (32), the relationship between the tilt angle

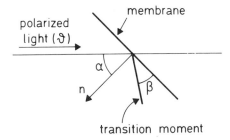

FIG. 5. Excitation with the polarized light. n is the normal to the membrane, θ is the angle between the electrical vector of the light and the plane of incidence (the plane formed by n and the light beam).

β between transition moment of the chromophore and the membrane is given by:

$$\frac{I(\theta)}{I(45°)} = 1 + \frac{3 \sin^2\beta - 1}{3 - \sin^2\beta} \cos 2\theta \qquad (2)$$

$I(\theta)$ is the photocurrent, determined at a certain angle θ of the polarized light. $I(45°)$ is the photocurrent for $\theta = 45°$. θ, $I(\theta)$, and $I(45°)$ are measurable quantities, so that β can be obtained from the slope plotting $I(\theta)/I(45°)$ versus $\cos 2\theta$ (Fig. 6).

A value of 25° was obtained for the tilt angle β (1), a value that is in good agreement with results of dichronic experiments (16,19).

APOMEMBRANE EXPERIMENTS

When a suspension of apomembrane is made to absorb to the bilayer membrane under the same conditions as those used in purple membrane experiments, only a small photosensitivity develops in a 3- to 4-hr time period. This residual photoeffect probably results from a small amount of unbleached bacteriorhodopsin remaining in the preparation. When all-*trans*-retinal is added to the aqueous

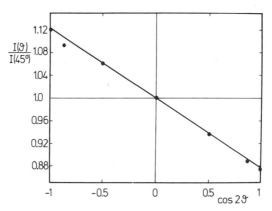

FIG. 6. Photoelectric effect with plane polarized light. $I(\theta)$ is the steady-state photocurrent I_∞ at low light intensity (0.05 mW/cm²), and θ is the angle between the electrical vector of the light and the plane of incidence (Fig. 5). The light was filtered through a Balzers K 4 filter ($\lambda_{max} \cong 550$ nm). To avoid light scattering in the cuvette, the light beam was focused and centered on the membrane plane. The diameter of the light beam was about 0.8 mm, whereas the membrane diameter was 1.8 mm. θ was varied by rotating the Polaroid filter that was interposed between light source and membrane. The aqueous solutions contained 0.1 M $MgCl_2$ and 0.5 mM Tris, pH 7.0. Gramicidin A, 30 nM, was added to the bacteriorhodopsin-free compartment. The level of the aqueous phases on both sides of the membrane was carefully controlled in order to keep the membrane planar. The experimental data were corrected for the residual polarization of the light source and the optical system as follows. The membrane was replaced by a bolometer and the bolometer signal $S(\theta)$ was measured as a function of the rotation angle (θ) of the Polaroid filter. The ratio $I(\theta)/I(45°)$ was corrected by multiplying the measured value of $I(\theta)/I(45°)$ by the factor $S(45°)/S(\theta)$. The correction was opposite to the polarization effect on the photocurrent and the deviation of $S(45°)/S(\theta)$ from unity was less than 0.06. $I(\theta)$ was symmetrical with respect to $\theta = 0$ within the limits of experimental error; the data points represent the average of $I(+\theta)$ and $I(-\theta)$.

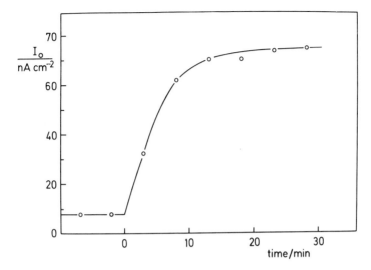

FIG. 7. Reactivation of photoactivity of the bleached membrane. At time $t = -3.5$ hr the protein was added to one aqueous phase to the final concentration of 5 μg/ml. The aqueous solution contained 0.1 M $MgCl_2$, 0.5 mM Tris, pH 7.0.

compartment containing the apomembrane, a pronounced photosensitivity develops within about 10 min as shown in Fig. 7. Addition of all-*trans*-retinal without prior addition of apomembrane has no effect. In agreement with previous findings using membrane suspensions (4,27), this experiment demonstrates that the photosensitivity of the bleached membrane is restored by addition of retinal. The reaction is much faster than the development of photosensitivity following the addition of a purple membrane suspension. This strongly suggests that apomembrane bound to the planar bilayer is reactivating *in situ*. It should be possible to use this technique, therefore, to explore effects of replacing retinal by its analogs.

KINETIC EXPERIMENTS

Following excitation by a laser flash, bacteriorhodopsin returns to its ground state through a photocycle (23) by passing through various spectroscopically identified intermediates. The photocycle can be considered as a sequence of monomolecular reactions shown in a simplified form in Fig. 8. The step from state K to M is probably responsible for proton uptake, whereas proton release may be attributed to the decay of the M intermediate. Fast photocurrent relaxation experiments on lipid bilayers covered with purple membrane should reveal how the proton transfer within the purple membrane is related to the photocycle.

Kinetic experiments were performed on the previously described sandwich-like structures. As can be seen from Figs. 1A, 4A, and 4B, the characteristic charging time of the capacitors formed by the two membranes in series is about 0.2 sec, indicating that one should be able to measure relaxations of the photo-

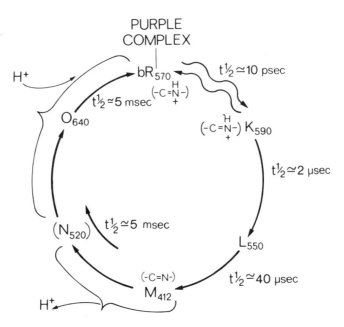

FIG. 8. Simplified photocycle of bacteriorhodopsin (13). The numbers on the intermediate produces indicate its absorption wavelength.

current at times shorter than 10 msec. At these times the capacitance of the underlying black lipid membrane acts as a short-circuit element. Kinetics of the light-induced charge displacement were measured by recording the short-circuit current on microsecond as well as millisecond time scales, following excitation of the bilayer purple membrane system with a 10 μsec laser flash (560-nm wavelength). Time resolution was limited to 0.5 μsec by the bandwidth of the current amplifier (for details see ref. 12). The time course of the photocurrent is shown in Fig. 9. The polarity of the current is as defined in Fig. 1A, that is, negative currents correspond to proton movement under steady-state illumination. During the first 10 μsec following the flash, the photocurrent exhibits a negative component. The current then rises toward positive values and finally decays toward zero. The finite slope of the falling phase of $I(t)$ reflects the time resolution of about 3 μsec of the amplifier in this particular experiment. The shape of $I(t)$ can be represented by a sum of 4 exponentials:

$$I(t) = \sum_{i=1}^{4} a_i \exp(-t/\tau_i) \tag{3}$$

Values of the time constants τ_i obtained from a computer fit are given in Table 1. The values obtained in D_2O solution were determined under similar conditions.

The evaluation of the data shown in Table 1 suggests that the four relaxation times can be assigned to intermediates in the photocycle. Interestingly, τ_2 was

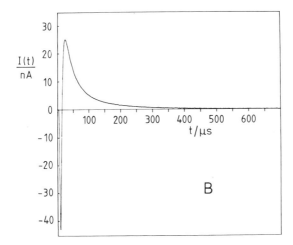

FIG. 9. Photocurrent $I(t)$ after a 10-msec laser flash at 575 nm. The curve has been directly reproduced from the original recorder plot. The polarity of the current is defined as indicated in Fig. 1., i.e., $I > 0$ corresponds to movement of positive charge toward the supporting membrane. The finite slope of the falling phase of $I(t)$ within the first 5 μsec is an artifact caused by the limited bandwidth of the amplifier. The signal represents an average over 3,000 single excitations. A qualitatively identical signal but superimposed with considerable noise is obtained in a single-flash experiment. The light intensity of 2 μJ corresponds (with an illuminated membrane area of 0.50 mm²) to 4 mJ/cm². The temperature was $T = 22°C$. Positive-peak currents at light saturation reached about 50 nA (~ 10 μA/cm²) at 22°C. For times > 10 μsec a highly accurate fit of the experimental $I(t)$ curve is obtained according to Eq. (3) with $\tau_1 = 3.9$ μsec, $\tau_2 = 21.0$ μsec, $\tau_3 = 64.4$ μsec, $\tau_4 = 910$ μsec, $a_1 = -708$ nA, $a_2 = 56.8$ nA, $a_3 = 17.6$ nA, and $a_4 = 1.23$ nA. The fitted curve (not shown) is almost indistinguishable by eye from the experimental curve for $t > 10$ μsec (apart from the noise components).

proposed by spectroscopic measurements to be in the photocycle but could not be described experimentally with absolute certainty (22). Our data have confirmed these results very well. Considering the amplitude of the 4 relaxation processes, it can be concluded that the process connected with τ_1 probably causes different charge movement processes than the following three processes. τ_1 can be assigned to K → L transition in the photocycle. The absence of an isotope effect on τ_1 and the negative sign of the amplitude suggests that this process is connected with a charge movement due to a conformational change of the protein and not with the movement of a proton or a deuterium ion. The other three relaxation processes with positive amplitude may be due to

TABLE 1. *Time constants τ_i at 25° of the photocurrent after excitation of the purple membrane lipid bilayer system with a 10-μsec, 560-nm laser flash*

	τ_1 (sec)	τ_2 (sec)	τ_3 (sec)	τ_4 (sec)
H_2O	1.25×10^{-6}	1.69×10^{-5}	0.58×10^{-4}	9.1×10^{-4}
D_2O	1.25×10^{-6}	4.3×10^{-5}	1.6×10^{-4}	1.56×10^{-3}

the proton movement within the purple membrane. The marked isotope effect of τ_2 and τ_3 is consistent with the notion that processes 2 and 3 are associated with proton dissociation from the binding site. Thus, τ_2 and τ_3 can be correlated with the L → M transition in the photocycle. The slowest component of the photocurrent transition has a time constant $\tau_4 = 950$ μsec. The M → N transition has been determined spectroscopically to be 1 msec under similar conditions. This process is probably involved in the rebinding of the proton to bacteriorhodopsin. This contention is consistent with the observation that τ_4 increases with the pH of the medium (12).

FLUCTUATION ANALYSIS OF PHOTOCURRENTS

Following excitation by a quanta of light, bacteriorhodopsin relaxes to its resting state. The end result of this process is the translocation of one or several protons across the purple membrane (34). The macroscopic photocurrent I arises from the summation of minute pulses of current $i(t)$ corresponding to the photon-induced net transfer of Q unitary charges (e.g., protons) across the purple membrane. For a purple membrane having a packing density of N bacteriorhodopsin molecules that are photoactivated k times per sec, the macroscopic photocurrent is given by

$$I = N k Q e \quad (4)$$

where e is the electrical change of a proton. Individual bacteriorhodopsin molecules are activated by light at random. This results in fluctuations of the photocurrent about its steady-state mean value I as, by chance, fewer or more of the bacteriorhodopsin molecules are activated. This phenomenon is analogous to the well-known shot noise generated by photodetectors (37). It can be characterized by its spectral density $S_i(f)$, which measures the intensity of current fluctuations having a frequency f. In the limit of low frequencies, that is, at frequencies below that of the relaxation features shown by bacteriorhodopsin, the spectral density of photocurrent fluctuations is given by the well-known formula for shot noise (29,37)

$$S_i = 2Q\, eI \quad (5)$$

The photocurrent I and the spectral density of its fluctuation S_i should be accessible to experimental measurement so that Q, the number of protons transported during one photocycle, may be determined directly.

$$Q = S_i / 2eI \quad (6)$$

Considerations such as these indicate that fluctuations of the photocurrent contain a wealth of information concerning the molecular events that occur during active proton transport.

Fluctuation measurements were carried out using the previously described sandwich-like structure formed by oriented purple membranes absorbed to a lipid bilayer membrane. The proton conductance of the underlying bilayer was carefully adjusted by the addition of carbonylcyanide-trifluoromethoxy-phenylhydrazone (FCCP) so that nearly all of the steady-state photocurrent could flow through the bilayer without being shunted by areas of membrane not covered by purple membrane. Furthermore, the buildup of proton concentration in the region near the membrane and in the spaces between the bilayer and the purple membrane was minimized by the addition of monensin (35), an electrically neutral exchanger of protons and cations (30). Typical photoresponse obtained under these circumstances is shown in Fig. 10. Note that the photocurrent is time invariant and large (0.25 μA/cm^2). Test pulses ($\pm$ 5 mV) were applied across the membrane, usually held at 0 mV, in order to measure the conductance of the system. Under the conditions of the experiment, the supporting bilayer is more conductive ($>$10 μS/cm^2) than the sandwich structure (1.2 μS/cm^2 in the dark). Thus, to a first approximation, the data of Fig. 10 show that illumination increases the conductance of the purple membrane from 1.2 μS/cm^2 to 5.2 μS/cm^2.

Illumination produces a photocurrent that fluctuates. These small but well-

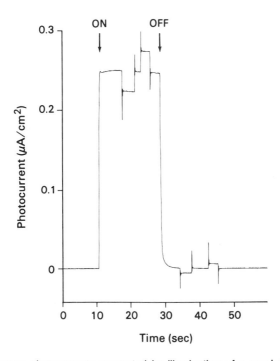

FIG. 10. Steady-state photocurrents generated by illumination of a sandwichlike structure. Positive current as defined in Fig. 1. Purple membranes were added to one side only of a bilayer formed from a dispersion of diphytanoylphosphatidylcholine and octadecylamine in decane (for details, see ref. 1). The 0.1 M NaCl aqueous solutions buffered to pH 5.6 with 0.01 M citrate also contained 4 μM FCCP and 40 μM monensin. The membrane potential was held at 0 mV except during the application of $\pm$ 5-mV test pulses used to determine membrane conductance. In the absence of purple membranes, photocurrents or conductance changes were absent.

defined fluctuations are most evident when purple membrane is present at both sides of the membrane. Photocurrents flowing in opposite directions across the bilayer membrane tend to cancel under these circumstances, thereby facilitating the observation of fluctuations. An experiment of this type is shown in Fig. 11. A pronounced decrease in current fluctuations is evident when illumination ceases. Figure 12 shows that the spectral density of the light-induced current fluctuations is practically frequency independent (white), as expected for shot noise generated by random photoactivation of bacteriorhodopsin molecules. Similar but less prominent fluctuations are present in photocurrents generated in experiments of the kind illustrated in Fig. 10, when purple membrane is applied to only one side of the bilayer membrane. Specifically, a light-induced noise of 3.8×10^{-25} A^2 sec/cm^2 is measured for the experiment of Fig. 10. Note that part of this noise may be contributed by excess thermal noise of the system generated by the increased membrane conductance observed on illumination. For the data of Fig. 10, this excess thermal noise can be calculated using the Nyquist formula $\Delta S = 4kT \Delta G$ where ΔG is the light-induced conductance change. For $\Delta G = 4.0$ $\mu S/cm^2$ $\Delta S = 6.5 \times 10^{-26}$ A^2 sec/cm^2. Thus the light-induced transport noise is 3.8×10^{-25} to $6.5 \times 10^{-26} = 3.1 \times 10^{-25}$ A^2 sec/cm^2.

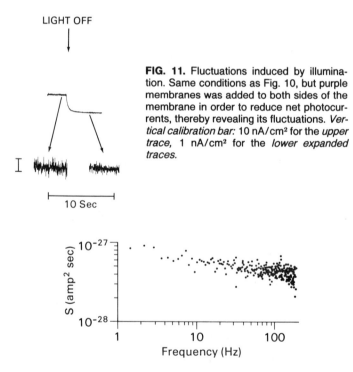

FIG. 11. Fluctuations induced by illumination. Same conditions as Fig. 10, but purple membranes was added to both sides of the membrane in order to reduce net photocurrents, thereby revealing its fluctuations. *Vertical calibration bar:* 10 nA/cm^2 for the *upper trace*, 1 nA/cm^2 for the *lower expanded traces.*

FIG. 12. Spectral density of the current fluctuations. Same experiment as in Fig. 11. The difference spectrum was obtained by subtracting the dark spectra from that observed in the presence of illumination.

The stoichiometry of the pump, that is, the number Q of photons translocated per photocycle, can be estimated using Eq. (6).

$$Q = 3.1 \times 10^{-25}/2 \times 1.6 \times 10^{-19} \times 0.25 \times 10^{-6} = 3.9$$

Note that the photocurrent I is probably underestimated in our measurements because of the shunting effects of lateral leakage pathways between patches of purple membranes. The value of 3.9 for Q is therefore an overestimate. It is likely that fewer than four protons are transported in a photocycle. Others, using different techniques, have estimated that $Q \sim 2$ or less (see ref. 1). The reasonableness of our estimate for Q is comforting. It reinforces confidence in the technically difficult fluctuation measurements. It is likely that further refinements of the technique will allow us to determine the value of Q with greater precision. Similar fluctuation measurements, applied to systems of active ion transport, may yield crucial information concerning the stoichiometery of active ion transport.

ACKNOWLEDGMENTS

This work was supported by the Deutsche Forschungsgemeinschaft and DHHS grant HL-24820.

REFERENCES

1. Bamberg, E., Apell, H.-J., Dencher, N.-A., Sperling, W., and Läuger, P. (1978): Photocurrents generated by bacteriorhodopsin on planar lipid membranes. *Biophys. Struct. Mech.*, 5:277–292.
2. Bamberg, E., Dencher, N.-A., Fahr, A., and Heyn, M. P. (1981): Transmembraneous incorporation of bacteriorhodopsin into lipid bilayer membranes. *Proc. Natl. Acad. Sci. USA*, 78:7502–7506.
3. Bayley, H., Huang, K.-S., Radhakrishnan, R., Ross, A. H., Takagaki, Y., and Khorana, H. G. (1981): Site of attachment of retinal in bacteriorhodopsin. *Proc. Natl. Acad. Sci. USA*, 78:2225–2229.
4. Becher, B., and Cassim, J. V. (1977): Effects of bleaching and regeneration of the purple membrane structure of Halobacterium halobium. *Biophys. J.*, 19:285–297.
5. Blok, M. C., Hellingwerf, K. I., and Van Dam, K. (1977): Reconstitution of bacteriorhodopsin in a Millipore filter system. *FEBS Lett.*, 76:45–50.
6. Blok, M. C., and Van Dam, K. (1978): Association of bacteriorhodopsin containing phospholipid vesicles with phospholipid impregnated Millipore filters. *Biochim. Biophys. Acta*, 507:48–61.
7. Campion, A., El-Sayed, M. A., and Terner, J. (1977): Resonance raman spectroscopy of bacteriorhodopsin in the microsecond time scale. *Biophys. J.*, 20:369–375.
8. Dancshazy, Z., and Karvaly, B. (1976): Incorporation of bacteriorhodopsin into a bilayer lipid membrane; a photoelectric spectroscopic study. *FEBS Lett.*, 72:136–138.
9. Dencher, N., and Wilms, M. (1975): Flash photometric experiments on the photochemical cycle of bacteriorhodopsin. *Biophys. Struct. Mech.*, 1:259–271.
10. Drachev, L. A., Kaulen, A. D., and Skulachev, V. P. (1978): Time resolution of intermediate steps in the bacteriorhodopsin-linked electrogenesis. *FEBS Lett.*, 87:161–167.
11. Drachev, L. A., Jasaitis, A. A., Kaulen, A. D., Kondrashin, A. A., Libermann, E. A., Nemecek, I. B., Ostroumov, S. A., Semenov, A. Yu., and Skulachev, V. P. (1974): Direct measurement of electric current generation by cytochrome oxidase, H^+-ATPase and bacteriorhodopsin. *Nature*, 249:321–324.
12. Fahr, A., Läuger, P., and Bamberg, E. (1981): Photocurrent kinetics of purple membrane sheets bound to planar lipid bilayer membranes. *J. Membr. Biol.*, 60:51–62.

13. Henderson, R. (1977): The purple membrane from Halobacterium halobium. *Annu. Rev. Biophys. Bioeng.*, 6:87–109.
14. Herrmann, T. R., and Rayfield, G. W. (1976): A measurement of the proton current generated by bacteriorhodopsin in black lipid membranes. *Biochim. Biophys. Acta*, 43:623–628.
15. Herrmann, T. R., and Rayfield, G. W. (1978): The electrical response to light of bacteriorhodopsin in planar membranes. *Biophys. J.*, 21:111–125.
16. Heyn, M. P., Cherry, R. J., and Müller, U. (1971): Transient and linear dichroism studies on bacteriorhodopsin. Determination of the orientation of the 568 nm all-trans retinal chromophore. *J. Mol. Biol.*, 117:607–620.
17. Hwang, S.-B., Korenbrot, J. I., and Stoeckenius, W. (1977): Proton transport by bacteriorhodopsin through an interface film. *J. Membr. Biol.*, 36:137–158.
18. Karvaly, B., and Dancshazy, Z. (1977): Bacteriorhodopsin: A molecular photoelectric regulator. Quenching of photovoltaic effect on BLM containing bacteriorhodopsin. *FEBS Lett.*, 76:36–40.
19. Korenstein, R., and Hess, B. (1978): Immobilization of bacteriorhodopsin and orientation of its transition moment in purple membrane. *FEBS Lett.*, 89:15–20.
20. Lemke, H.-D., and Oesterhelt, D. (1981): Lysine 216 is a binding site of the retinyl moiety in bacteriorhodopsin. *FEBS Lett.*, 128:255–260.
21. Lewis, A., Spoonhower, J., Bogomolni, R. A., Lozier, R. H., and Stoeckenius, W. (1974): Turnable laser resonance raman spectroscopy of bacteriorhodopsin. *Proc. Natl. Acad. Sci. USA*, 71:4462–4466.
22. Lewis, A. (1978): The molecular mechanism of excitation in visual transduction and bacteriorhodopsin. *Proc. Natl. Acad. Sci. USA*, 76:549–552.
23. Lozier, R. H., Bogomolni, R. A., and Stoeckenius, W. (1975): Bacteriorhodopsin: a light driven proton pump in Halobacterium halobium. *Biophys. J.*, 15:955–962.
24. Lozier, R. H., Niederberger, W., Bogomolni, R. A., Hwang, S.-B., and Stoeckenius, W. (1976): Kinetics and stoichiometry of light induced proton release and uptake from purple membrane fragments, Halobacterium halobium cell envelopes, and phospholipid vesicles containing oriented purple membranes. *Biochim. Biophys. Acta*, 440:545–556.
25. Oesterhelt, D., and Stoeckenius, W. (1973): Functions of a new photoreceptor membrane. *Proc. Natl. Acad. Sci. USA*, 70:2853–2857.
26. Oesterhelt, D., and Hess, B. (1973): Reversible photolysis of the purple complex in the purple membrane of Halobacterium halobium. *Eur. J. Biochem.*, 37:316–326.
27. Oesterhelt, D., Schuhmann, L., and Gruber, H. (1974): Light dependent reaction of bacteriorhodopsin with hydroxylamine in cell suspensions of Halobacterium halobium: Demonstration of an apomembrane. *FEBS Lett.*, 44:257–261.
28. Oesterhelt, D. (1976): Isoprenoids and bacteriorhodopsin in halobacteria. *Prog. Mol. Subcell. Biol.*, 4:133–166.
29. Papoulis, A. (1965): *Probability, Random Variables and Stochastic Processes*. McGraw-Hill, New York.
30. Pressman, B. C., and deGuzman, N. (1975): Biological application of ionophores: Theory and practice. *Ann NY Acad. Sci.*, 264:373–385.
31. Shieh, P., and Packer, L. (1976): Photoinduced potentials across a polymer stabilized planar membrane in the presence of bacteriorhodopsin. *Biochem. Biophys. Res. Commun.*, 71:603–609.
32. Steinemann, A., Stark, G., and Läuger, P. (1972): Orientation of the porphyrin ring in artificial chlorophyll membranes. *J. Membr. Biol.*, 5:177–194.
33. Stoeckenius, W., Lozier, R. H., and Bogomolni, R. A. (1979): Bacteriorhodopsin and the purple membrane of halobacteria. *Biochim. Biophys. Acta*, 505:215–278.
34. Stoeckenius, W., and Bogomolni, R. A. (1982): Bacteriorhodopsin and related pigments of halobacteria. *Annu Rev. Biochem.*, 52:587–616.
35. Szabo, G., and Bamberg, E. (1981): Fluctuations of the photocurrent associated with light induced proton transport in purple membranes. In: *VII International Biophysics Congress*, p. 146 (abstr.). Mexico City, Mexico.
36. Trissl, H.-W., and Montal, M. (1977): Electrical demonstration of rapid light-induced conformational changes in bacteriorhodopsin. *Nature*, 266:655–657.
37. van der Ziel, A. (1977): Noise in solid state devices and lasers. In: *Electrical Noise: Fundamentals and Sources*, edited by M. S. Gupta, pp. 237–265. IEEE Press, New York.

Subject Index

Subject Index

Active transport; *see also* Electrogenic
  pump/ion transport; Primary active
  transport; Secondary active transport
  in cardiac muscle cells, 181–189
  characteristics of, 51, 307
  electrochemical gradients in, 162–163,
    181–182, 215
  energetic considerations in, 50, 54–55
  equilibrium conditions in, 51–55
  ion flux in, 105, 107; *see also specific ions
    by name*
  kinetic favorability in, 57–59
  negative free-energy change in, 50
  primary vs. secondary, 49, 93–94
  by protons, 307
  rate of, 54–59, 62–67; *see also* Solute
    transport rate
Adenosine diphosphate, 4
Adenosine triphosphatase (ATPase)
  calcium-activated, 130, 139
  chloride ion stimulation of, 337–338
  hydrogen activated, *see* Proton-activated
    ATPase
  mitochondrial, 340
  in plasma membrane vs. tonoplasts,
    339–340
  proton-activated, *see* Proton-activated
    ATPase
  sodium-potassium activated, *see*
    Sodium + potassium-activated
    ATPase
Adenosine triphosphate (ATP)
  in calcium-calcium exchange, 135–138
  in calcium uncoupled extrusion, 138–139,
    142
  chromaffin ghost permeability to, 77
  chromaffin granule membrane permeability
    to, 74, 75
  proton-activated ATPase responses to,
    79–80
  in sodium-calcium exchange activation,
    133–135, 149, 151–152, 157–158
  in sodium-calcium exchange
    dephosphorylation, 153–154
ATP-dependent membrane transport
  chloride ion (Cl^-) effects on, 334–339
  in plant root microsomal preparations,
    332, 334–338
  in squid giant axons, 40

ATP hydrolysis
  chemiosmotic mechanism role of, 17, 72
  free energy of, 3–4, 17
  in plant cells, 4
  in plant microsomal vesicles, 332
  as primary active transport mechanism, 49,
    72
  in proton pump energy transfer, 315
ATP synthesis, in chemiosmotic theory, 71
Ammonium-hydrogen ion exchange, 57
Amphiuma red cell membrane, 36
Aplysia neurons
  electrogenic pump effects in, 264
  membrane resistance in, 254, 257, 265
  vs. squid, 257
  temperature sensitivity of, 254–257
Artificial vesicles, *see* Vesicles

Bacteriorhodopsin
  as electrogenic pump, 381–393
  molecular characteristics of, 381
  photocycle developments in, 387–390
  stationary photocurrents in, 382
Balanus nubilis, 365–370; *see also* Barnacle
  muscle fibers
Barium ions, in sodium-potassium exchange
  determination, 40–41
Barnacle muscle fibers
  ATP-driven calcium extrusion in,
    138–139, 365–370
  ATP role in sodium-calcium exchange in,
    134–135
  sodium-calcium exchange in, 134–135,
    365–370; *see also* Sodium-calcium
    exchange
Basolateral membrane, 7
Bilayer lipid membranes; *see also* Vesicles
  bacteriorhodopsin in, 381–393
  electrogenicity in, 35–36
  purple membrane attachment to, 381,
    382–383
  sodium-potassium exchange ratio in, 40
Biological amines, chemiosmotic coupling
  and, 71–90
Biological potentials
  concept and theory of, 105–106; *see also*
    Membrane potentials
  Goldman-Hodgkin-Katz equation for, 106

395

Calcineurin, 149, 154–155, 157
Calcium-binding proteins, 129
Calcium-calcium exchange mode
  alkali cation activation of, 138
  calcium ion translocation by, 130–131, 135–138
  electroneutrality of, 135
  sodium-calcium exchange modulation by, 130–131, 135–138, 372
  sodium ion inhibition of, 135, 137–138
Calcium-dependent ATPase
  in calcium extrusion, 130–131
  in control of calcium ion transport, vs. sodium-calcium exchange, 40, 139–142
  in red blood cells, 130
  sodium-potassium exchange system role of, 130
Calcium ion(s)
  ATP in uncoupled extrusion of, 138–139, 142
  calcium-calcium exchange translocation of, 130–131, 135–138
  across cardiac sarcolemma, 239–249; see also Sarcolemmal vesicles
  carrier reversal potential in extrusion of, 167
  chromaffin granule membrane permeability to, 74, 75
  cytoplasmic regulation of, 129–130
  electrochemical gradient of, in sodium-calcium exchange, 131–132, 162–167, 368–370
  extrusion of, in sodium-calcium exchange, 130, 138–139, 142, 144–145, 162–163, 166–167, 366–370, 376–378
  influx mode/mechanisms, in sodium-calcium exchange, 130, 131, 162–170
  intracellular, 129, 140
  membrane bound, in sodium-calcium exchange, 149, 151–153, 158
  membrane potential effects on movement by, 162–170, 244–249, 374, 379
  membrane potential responses to, in sodium-calcium exchange, 167–170, 374, 379
  organelles modulating cytoplasmic, 129–130
  in sarcolemmal vesicles, see Sarcolemmal vesicles
  as second messenger, 129
  signal-to-noise ratio in influx of, 129
  smooth-muscle slow-wave response to, 275–283
  sodium-calcium exchange transport of, 130–131, 138–140, 142, 144–145, 161–167, 239–249, 277–283, 366–370, 376–378; see also Sodium-calcium exchange
  sodium ion inhibition of movement by, 135–139, 211–212
Calmodulin, 129, 130, 149–154
Canine red blood cells, sodium-calcium exchange paradox in, 131
Cardiac action potential, 163
  characterization of, problems in, 181, 215
  internal sodium stimulation by, 225–226
  sodium-potassium pump current and, 216, 224–225
Cardiac muscle cells
  ATP-driven calcium regulation in, 139, 142
  calcium ion in, 138–140, 161–166, 170–172
  calcium/sodium electrochemical gradient coupling in, 140, 161–167
  contractility of, and internal sodium concentration, 174–177
  electrochemical characteristics of cultured, 181–191
  membrane potential changes in, 194–196, 197–201, 218–220
  polystrand tissue culture preparations of, 183–189
  sodium-calcium exchange as electrogenic in, 122–124, 140, 149–158, 161–167, 171–177, 187–188, 239, 244–249, 273–280, 366
  sodium + potassium-activated ATPase in, 18–20
  sodium-potassium pump as electrogenic in, 194, 215–216, 224, 372
  sodium pump as electrogenic in, 25–26, 193–212, 218–220
  steady-state ion-flux measurements in, 184–185
  transmembrane ion transport in, 181–189
Cardioactive steroids
  sodium + potassium-activated ATPase modulation by, 267
  sodium-potassium pump stimulation by, 222
  sodium pump inhibition by, 196–197, 208, 212, 220–221, 255–256, 267
Carrier reversal potential, 167
Catecholamines, chromaffin granule uptake of, 72–73, 80–90
Cationic amine species, 88
Cell membrane conductance, see Membrane conductance
Cell membrane potential, see Membrane potential
Cell membrane translocation
  active, see Active transport; Electrogenic ion transport
  coupled active, see Secondary active transport

SUBJECT INDEX 397

coupled flow in, 103, 161
coupling ratio in, 107, 165, 166, 307
electrochemical driving force in, *see*
  Electrochemical driving force
electrochemical gradients in, *see*
  Electrochemical gradients
electrodiffusional fluxes in, 105–106, 108,
  181, 182
as electrogenic, *see* Electrogenic ion
  transport
ion flux equation for, 105
leakage conductance in, 309–315
leakage routes in, 95
positive direction for, 107
as rheogenic, 93–103; *see also* Rheogenic
  transport
secondary active transport in, *see*
  Secondary active transport
substrate/ion gradient intraconversion in,
  161
velocity effects in, 95
Cesium ion, 201, 300–301, 358–359
Chemiosmotic coupling hypothesis, 71–72,
  85, 323, 331
Chloride ion
  ATPase activity stimulation by, 337–338
  ATP-dependent proton transport and, 333,
    334–338
  in cardiac muscle cells, 185
  chromaffin granule permeability to, 74–75
  electrical neutrality of, 105–106, 119, 185
  electrogenic equation for pumping, 112,
    119–120
  intestinal slow-wave activity and, 275
  steady-state flux value, 185
  nerve cell current, 120
  plasma membrane vs. vesicle activation by,
    334
  proton transport stimulation by, 333–339
Chord conductance
  in animal vs. plant cell pumps, 6
  definition of, 5
  in sodium pump vs. cell membrane, 6–7
Chromaffin ghosts/vesicles
  catecholamine uptake by, 86–88
  chromaffin granules compared with, 77
  electrochemical membrane potential in,
    86–87
  pH development in, 77–78
  preparation of, 77
Chromaffin granules
  buffering capacity of, 74
  calcium ion uptake by, 74–75
  catecholamine uptake by, 73–74, 81–90
  chromaffin ghosts compared with, 77
  diagrammatic characterization of, 72
  electrochemical proton gradient (ΔpH)
    across, 74–75
  membrane-associated proteins of, 73
  permeability of, 74
  pH gradient in catecholamine uptake by,
    85–88
  proton-activated ATPase modulation of,
    78–80, 339–340
  transmembrane potential across, 74–75
Chromogranins, 73
Cold receptor, 267
Constant field equation, 106; *see also*
  Goldman-Hodgkin-Katz equation
Cotransport systems
  anion involvement in, 99–100
  binding-unbinding processes in, 96–97
  carrier translocation as rate-limiting in, 97
  cation participation in, 99
  channel mechanisms as colimiting in, 98
  glide symmetry in, 96
  incomplete complexes in, 95
  leakage pathways in, 96
  mirror symmetry in, 96
  pH lowering and, 99–100
  potassium ion effect in, 93, 101
  proton involvement in, 99
  rate-limiting step alternatives in, 96–97
  reaction sequences in, 96
  rheogenicity of, 93
  sodium ion-dependent, 93, 94, 98
  substrate-positive rheogenicity in, 101
  transport effective routes in, 95
Countertransport, 93
  in sodium-calcium exchange, 239
  translocation routes effective in, 95
Coupled flow
  as electrogenic, 103
  equation for, 103
Coupling ratio, 107, 165–166, 367, 370

Diffusion, passive uncoupled, 50

Electric charge
  in cotransport, 93
  in rheogenic transport, 94
Electrical neutrality
  of chloride ions, 105–106, 119, 185
  in membrane potential calculations,
    105–106, 108–109, 120
Electrical potential difference
  in cotransport rate-limiting steps, 97–98
  in positive rheogenic vs. electrogenic
    processes, 101–103
  in rheogenic transport activation, 94, 101
Electrical work, in cell membrane activity,
  36–37
Electrochemical driving force
  active transport rate vs., 54–55
  external cation effects and, 57–59
  sodium efflux vs. external cations and, 59
  solute transport rate vs., 53–54, 67–68

Electrochemical driving force *(cont.)*
  substrate binding site saturability vs., 56–67
  transportable allosteric solute as modifying, 59–61
Electrochemical gradients
  cardiac cell membrane types, 162–163, 181–182
  in embryonic cultured heart cells, 183, 184
  of protons, *see* Proton gradients
  in sodium-calcium exchanges, 131–132, 161–167, 368–370
Electrodiffusional fluxes, 105–106, 108, 181, 182
Electrogenic current, 3
Electrogenic pump/ion transport
  active transport as, 51, 307; *see also* Active transport
  ATP role in, 1–4, 17, 42, 72
  bacteriorhodopsin as model for, 381–393
  in bilayer lipid membranes, 35–36
  calcium ion role in, 162–163
  in cardiac muscle cell membranes, *see* Cardiac muscle cells
  chemiosmotic coupling hypothesis of, 71–72, 85, 323, 331
  chord conductance characteristics, 5–7
  classic type, characterized, 2–4
  current-voltage relationships for, 7–14, 345
  electromotive force in, 2–6
  equivalent circuit for, 2–3
  in *Escherichia coli*, 323–329
  in higher plants, 331–341
  in human red blood cells, 41, 287–292
  hyperpolarization and, *see* Hyperpolarization
  in intestinal smooth muscle cells, 271–283
  membrane conductance and, *see* Membrane conductance
  membrane potential in, *see* Membrane potential
  neuron excitability regulation by, 253–268
  pacemaker potential generation by, 272, 277, 282, 283
  in pancreatic beta cell membranes, 295–305
  physiological criteria for, in cultured cardiac muscle cells, 181–189
  potassium external depletion and, 33–35, 42–43, 194, 197–201, 208–216, 255, 272–275
  as primary active transport, 1–2, 49–51, 93–94
  proton electrochemical gradient in, 71, 161–162; *see also* Proton gradient
  proton pump as, *see* Proton pump
  rheogenic transport and, 93–103; *see also* Rheogenic transport
  secondary active transport and, *see* Secondary active transport
  slope conductance in, 8, 25–26
  smooth muscle, slow-wave activity and, 271–283; *see also* Smooth muscle(s)
  sodium-calcium exchange as, *see* Sodium-calcium exchange
  thermodynamic force applications to, 28–30
  voltage dependence in, 4–5, 7–8, 14, 21–25; *see also* Membrane currents; Membrane current-voltage relations
Electromotive force, 2–6
Electroneutral ion transport
  in calcium-calcium exchange, 135
  in cardiac cell membrane, 181, 182
  chloride ion and, 105–106, 119, 185
  equation for, 109
Equivalent circuits, in electrogenic transport, 2–3, 7
Erythrocytes, *see* Red blood cells
Escherichia coli
  beta-galactoside transport in, 323–329
  electrogenic transport in, 323–329
  lac carrier protein in, 323–329

Free energy change
  in passive diffusion, noncoupled, 50–51
  in secondary active transport, 51
Frog skin
  sodium/potassium coupling determination for, 40–41
  two-membrane hypothesis of, 40

β-Galactosides, 323
Glucose
  membrane potential response to, in pancreatic beta cells, 296–304
  potential dependence of sodium-dependent vs. sodium-independent, 62–64
  sodium cotransport of, *see* Sodium-glucose cotransport
Goldman-Hodgkin-Katz equation, 106

Halobacterium halobium, 381
Heart cells, *see* Cardiac muscle cells
Helix aspera, 353–362
Hevea brasiliensis, 333
Hydrogen-activated ATPase, *see* Proton-activated ATPase
Hydrogen ion, *see* Proton
Hyperpolarization, posttetanic, 33–35, 253

Ion transport, *see* Transmembrane ion transport

lac carrier protein, 323–329
  amino acid residue sequence for, 328
  antibody-induced inhibition of, 327–328

bilayer lipid membrane binding by, 326–327
hydropathic profile of, 328–329
lactose translocation by, 324–326
molecular structure of, 326, 328–329
  dual forms indicated for, 326–327
molecular weight of, 326, 327–328
lac permease, 323
lac y gene, 323, 324, 326
Lactose translocation, 324–326
Leakage conductance, in proton pump, 309–315
Leakage pathways, in secondary active transport, 95
Liposomes, 35–36; *see also* Bilayer lipid vesicles
Lithium-hydrogen ion exchange, 57
Lithium ions
  in calcium-calcium exchange, 138
  in sodium pump current reactivation, 201
  in sodium translocation, 57–59

Magnesium ion
  chromaffin granule permeability to, 74, 75
  in sodium-calcium exchange reactivation, 151–153, 157
Manganese ion, in calcium efflux reduction, 139
Membrane conductance
  electrogenic pump effects on, 8
  external calcium effects on, 377–379
  external sodium effects on, 366
  measurement of, in *Xenopus* blastomeres, 345–351
  pump chord conductance and, 5–7
  in sodium-calcium exchange, 366, 377–379
  total, 3
Membrane currents
  extracellular sodium effects on, in sodium-calcium exchange, 366–367, 373–380
  in intestinal smooth muscle activity, 272–282
  pacemaker activity and, 277
  sodium-potassium pump generation of, 22–24, 36–38, 44–45, 195–198, 215–216
Membrane current-voltage relations
  leakage conductance in, 309–315
  measurement of, 308–311, 345–351
  in *Neurospora*, 308–320
  pH change effects on, 311–315
  in proton pump, 308–320
  in *Xenopus* blastomeres, 345–351
Membrane potential
  ATP-dependent, in corn root vesicles, 334–335
  in binding of nontranslocated substances, 65–67

calcium ion movement effects on, 167–170, 374, 379
in catecholamine uptake by chromaffin ghosts, 86–87
chloride ion current equilibrium and, 108, 112, 119–120, 334–335
electrical neutrality in, 105–106, 108–109, 120
electrochemical, 86–87, 140, 161–170
electrochemical driving force and, 62–67
in electrodiffusional fluxes, calculations for, 105–110
electrogenic pump current and, 3–6, 8, 36–39, 108
  calculations for, 108, 110–115
electromotive force vs., 2–6
in excitable cells, 106, 122
glucose effects on, 296–304
Goldman-Hodgkin-Katz equation and, 106, 112
in human red blood cell membranes, 287–292
hyperpolarization of, 22–24, 36–38, 45–46
lactose transmembrane movement and, 324, 325
measurement of, 287–292, 345–351
in neuronal cells, 255, 265–267
in *Neurospora*, 307–320
in pancreatic beta cell membranes, modulation of, 295–305
pH change and, 38, 311–315
potassium depletion in hyperpolarization of, 106, 108, 183–184, 194–201, 208–211, 255, 296–301
proton current effects on, 37, 45–46, 115–119, 307–320
resting, 308, 311–312
sodium-calcium exchange sensitivity to, 167–170, 244–249, 374
sodium-calcium pump electrogenicity vs. membrane conductance in generation of, 374
in sodium electrochemical gradient changes, 162–165
in sodium-glucose cotransport, 62–65
sodium intracellular concentration and, 43–44, 201–203, 301–304
sodium-potassium-activated ATPase reaction and, 17–19, 267
in sodium-potassium ion transport, 22–24, 36–38, 45–46
  voltage sensitivity and, 42–44
sodium pump reactivation effects on, 193–212
steady-state equation for, 6–7, 107
temperature sensitivity of
  in neurons, 255–257
  in *Aplysia* vs. squid, 257

Membrane potential (cont.)
  in mammals vs. invertebrates, 257
  in *Xenopus* blastomeres, 345–351
Membrane resistance
  in human red blood cells, calculations for, 287–288
  in neuron cell bodies vs. axons, 257
Microsomal vesicles; *see also* Bilayer lipid membranes; Chromaffin ghosts
  ATPase activity in, 332–334
  ATP-dependent proton transport in, 331–341
  ATP hydrolysis site in, 332
  chloride effects on proton transport in, 333–338
  chloride ion, two-route permeation of, 337
  corn root vs. other types, 333–334
  proton transport in, 332–334, 338
  vs. tonoplasts, 334, 339–341
Mitochondria
  ATPase in, 340
  calcium ion release by, 129–130
  permeability of, vs. chromaffin granules, 74

Neurons/neuronal cells
  cardiac glucoside effects on, 255–256
  electrogenic pump activity in excitability of, 253–268
  membrane resistance in cell body (soma) vs. axons of, 257
  pacemaker discharge patterns in, 255–256
  posttetanic hyperpolarization in, 34, 253
  potassium ion concentration effects in, 255–256, 259, 263
  sodium-calcium exchange depolarization in, 162–163
  sodium pump current measurement in snail, 353–362
  temperature sensitivity of, 254–263, 267–268
Neurospora crassa
  electrogenic pump reversal in, 5, 37
  membrane current-voltage analysis in, 308–315
  membrane potential in, 307–320

Pacemaker activity
  in intestinal smooth muscle cells, 272, 277, 282, 283
  in neurons, 254–256
  potassium ion absence and, 255
  sodium pump action and, 255, 268
  temperature sensitivity of, 255–256
Pancreatic beta cells
  glucose effects on, 296–304
  membrane potential modulation in, 295–305

sodium ion intracellular hyperpolarization in, 301–304
  sodium pump electrogenicity in, 295–305
Parallel positive transfer effect, in proton-membrane binding, 313
Patlak-Läuger type gated channel, 97–99
pH
  ATPase activity and, 79
  in catecholamine transport, 88–89
  in chromaffin granules, 74–75, 79
  intracellular regulation of, in snail neurons, 353
  in lactose translocation, 324–325
  measurement of intravesicular, 74–75
  membrane potential response to, 38, 311–315
  proton pump responses to, extracellular, 311–313
  intracellular, 313–315
pH gradient
  ATPase activity response to, 80
  in catecholamine uptake by chromaffin granules, 81, 84–88
  characterized, 71
  external cation availability and, 57–58
Photocurrents, in bacteriorhodopsin, 381–393
Photocycles
  in bacteriorhodopsin, 387–390
  in purple membrane, 381
Plant cell membranes
  electrogenic pump potential in, 4,6; *see also* Proton pump
  pump conductance in, 6
Plant cells, experimental current measurement in, 7
Plasmalemma, 332
Phosphoenzyme, interconversion of, 42–45
Phosphorylase phosphatase, in sarcolemmal sodium-calcium exchange, 151–154, 158
Polystrand, characterized, 183
Posttetanic hyperpolarization, 33–35, 253
Potassium ion
  ATPase activation by, in plasma membrane vs. microsomal vesicles, 334
  in calcium-calcium exchange, 138
  chromaffin granule permeability to, 74, 75
  conformational change and release of, 42–44
  cotransport effects of, 93, 101
  coupling/exchange ratio with sodium, *see* Sodium-potassium exchange/ion transport, stoichiometry
  equilibrium potential negativity of, and GHK equation, 106, 108, 255
  extracellular, in sodium pump current reactivation, 33–35, 194–201,

208–211, 255
  intestinal slow wave activity on depletion
    of, 272
  membrane potential hyperpolarization on
    depletion of, 106, 108, 183–184,
    194–201, 210–211, 255, 293–301
Primary active transport; see also Active
    transport; Electrogenic pump/ion
    transport
  as electrogenic, 93–94
  free energy change equation for, 50, 51
  secondary active transport compared with,
    43, 93–94
Protein kinase, in sodium-calcium exchange
    regulation, 149–158
Proton-activated ATPase
  ATP effects on, 79–80
  in catecholamine uptake, 81, 85–86
  chloride ion stimulation of, 337–339
  in chromaffin granule membranes, 72–80,
    85–86
  in corn root vs. other type vesicle
    membranes, 333–334
  inhibition of, in chromaffin granules,
    79–80
  solubilization and reconstitution of,
    340–342
  in tonoplasts vs. plasma membranes, 334,
    339–341
Proton gradient
  in catecholamine uptake by chromaffin
    granules, 74, 81–85
  in electrogenic transport, 161, 162
  expression for, 71, 77
  in galactoside translocation, 323
  in lac carrier protein kinetics, 326
  measurement of, in chromaffin granule
    membrane, 74–75
  proton-activated ATPase effects on, 77, 80
Proton pump/ion transport
  active transport in, 307
  ATP-dependent, 332–333
  ATP hydrolysis in, 315
  in bacteriorhodopsin, as light-driven,
    381–393
  chloride ion stimulation of, 333, 334–339
  current densities in, 6, 308
  energy transfer localization in, 315–317
  experimental current measurement in, 7,
    308–309
  leakage conductance in, 309–315
  membrane current-voltage relations in, 8,
    308–320
  metabolic inhibition effects on, 315–318
  in microsomal vesicles, 332–334
  in Neurospora, as electrogenic, 307–320
  parallel positive transfer effects and, 313
  pH effects on, extracellular, 311–312
    intracellular, 41–42, 313–315

in purple membranes, 381–393
  in rheogenic cotransport, 93
  tonoplasts and, 332, 333–334
  two-state carrier model for, 310–311
Pump chord conductance, 5–7
Pump current
  characterized, 115
  electrogenic change in membrane potential
    vs., 115–119, 374
  measurement of
    in Neurospora, 308–309
    in Xenopus blastomeres, 345–351
  reversal potential and, 117–118
Purkinje fibers
  canine, preparation of, 216–218
  current abolition by cardioactive steriods in,
    197
  extracellular potassium effects on size of,
    235–236
  membrane current generation in, 195–201,
    215–223
  membrane potential response to potassium
    exclusion in, 218–220
  sheep cardiac, preparation of, 374
Purple membrane
  action spectrum vs. absorption spectrum
    in, 382
  in bacteriorhodopsin, 381–393; see also
    Bacteriorhodopsin
  bilayer lipid attachment, 382–383
  chromophore transient moment in,
    385–386
  photocurrents in
    activated, 390–393
    stationary, 382–384
  photoelectric responses by, 381, 382–383

Red blood cell membrane
  anion influx via human, 41
  electrical resistance measurement in
    human, 287–288
  membrane potential measurement in,
    287–289
  preparation of human, 288
  sodium pump characteristics in human, 39,
    41, 193, 287–292
Renal microvillus membrane vesicles
  secondary active transport via, 49–68
  sodium-hydrogen exchange via, 49–68
  sodium ion efflux response in, 57–59
Resting potential, 105, 121
Reverse potential
  in electrogenic pump current stoichiometry
    computation, 4–5
  in sodium pump current, 117–118
  voltage-dependent, 4–5, 34–35
Rheogenic transport
  characterized, 2, 7, 14, 93–94
  in cotransport systems, 93–103

Rheogenic transport *(cont.)*
  electric potential difference and, 94–98
  verification in, 94

Sarcolemma/sarcolemmal vesicles
  calcium transport in, 150, 239–248, 372
  isolation and preparation of, 150, 239–240, 243
  phosphorylation-dephosphorylation of, 150–151
  sodium-calcium exchange in, 149–158, 240–249, 372
Sarcoplasmic reticulum, calcium release by, 130
Secondary active transport; *see also* Active transport
  allosteric modifiers in, 59–61
  as carrier-mediated transport mechanism, 64
  characteristics of, 49, 50–54, 64, 67–68, 93
  cotransport as, 95
  countertransport as, 95
  electric charge associations with, 93
  equilibrium conditions in, 51–55, 67–68
  external cation effects on, 57–59
  free-energy change equation for, 50
  inward vs. outward flow potential in, 51–52
  leakage routes in, 95
  rates of, and electrochemical driving forces, 67–68
  via renal microvillus membrane vesicles, 52–59
  as rheogenic, 93–103
  saturability of substrate binding in, 56–57
  sodium-calcium exchange in cardiac cells as, 177–178
  sodium ion efflux via, 57–61
  sodium-proton exchange via, 49–68
  supplemental driving force in, 51
Skeletal muscle fibers, membrane potential calculations in, 110–115
Slope conductance, 5, 8, 25–26
Smooth muscle, intestinal
  calcium ion concentration effects on, 275–283
  circular and longitudinal layer differences in, 271–272
  membrane current modulation in, 272–277, 282
  pacemaker activity in, 272, 277, 283
  slow-wave depolarization in, 271–275
  sodium-potassium pump modulation of, 272–275, 282
Snail neurons, sodium pump current measurement in, 353–362
Sodium-ammonium exchange, 58–59

Sodium-calcium exchange
  ATP in, 133–135
  as ATP-dependent second pathway, 40
  in barnacle muscle fibers, 134–135, 365–370
  calcineurin in deactivation of, 149, 154–155, 157
  calcium-calcium exchange effects in, 130, 131, 135–138, 372
  calcium-dependent ATPase role in, 130
  calcium external concentration and, 376–379
  calcium extrusion mode in, 130–131, 144–145, 166–167, 365–370
  calcium influx mode in, 130, 131, 162–172
  calcium transport control by, 140–142, 161–179
  in cardiac muscle cells, 140, 161–169, 171–177
  in cardiac muscle contractility, 174–177, 373
  carrier reversal potential in, 165–166
  coupling ratio in, 165–166, 370
  in dog red blood cells, as paradoxical, 131
  dual, fast and slow, components of, 240–248
  electrochemical gradient changes in, 161–167
  electrochemical gradient coupling in, 131–132
  as electrogenic, 122–124, 161–167, 187–188, 239, 244–249, 366, 373–380
  forward-reverse modulations in, 130, 131, 144, 163–164
  inhibition and reactivation of, 149–158
  membrane current responses to, 366–367, 373–380
  membrane potential responses to, 167–170, 244–245, 248, 374
  in nerve terminal depolarization, 162–163
  in pancreatic beta cell membranes, 302
  phosphorylation-dephosphorylation regulation of, 133–135, 149–158
  sarcolemma in cardiac regulation of, 373
  in sarcolemmal vesicles, 149–158, 239–249, 372
  as secondary active transport, 177–178
  sodium electrochemical gradient in, 161–164
  in squid giant axons, 132–135, 162
  stoichiometry of, 130–131, 142–144, 163–166, 239–240, 365, 367, 370, 379
  temperature effects on, 241–243, 245
  transient inward currents in, 369–370
  trypsin proteolysis activation of, 151, 155–157

SUBJECT INDEX 403

as two-directional, forward and reverse, 130–131, 144, 163–164, 239, 272, 365–370
  voltage sensitivity of, 177–178, 367–368
Sodium-dependent countertransport, 93, 94, 98
Sodium-glucose cotransport
  coupling ratios for, 52–55
  as electrogenic, 62–63
  general carrier model of, 64
  glucose gradient calculation for, 52
  glucose uptake rate and sodium dependency, 62–67
  membrane potential variability in, 62–67
  via renal microvillus membrane vesicles, 49–50, 54–67
  as secondary active transport, 52–53
Sodium-hydrogen exchange, *see* Sodium-proton exchange
Sodium-independent glucose transport, 62–67
Sodium ion(s)
  calcium transport inhibition by, 135–139, 211–212
  cardiac action-potential influx stimulation, 225–226
  chromaffin granule permeability to, 74–75
  electrochemical driving forces in extrusion of, 59
  electrochemical gradient modification in transport of, 161–164
  external cation effects on, 57–69
  external concentration of, in intestinal slow-wave activity, 275
  internal pH effects on, in renal microvillus vesicles, 59–61
  membrane current and extrusion of, 374–376
  membrane potential responses to intracellular, 43–44, 201–203, 301–304
  in pancreatic beta cell hyperpolarization, 301–304
  in rheogenic cotransport systems, 93, 94, 98
  steady-state flux values for, in heart cells, 185–186
Sodium + potassium-activated ATPase
  ATP hydrolysis by, 17
  in artificial vesicle pump electrogenicity, 35–36
  cardioactive steroid effects on, 267
  current-voltage relations in ion translocation by, 20–22
  ion translocation step possibilities for plasmalemmal, 18, 20–21
  membrane potential effects on, 17–19, 267
  reaction scheme for, in sodium pump, 18–19

sodium-dependent calcium two-phase uptake and, 241–242
sodium-potassium exchange ratio determination by, 40
voltage-dependent steps in ion activation by, 20–25
Sodium-potassium pump/ion transport; *see also* Sodium pump/ion transport
  calcium levels and, 40, 139–142, 277–283
  cardiac action potential repolarization and, 224–230
  in cardiac muscle cells, 193–212, 215–236, 372
  cardioactive steroid effects on, *see* Sodium pump/ion transport
  charge translocation model of, 9
  conformational change voltage sensitivity in, 42–44
  current generated by, in Purkinje fibers, 215–223
  electrogenicity characteristics of, 193–212, 215–236, 287–292
  in embryonic cultured heart cells, 185–187
  exchange ratio determination in, 39–42; *see also* stoichiometry, *following*
  in human red blood cell membranes, 287–292
  hyperpolarization of, 22–24, 36–38, 44–45, 193–194, 218–225
  intestinal slow-wave activity and, 271–283
  membrane current activation by, 215–216
  in pancreatic beta cell membranes, 293–305
  phosphorylation-dephosphorylation, 42–44
  rate changes for, experimental, 216–236
  sodium ion internal increase activation by, 225–226, 287–292
  steady-state chord conductance in, 6–7
  steady-state equation for, 107
  stoichiometry ratio characteristics of, 6–7, 39–42, 107, 186–187, 193–194, 204, 210, 215
  transient outward overshoot by, 196–197, 208, 219–221
  transmembrane ion gradient maintenance by, 215
  voltage-sensitive step determination in, 43–45
Sodium-proton exchange
  external cation alternative effects on, 57–59
  internal pH effects on sodium extrusion in, 59–61
  in renal microvillus membrane vesicles, 52–57
  as secondary active transport, 49–68
  sodium ion influx rate and saturation in, 56
  transport rate vs. electrochemical driving force in, 67

Sodium pump/ion transport; see also
  Sodium-potassium pump/ion transport
  ATP hydrolysis reaction and, 3–5, 17, 42, 72
  anion transport via, 41–42
  in artificial lipid vesicles, 35–36, 40
  cardioactive steroid inhibition of, 196–197, 208, 212, 220–222, 255–256, 267
  chord conductance in, 5–7
  current densities, 6
  current measurement in, 193–212, 353–362
  current-voltage relationships in, 7–11, 115–119, 345–351
  electrodiffusion in, calculation of, 105–110
  electrogenicity of, 4–17, 33–36, 110–115, 193–212, 255
  membrane potential responses to, 22–24, 36–38, 44–45, 111–113, 115–119, 195–198, 218–221, 253–255, 265–268, 287–292, 374
    calculations for, 107–110
  outward current generation mechanism of, 39–45
  pH sensitivity of, 41–42
  posttetanic hyperpolarization and, 333–335
  potassium external depletion effects on, 195–201, 208–211, 255
  proton transport and, 41–42
  sodium extrusion by, 203–204, 374–376
  sodium intracellular activity and electrogenicity of, 202–203, 205–208, 211–212
  temperature sensitivity of, 255–263, 267–268
Solute transport rate
  electrochemical driving force and, 54–56, 67–68
  external alternative cation effects on, 57–59
  membrane potential variability in, 62–67
  substrate binding-site saturability and, 56
  transportable allosteric solute effects on, 59–61

Tonoplasts, 332
  proton-activated ATPase vs. plasma membrane in, 334, 339–341
  proton transport indicated for, 332–338
Total membrane conductance, characterized, 3
Transmembrane potential
  ATPase activity and, 80
  in catecholamine uptake by chromaffin granules and ghosts, 85–88
  measurement of, in chromaffin granules, 74–76
True coupling, 93
Trypsin, in sodium-calcium exchange activation, 151, 155–157

Velocity effect, in translocation, 95
Vesicles, artificial, see Bilayer lipid membranes; Chromaffin ghosts/vesicles; Microsomal vesicles; Renal microvillus membrane vesicles; Sarcolemmal vesicles

Xenopus laevis, 345–351
  blastosphere preparation of, 346
  membrane conductance response in, 345–351

Zwitterionic amine species, 88